Basic Ship Theory

Basic Ship Theory

K.J. Rawson
MSc, FEng, RCNC, FRINA, WhSch

E.C. Tupper
Bsc, CEng, RCNC, FRINA, WhSch

Fourth edition
Volume 1
Chapters 1 to 9
Hydrostatics and Strength

Addison Wesley Longman Limited,
Edinburgh Gate, Harlow
Essex CM20 2JE, England
and Associated Companies throughout the world.

Copublished in the United States with
John Wiley & Sons, Inc., 605 Third Avenue, New York, NY 10158

First published 1968
Second edition 1976 (in two volumes)
Third edition 1983
Sixth impression 1989
Fourth edition 1994
Reprinted 1996

British Library Cataloguing in Publication Data
A catalogue entry for this title is available from the British Library

ISBN 0-582-21922-1

Rawson, K. J.
Basic ship theory.
p. cm.
Includes bibliographical references and indexes.
Contents: v. 1. Hydrostatics and strength – v. 2. Ship dynamics and design.
1. Naval architecture. I. Tupper, E. C. II. Title.
VM156.R37 1994 623.8'1 94-9031

ISBN 0-470-23388-5 (pbk.: v. 1) (USA only)
ISBN 0-470-23429-6 (pbk.: v. 2) (USA only)

Set by 4 in 10½/12pt Times
Produced through Longman Malaysia, TCP

Contents

Volume 2

Preface to the fourth edition

The fundamental concepts of naval architecture, which the authors set out when Basic Ship Theory was first published in 1968, remain as valid as ever. However, many advances have been made in our understanding of the environment in which ships operate and in their behaviour in that environment. More recent developments of theory and application have been included, some of it emanating from the steady work of the International Maritime Organisation and, unhappily, inspired by continuing losses at sea. Prescriptive regulation is seen to be increasingly over-complicated and not adequately relevant and licensing agencies, working on the basis of acceptable limits in the behaviour of ships, cannot be far away. It remains important, therefore, to strive to understand better the behaviour of ships, to predict the limits of their performances and consequently to assure their safety. Safety management must surely soon be normal practice in the shipping industry as it is in other industries.

Besides advances in knowledge, the power and relative cheapness of computers has now significantly changed the way in which a naval architect goes about his or her work. Many computer based solutions to analytical problems are now readily available. It is vital that practitioners avoid a blind application of such programs. They must examine knowledgeably the relevance of the programs they use to their problems on which people's lives may depend.

In this edition, we have maintained the objective of conveying principles and understanding to help student and practitioner to avoid that superficiality which derives from an inadequate appreciation of a computer's strengths and weaknesses.

To make way for new developments, it has been necessary to remove some of the earlier material. Most of this, while useful in the past for obtaining a simple view of what was really a complex problem, has now been superseded by better methods, However, as naval architects, we are glad to acknowledge the work of those who originated such development of our understanding.

We have extended later chapters to embrace exciting new developments in maritime transport and have brought up to date the standards upon which ship design is dependent. Note that such standards change continually and should be checked from the reference material.

Our thanks are extended to many friends who have advised us.

June 1993 KJR ECT

Preface to the third edition

This new edition provides an opportunity to up-date the subject matter in line with modern practice and thinking. The most significant changes are not surprisingly to be found in those chapters dealing with structures, hydrodynamics and design. It is in these areas that the increasing power of computer systems is making itself felt and providing the naval architect with an ability to respond to the measures of modern day economies. Nevertheless, it is only possible in a book of this type to provide the student with a general understanding of such developments and to set them within the context of the basic grounding in ship theory which the book primarily aims to provide. This is inevitable, as many of the new developments require an advanced theoretical knowledge while others are made possible by the ability of the computer to manipulate large volumes of arithmetical data. It is hoped that on the one hand this book will whet the student's appetite for more detailed study and on the other enable him to apply the computer intelligently.

Following on from the above, the authors have judged it desirable to retain some elements of the subject which, whilst they are seldom met in practice, do provide the student with an insight into the fundamental principles involved. The aim is to provide a real understanding of the underlying concepts. All too often these concepts are not so apparent in the more advanced procedures now adopted in the design office, which do not necessarily therefore provide the best path to a proper understanding of the subject.

After careful consideration it was decided to retain the balance between Imperial and SI units generally as in the Second Edition. This is felt to reflect the balance within the profession taken globally.

Preface to the second edition

Knowledge and practice advances today with such speed that textbooks must be updated with increasing regularity. Basic Ship Theory was first published at a time when SI units were to be introduced but before any clear guidance upon preferred units or values had been prepared—indeed, debate upon this subject is far from dead. Fortunately, we were able to anticipate correctly how the profession would move and changes in the second edition on this score are few. We have extended the use of metric SI but have felt that it is not yet time to abandon the old units totally; consequently, old and new units have been retained—a state of affairs which practising naval architects will have to face for many years yet. Ships and shipping are, after all, international and enduring!

It is a good deal more difficult to incorporate in a book of this type the very rapid advances in theory which have been encouraged by the development of the digital computer. So many of these advances depend upon a mathematical technique which cannot reasonably be yet expected in students at this stage of their learning. Yet the exclusion of this basic work would render the book incomplete. The authors have therefore incorporated a superficial description of such material as strip theory, finite element technique, structural dynamics, computing techniques, etc., so that the students will be aware of its existence and the way in which they are forced to rely upon an increasing array of specialists until they are themselves able to study these topics in detail.

It has been decided to issue the second edition in two volumes, the first volume dealing with statics, structure and environment and the second dealing with hydrodynamics, ship dynamics and design. In this way, it is hoped that the swelling number of people, inspired perhaps by the challenge of offshore engineering and maritime affairs, will be encouraged to establish their interest in the sea with some basic knowledge of its technology and science.

September 1974

KJR ECT

Foreword to the first edition

By Sir Alfred J. Sims, K.C.B., O.B.E., D.Sc., R.C.N.C., formerly Director General Ships, Ministry of Defence, Principal Technical Adviser to the Navy Board and President of the Royal Institution of Naval Architects

The textbooks available in theoretical naval architecture are few. The one with which I have had some association in later years—Attwood's *Theoretical Naval Architecture*—has stood the profession in good stead since the beginning of this century.

For some time I have looked forward to a new book in this field, encompassing the many important developments in the subject and the modern aids to good ship design.

Now, two members of the Royal Corps of Naval Constructors have undertaken this task with enthusiasm and competence. Both authors have recently been in the forefront of the design of complex warships. Both are well able to expound the latest theories and to be able to select the most useful material for students of all ages.

The authors have taken care to cover the merchant ship field as well as warships and the book will have wide appeal throughout the profession on this account.

To new students, I commend the book in the confidence that it will be helpful and stimulating. To those already in the profession, or well advanced in their studies, I commend the work as a valuable book of reference.

December 1966 ALFRED J. SIMS

Acknowledgements

We are grateful to the following for permission to reproduce copyright material:

Editor, Naval Architect for our Figs. 6.33 & 6.35 after Figs. 3 & 2 (Faulkner & Saddon, Ref. 7); Society of Naval Architects & Marine Engineers for our Figs. 7.36 & 7.37 after Figs. 1b & 2 (Dow, Hugill & Clarke, Ref. 9); National Physical Laboratory and N. Hogben for our Fig. 9.20, Tables 9.10, 9.11 & 9.12 after Fig. p.2, Tables pp. 6, 13 & 68 (Hogben & Lumb, Ref. 20); Human Factors Research for our Fig. 9.26 (McCauley *et al.*, Ref. 16); Secretariat of ISO/TC for our Fig. 9.30 based on Tables 1 & 2, Fig. 2a & b (ISO/DP 2631, Ref. 14).

Introduction

Naval architecture is a rapidly developing subject. Advances in other scientific and technical subjects, development of new design techniques and the ubiquitous computer all enable the ship designer to carry out his work with greater precision and confidence. New ideas and methods are complementing or replacing old concepts.

Not unnaturally the major changes are taking place at the more advanced levels. Nevertheless, they influence the method of presentation of the more basic ideas and the emphasis placed on various aspects of the subject. At the same time, in common with other professions, it is desirable that students be presented with concepts more advanced than have been available in general textbooks to date.

This book aims at providing an introduction to ship theory—a term embracing theoretical naval architecture and other specialized theory used by those whose work is associated with ships. Besides extending the scope of earlier textbooks in this way, the conventional bounds of the subject have been extended. Care has been taken to avoid duplicating, as far as possible, work the student will be doing in other subjects as part of his degree or engineering or technician courses; indeed, it is necessary to assume that his knowledge in all subjects advances with his progress through this book. It is hoped that this book will encourage more people to interest themselves and advance their knowledge in ships. The authors have tried to stimulate and hold the interest of the student by a careful arrangement of the subject matter of the book. Chapter 1 and the opening paragraphs of each succeeding chapter have been presented in somewhat lyrical terms in the hope that the authors will convey to the students some of the enthusiasm which they feel themselves for this fascinating subject. Naval architects need never fear that they will have, during their careers, to face the same problems day after day. Before they reach the Board Room to direct basic thought and policy, they will spend many years on as wide a variety of sciences as are touched upon by any profession.

Before embarking on the book proper it is necessary to comment on the units employed, and the use made of examples and references.

UNITS

In May 1965, the Government announced its view that British Industry should move to the use of the metric system. In addition, at the present time, a newly rationalized set of metric units is coming into international use following endorsement by the International Organization for Standardization. It is the Système International d'Unités (SI)—and the UK is able to adopt it from the outset. It replaces also the British Standard Specification 1991 for symbols and terminology.

The adoption of SI units has been patchy in many countries while some have yet to change from their traditional positions. A change from inches to metres is simple and has been widely effected; the rejection of tonf in favour of meganewton has proceeded slowly. One of the reasons for this is the retention of kg as the unit of mass and its ready—and erroneous—acceptance by trade and industry as the unit of force because weighing devices are dependent upon springs which respond to force. Students beware!

In the following notes, the SI system of units is presented briefly; a fuller treatment appears in Ref. 1. To assist the change to the metric system in the United Kingdom and to meet the needs of the readers in other countries, this book has been written in terms of British and metric units. Examples in both British and metric units are included both as worked examples and set problems and, wherever appropriate, basic data is presented in terms of both sets of units.

The SI is a rationalized selection of units in the metric system. It is a coherent system, i.e. the product or quotient of any two unit quantities in the system is the unit of the resultant quantity. The basic units are as follows:

Quantity	Name of unit	Unit symbol
Length	metre	m
Mass	kilogramme	kg
Time	second	s
Electric current	ampere	A
Thermodynamic temperature	kelvin	K
Luminous intensity	candela	cd
Amount of substance	mole	mol
Plane angle	radian	rad
Solid angle	steradian	sr

Special names have been adopted for some of the derived SI units and these are listed below together with their unit symbols:

Physical quantity	SI unit	Unit symbol
Force	newton	N = kg m/s^2
Work, energy	joule	J = N m
Power	watt	W = J/s
Electric charge	coulomb	C = A s
Electric potential	volt	V = W/A
Electric capacitance	farad	F = A s/V

(*continued on p. xiv*)

Electric resistance	ohm	Ω = V/A
Frequency	hertz	Hz = s^{-1}
Illuminance	lux	lx = lm/m^2
Self inductance	henry	H = V s/A
Luminous flux	lumen	lm = cd sr
Pressure, stress	pascal	Pa = N/m^2
	megapascal	MPa = N/mm^2
	bar	bar = 10^5 Pa
Electrical conductance	siemens	S = $1/\Omega$
Magnetic flux	weber	Wb = V s
Magnetic flux density	tesla	T = Wb/m^2

In the following two tables are listed other derived units and the equivalent values of some UK units respectively:

Physical quantity	SI unit	Unit symbol
Area	square metre	m^2
Volume	cubic metre	m^3
Density	kilogramme per cubic metre	kg/m^3
Velocity	metre per second	m/s
Angular velocity	radian per second	rad/s
Acceleration	metre per second squared	m/s^2
Angular acceleration	radian per second squared	rad/s^2
Pressure, Stress	newton per square metre	N/m^2
Surface tension	newton per metre	N/m
Dynamic viscosity	newton second per metre squared	$N s/m^2$
Kinematic viscosity	metre squared per second	m^2/s
Thermal conductivity	watt per metre Kelvin	W/(mK)

Quantity	UK unit	Equivalent SI units
Length	1 yd	0·9144 m
	1 ft	0·3048 m
	1 in	0·0254 m
	1 mile	1609·344 m
	1 nautical mile (UK)	1853·18 m
	1 nautical mile (International)	1852 m
Area	1 in^2	$645{\cdot}16 \times 10^{-6}$ m^2
	1 ft^2	0·092903 m^2
	1 yd^2	0·836127 m^2
	1 $mile^2$	$2{\cdot}58999 \times 10^6$ m^2
Volume	1 in^3	$16{\cdot}3871 \times 10^{-6}$ m^3
	1 ft^3	0·0283168 m^3
	1 UK gal	0·004546092 m^3 = 4·546092 litres
Velocity	1 ft/s	0·3048 m/s
	1 mile/hr	0·44704 m/s; 1·60934 km/hr
	1 knot (UK)	0·51477 m/s; 1·85318 km/hr
	1 knot (International)	0·51444 m/s; 1·852 km/hr

(*continued on p. xv*)

(continued from p. xiv)

Standard acceleration, g	32·174 ft/s²	9·80665 m/s²
Mass	1 lb	0·45359237 kg
	1 ton	1016·05 kg = 1·01605 tonnes
Mass density	1 lb/in³	27·6799 × 10³ kg/m³
	1 lb/ft³	16·0185 kg/m³
Force	1 pdl	0·138255 N
	1 lbf	4·44822 N
Pressure	1 lbf/in²	6894·76 N/m²
Stress	1 tonf/in²	15·4443 × 10⁶ N/m²
		15.4443 MPa or N/mm²
Energy	1 ft pdl	0·0421401 J
	1 ft lbf	1·35582 J
	1 cal	4·1868 J
	1 Btu	1055·06 J
Power	1 hp	745·700 W
Temperature	1 Rankine unit	5/9 Kelvin unit
	1 Fahrenheit unit	5/9 Celsius unit

Note that, while multiples of the denominators are preferred, the engineering industry has generally adopted N/mm² for stress instead of MN/m² which has, of course, the same numerical value and are the same as MPa.

Prefixes to denote multiples and sub-multiples to be affixed to the names of units are:

Factor by which the unit is multiplied	Prefix	Symbol
1 000 000 000 000 = 10^{12}	tera	T
1 000 000 000 = 10^{9}	giga	G
1 000 000 = 10^{6}	mega	M
1 000 = 10^{3}	kilo	k
100 = 10^{2}	hecto	h
10 = 10^{1}	deca	da
0·1 = 10^{-1}	deci	d
0·01 = 10^{-2}	centi	c
0·001 = 10^{-3}	milli	m
0·000 001 = 10^{-6}	micro	μ
0·000 000 001 = 10^{-9}	nano	n
0·000 000 000 001 = 10^{-12}	pico	p
0·000 000 000 000 001 = 10^{-15}	femto	f
0·000 000 000 000 000 001 = 10^{-18}	atto	a

We list, finally, some preferred metric values (values preferred for density of fresh and salt water are based on a temperature of 15°C(59°F)).

Item	Accepted British figure	Direct metric equivalent	Preferred SI value
Gravity, g	32·17 ft/s^2	9·80665 m/s^2	9·807 m/s^2
Mass density salt water	64 lb/ft^3 35 ft^3/ton	1·0252 tonne/m^3 0·9754 m^3/tonne	1·025 tonne/m^3 0·975 m^3/tonne
Mass density fresh water	62·2 lb/ft^3 36 ft^3/ton	0·9964 tonne/m^3 1·0033 m^3/tonne	1·0 tonne/m^3 1·0 m^3/tonne
Young's modulus, E (steel)	13,500 tonf/in^2	2·0855 × 10^7 N/cm^2	209 GN/m^2 or GPa
Atmospheric pressure	14·7 lbf/in^2	101,353 N/m^2 10·1353 N/cm^2	10^5 N/m^2 or Pa or 1·0 bar
TPI (salt water)	$\frac{A_W}{420}$ tonf/in A_W (ft^2)	1·025 A_W (tonnef/m) A_w (m^2)	1·025 A_W tonnef/m
NPC	A_w (m^2)	100·52 A_W (N/cm)	
NPM		10,052 A_W (N/m)	10^4 A_W (N/m)
MCT 1″ (salt water) (Units of tonf and feet)	$\frac{\Delta \overline{GM}_L}{12L} \frac{\text{tonf ft}}{\text{in}}$		
One metre trim moment, (Δ in MN or $\frac{\text{tonnef m}}{\text{m}}$, Δ in tonnef)		$\frac{\Delta \overline{GM}_L}{L}\left(\frac{\text{MN m}}{\text{m}}\right)$	$\frac{\Delta \overline{GM}_L}{L}\left(\frac{\text{MN m}}{\text{m}}\right)$
Force displacement Δ	1 tonf	1·01605 tonnef 9964·02N	1·016 tonnef 9964 N
Mass displacement Σ	1 ton	1·01605 tonne	1·016 tonne
Weight density:			
Salt water			0·01 MN/m^3
Fresh water			0·0098 MN/m^3
Specific volume:			
Salt water			99·5 m^3/MN
Fresh water			102·0 m^3/MN

Of particular significance to the naval architect are the units used for displacement, density and stress. The force displacement Δ, under the SI scheme must be expressed in terms of newtons. In practice the meganewton (MN) is a more convenient unit and 1 MN is approximately equivalent to 100 tonf (100·44 more exactly). This new unit will eliminate any confusion between the displacement and tonnage of a ship but, to assist students in making the change from one system of units to another, the authors have additionally introduced the tonnef (and, correspondingly, the tonne for mass measurement) as explained more fully in Chapter 3.

Many preferred values are used by naval architects, e.g. 36 for the number of cubic feet in a ton of fresh water, although the precise figure is slightly different. Conversion to the metric system will involve the adoption of new preferred values which are not precisely equivalent to the old. The authors have adopted those shown in the table immediately above.

EXAMPLES

A number of worked examples has been included in the text of most chapters to illustrate the application of the principles enunciated therein. Some are relatively short but others involve lengthy computations. They have been deliberately chosen to help educate the student in the subject of naval architecture, and the authors have not been unduly influenced by the thought that examination questions often involve about 30 minutes' work.

In the problems set at the end of each chapter, the aim has been adequately to cover the subject matter, avoiding, as far as possible, examples involving mere arithmetic substitution in standard formulae. The opportunity has also been taken to extend the scope of the chapter in some instances. Many of these are or are based closely on actual examination questions. Others have been introduced to lead to a better understanding of the subject without direct concern for time of execution. All have been made as realistic as possible.

REFERENCES

References are listed at the end of each chapter in order to acknowledge the source of information presented and to enable the student to study in more detail any particular aspect of the subject. The references are not exhaustive but most provide further guidance on recommended reading. Where information has been presented or used in more than one paper or book an attempt has been made to select the one most likely to be accessible to the student. A large number of references have been made to the transactions of the Royal Institution of Naval Architects as these are widely available. The authors apologize if this should, in any instance, appear to credit a development to anyone other than the true originator but would point out that some students may not have as ready access to scientific journals as research workers often enjoy.

The authors also seek indulgence where limitations in the space available in or scope of this volume have meant that the arguments of other authors have not been fully covered. Indeed, in some cases, it has been necessary to forsake mathematical precision in order to present the essential elements of an idea without confusing the student with the many qualifying conditions often involved.

The following abbreviations have been used in the references:

RINA	Royal Institution of Naval Architects (INA prior to 1960).
NA	*The Naval Architect Journal*, RINA.
SNAME	The Society of Naval Architects and Marine Engineers.
SNAJ	Society of Naval Architects of Japan.
IESS	Institution of Engineers and Shipbuilders of Scotland.

IMarE	Institute of Marine Engineers.
IMechE	Institution of Mechanical Engineers.
HMSO	Her Majesty's Stationery Office.
BSI	British Standards Institution.
DTMB	David Taylor Model Basin, Washington.

The letter *T* before a set of the above initials represents the transactions, and the letter *J* the journal, of that body.

Over the years many of the organizations referred to in this book have been subject to regrouping and/or change of name. When reference is made to these the title used is that at the time the work under discussion was carried out. It is not possible here to cover all renamings but the following examples will assist the reader.

The Admiralty Experiment Works (AEW) and Naval Construction Research Establishment (NCRE) became in sequence part of the Admiralty Marine Technology Establishment (AMTE) then of the Admiralty Research Establishment (ARE) and now the Defence Research Agency (DRA). The Ship Division of the National Physical Laboratory (NPL) became the National Maritime Institute (NMI) and then joined with the British Ship Research Association (BSRA) to become British Maritime Technology (BMT). The David Taylor Model Basin (DTMB) in Washington became the David Taylor Naval Ship R & D Centre (DTNSRDC) then the David Taylor Research Centre (DTRC) and has now reverted to DTMB.

MARINE SAFETY

The statutory authority on ship safety in the UK is the Department of Transport (DTp). Executive authority for marine safety was invested in 1994 in the Marine Safety Agency (MSA) created from the former Surveyor General's organization of the DTp. Responsibility for the safety of offshore structures was transferred in 1991 from the Department of Energy to the Health and Safety Executive following the *Piper Alpha* disaster.

References and further reading

1. *SI units and recommendations for the use of their multiples*, British Standard 5555, 1981.
2. *Metric Units for use by the Ministry of Defence*, Defence Standard 00-11, 1978.
3. Symposium on Education and Training for Naval Architecture and Ocean Engineering, *RINA*, 1976.
4. Guides to the preparation of papers, *RINA* and *IMechE*.

Symbols and nomenclature

GENERAL

a	linear acceleration
A	area in general
B	breadth in general
D, d	diameter in general
E	energy in general
F	force in general
g	acceleration due to gravity
h	depth or pressure head in general
h_W, ζ_W	height of wave, crest to trough
H	total head, Bernoulli
L	length in general
L_W, λ	wave-length
m	mass
n	rate of revolution
p	pressure intensity
p_V	vapour pressure of water
p_∞	ambient pressure at infinity
P	power in general
q	stagnation pressure
Q	rate of flow
r, R	radius in general
s	length along path
t	time in general
$t°$	temperature in general
T	period of time for a complete cycle
u	reciprocal weight density, specific volume,
u, v, w	velocity components in direction of x-, y-, z-axes
U, V	linear velocity
w	weight density
W	weight in general
x, y, z	body axes and Cartesian co-ordinates Right-hand system fixed in the body, z-axis vertically down, x-axis forward. Origin at c.g.
x_0, y_0, z_0	fixed axes Right-hand orthogonal system nominally fixed in space, z_0-axis vertically down, x_0-axis in the general direction of the initial motion.
α	angular acceleration
γ	specific gravity
Γ	circulation
δ	thickness of boundary layer in general
θ	angle of pitch
μ	coefficient of dynamic viscosity
ν	coefficient of kinematic viscosity
ρ	mass density
ϕ	angle of roll, heel or list
χ	angle of yaw
ω	angular velocity or circular frequency
∇	volume in general

GEOMETRY OF SHIP

A_M	midship section area
A_W	waterplane area
A_X	maximum transverse section area
B	beam or moulded breadth
$\overline{BM}$	metacentre above centre of buoyancy
C_B	block coefficient
C_M	midship section coefficient
C_P	longitudinal prismatic coefficient
C_{VP}	vertical prismatic coefficient
C_{WP}	coefficient of fineness of waterplane
D	depth of ship
F	freeboard
$\overline{GM}$	transverse metacentric height
$\overline{GM}_L$	longitudinal metacentric height
I_L	longitudinal moment of inertia of waterplane about CF
I_P	polar moment of inertia
I_T	transverse moment of inertia
L	length of ship—generally between perps
L_{OA}	length overall
L_{PP}	length between perps
L_{WL}	length of waterline in general
S	wetted surface
T	draught
Δ	displacement force
λ	scale ratio—ship/model dimension
∇	displacement volume
Σ	displacement mass

PROPELLER GEOMETRY

A_D	developed blade area
A_E	expanded area
A_O	disc area
A_P	projected blade area
b	span of aerofoil or hydrofoil
c	chord length
d	boss or hub diameter
D	diameter of propeller
f_M	camber
P	propeller pitch in general
R	propeller radius
t	thickness of aerofoil
Z	number of blades of propeller
α	angle of attack
ϕ	pitch angle of screw propeller

RESISTANCE AND PROPULSION

a	resistance augment fraction
C_D	drag coeff.
C_L	lift coeff.
C_T	specific total resistance coeff.
C_W	specific wave-making resistance coeff.
D	drag force
F_n	Froude number
I	idle resistance
J	advance number of propeller
K_Q	torque coeff.
K_T	thrust coeff.
L	lift force

P_D	delivered power at propeller
P_E	effective power
P_I	indicated power
P_S	shaft power
P_T	thrust power
Q	torque
R	resistance in general
R_n	Reynolds number
R_F	frictional resistance
R_R	residuary resistance
R_T	total resistance
R_W	wave-making resistance
s_A	apparent slip ratio
t	thrust deduction fraction
T	thrust
U	velocity of a fluid
U_∞	velocity of an undisturbed flow
V	speed of ship
V_A	speed of advance of propeller
w	Taylor wake fraction in general
w_F	Froude wake fraction
W_n	Weber number
β	appendage scale effect factor
β	advance angle of a propeller blade section
δ	Taylor's advance coeff.
η	efficiency in general
η_B	propeller efficiency behind ship
η_D	quasi propulsive coefficient
η_H	hull eff.
η_O	propeller eff. in open water
η_R	relative rotative efficiency
σ	cavitation number

SEAKEEPING

c	wave velocity
f	frequency
f_E	frequency of encounter
I_{xx}, I_{yy}, I_{zz}	real moments of inertia
I_{xy}, I_{xz}, I_{yz}	real products of inertia
k	radius of gyration
m_n	spectrum moment where n is an integer
M_L	horizontal wave bending moment
M_T	torsional wave bending moment
M_V	vertical wave bending moment
s	relative vertical motion of bow with respect to wave surface
$S_\zeta(\omega)$, $S_\theta(\omega)$, etc.	one-dimensional spectral density
$S_\zeta(\omega, \mu)$, $S_\theta(\omega, \mu)$, etc.	two-dimensional spectral density
T	wave period
T_E	period of encounter
T_z	natural period in smooth water for heaving
T_θ	natural period in smooth water for pitching
T_ϕ	natural period in smooth water for rolling
$Y_{\theta\zeta}(\omega)$	response amplitude operator—pitch
$Y_{\phi\zeta}(\omega)$	response amplitude operator—roll
$Y_{\chi\zeta}(\omega)$	response amplitude operator—yaw
β	leeway or drift angle
δ_R	rudder angle
ε	phase angle between any two harmonic motions
ζ	instantaneous wave elevation
ζ_A	wave amplitude

ζ_w	wave height, crest to trough
θ	pitch angle
θ_A	pitch amplitude
κ	wave number
ω_E	frequency of encounter
Λ	tuning factor

MANOEUVRABILITY

A_C	area under cut-up
A_R	area of rudder
b	span of hydrofoil
c	chord of hydrofoil
K, M, N	moment components on body relative to body axes
O	origin of body axes
p, q, r	components of angular velocity relative to body axes
X, Y, Z	force components on body
α	angle of attack
β	drift angle
δ_R	rudder angle
χ	heading angle
ω_C	steady rate of turn

STRENGTH

a	length of plate
b	breadth of plate
C	modulus of rigidity
ε	linear strain
E	modulus of elasticity, Young's modulus
σ	direct stress
σ_y	yield stress
g	acceleration due to gravity
I	planar second moment of area
J	polar second moment of area
j	stress concentration factor
k	radius of gyration
K	bulk modulus
l	length of member
L	length
M	bending moment
M_p	plastic moment
M_{AB}	bending moment at A in member AB
m	mass
P	direct load, externally applied
P_E	Euler collapse load
p	distributed direct load (area distribution), pressure
p'	distributed direct load (line distribution)
τ	shear stress
r	radius
S	internal shear force
s	distance along a curve
T	applied torque
t	thickness, time
U	strain energy
W	weight, external load
y	lever in bending
δ	deflection, permanent set, elemental (when associated with element of breadth, e.g. δb)
ρ	mass density
ν	Poisson's ratio
θ	slope

NOTES

(*a*) A distance between two points is represented by a bar over the letters defining the two points, e.g. $\overline{\mathrm{GM}}$ is the distance between G and M.
(*b*) When a quantity is to be expressed in non-dimensional form it is denoted by the use of the prime $'$. Unless otherwise specified, the non-dimensionalizing factor is a function of p, L and V, e.g. $m' = m/\frac{1}{2}\rho L^3$, $x' = x/\frac{1}{2}\rho L^2 V^2$, $L' = L/\frac{1}{2}\rho L^3 V^2$.
(*c*) A lower case subscript is used to denote the denominator of a partial derivative, e.g. $Y_u = \partial Y/\partial u$.
(*d*) For derivatives with respect to time the dot notation is used, e.g. $\dot{x} = \mathrm{d}x/\mathrm{d}t$.

1 Art or science?

Many thousands of years ago when man became intelligent and adventurous, those tribes who lived near the sea ventured on to it. They built rafts or hollowed out tree trunks and soon experienced the thrill of moving across the water, propelled by tide or wind or device. They experienced, too, the first sea disasters; their boats sank or broke, capsized or rotted and lives were lost. It was natural that those builders of boats which were adjudged more successful than others, received the acclaim of their fellow men and were soon regarded as craftsmen. The intelligent craftsman observed perhaps, that capsizing was less frequent when using two trunks joined together or when an outrigger was fixed, or that he could manoeuvre better with a rudder in a suitable position. His tools were trial and error and his stimulus was the pride in his craft. He was the first naval architect.

The craftsmen's expertise developed as it was passed down the generations: the Greeks built their triremes and the Romans their galleys; the Vikings produced their beautiful craft to carry soldiers through heavy seas and on to the beaches (Ref. 12). Several hundred years later, the craftsmen were designing and building great square rigged ships for trade and war and relying still on knowledge passed down through the generations and guarded by extreme secrecy. Still, they learned by trial and error because they had as yet no other tools and the disasters at sea persisted.

The need for a scientific approach must have been felt many hundreds of years before it was possible and it was not possible until relatively recently, despite the corner stone laid by Archimedes two thousand years ago. Until the middle of the eighteenth century the design and building of ships was wholly a craft and it was not, in Britain, until the second half of the nineteenth century that science affected ships appreciably.

Isaac Newton and other great mathematicians of the seventeenth century laid the foundations for so many sciences and naval architecture was no exception. Without any doubt, however, the father of naval architecture was Pierre Bouguer who published in 1746, *Traité du Navire* (Ref. 1). In this book, Bouguer laid the foundations of many aspects of naval architecture which were developed later in the eighteenth century by Bernoulli, Euler and Santacilla (Ref. 2). Lagrange and many others made contributions but the other outstanding figure of that century was the Swede, Frederick Chapman who pioneered work on ship resistance which led up to the great work of William Froude a hundred years later. A scientific approach to naval architecture was encouraged more on the continent than in Britain where it remained until the 1850's, a craft surrounded by pride and secrecy. On 19 May 1666, Samuel Pepys wrote of a Mr Deane,

> And then he fell to explain to me his manner of casting the draught of water which a ship will draw before-hand; which is a secret the King and all admire in him, and he is the first that hath come to any certainty before-hand of foretelling the draught of water of a ship before she be launched. (Ref. 3.)

The second half of the nineteenth century, however, produced Scott Russell, Rankine and Froude and the development of the science, and dissemination of knowledge in Britain was rapid. This development is described in Ref. 4.

NAVAL ARCHITECTURE TODAY

It would be quite wrong to say that the art and craft built up over many thousands of years has been wholly replaced by a science. The need for a scientific approach was felt, first, because the art had proved inadequate to halt the disasters at sea or to guarantee to the merchant that he was getting the best for his money. Science has contributed much to alleviate these shortcomings but it continues to require the injection of experience of successful practice. Science produces the correct basis for comparison of ships but the exact value of the criteria which determine their performances must, as in other branches of engineering, continue to be dictated by previous successful practice, i.e. like most engineering, this is largely a comparative science. Where the scientific tool is less precise than one could wish, it must be heavily overlaid with craft; where a precise tool is developed, the craft must be discarded. Because complex problems encourage dogma, this has not always been easy.

The question, 'Art or Science?' is therefore loaded since it presupposes a choice. Naval architecture is art and science.

Basically, naval architecture is concerned with ship safety, ship performance and ship geometry, although these are not exclusive divisions.

With ship safety, the naval architect is concerned that the ship does not capsize in a seaway, or when damaged or even when maltreated. He must ensure that the ship is sufficiently strong so that it does not break up or fracture locally to let the water in. He must ensure that the crew has a good chance of survival if the ship does let water in through accident or enemy action.

The performance of the ship is dictated by the needs of trade or war. The required amount of cargo must be carried to the places which the owner specifies in the right condition and in the most economical manner; the warship must carry the maximum hitting power of the right sort and an efficient crew to the remote parts of the world. Size, tonnage, deadweight, endurance, speed, life, resistance, methods of propulsion, manoeuvrability and many other features must be matched to provide the right primary performance at the right cost. Over 90 per cent of the world's trade is still carried by sea.

Ship geometry concerns the correct interrelation of compartments which the architect of a house considers on a smaller scale. In an aircraft carrier, the naval architect has 2000 rooms to relate, one with another. He must provide up to fifty different piping and ducting systems to all parts of the ship. He must provide comfort for the crew and facilities to enable each member to perform his correct function. The ship must load and unload in harbour with the utmost speed and perhaps replenish at sea. The architecture of the ship must be such that it can be economically built, and aesthetically pleasing. The naval architect is being held increasingly responsible for ensuring that the environmental impact of his product is minimal both in normal operation and following any foreseeable accident. He has a duty to the public at large for the safety of marine transport. In common with other professionals the naval architect is expected to abide by a stringent code of conduct (Ref. 14).

It must be clear that naval architecture involves complex compromises of many of these features. The art is, perhaps, the blending in the right proportions. There can be few other pursuits which draw on such a variety of sciences to blend them into an acceptable whole. There can be few pursuits as fascinating.

SHIPS

Ships are designed to meet the requirements of owners or of war and their features are dictated by these requirements. The purpose of a merchant ship has been described (Ref. 5) as conveying passengers or cargo from one port to another in the most efficient manner. This was interpreted by the owners of *Cutty Sark* as the conveyance of relatively small quantities of tea in the shortest possible time, because this was what the tea market demanded at that time. The market might well have required twice the quantity of tea per voyage in a voyage of twice the length of time, when a fundamentally different design of ship would have resulted. The economics of any particular market have a profound effect on merchant ship design. Thus, the change in the oil market following the second world war resulted in the disappearance of the 12,000 tonf deadweight tankers and the appearance of the 400,000 tonf deadweight supertankers. The economics of the trading of the ship itself have an effect on its design; the desire, for example, for small tonnage (and therefore small harbour dues) with large cargo-carrying capacity brought about the three island and shelter deck ships where cargo could be stowed in spaces not counted towards the tonnage on which insurance rates and harbour dues were based. Such trends have not always been compatible with safety and requirements of safety now also vitally influence ship design. Specialized demands of trade have produced the great passenger liners and bulk carriers, the natural-gas carriers, the trawlers and many other interesting ships. Indeed, the trend is towards more and more specialization in merchant ship design (see Chapter 16).

Specialization applies equally to warships. Basically, the warship is designed to meet a country's defence policy. Because the design and building of warships takes several years, it is an advantage if a particular defence policy persists for at least ten years and the task of long term defence planning is an onerous and responsible one. The Defence Staff interprets the general Government policy into the needs for meeting particular threats in particular parts of the world and the scientists and technologists produce weapons for defensive and offensive use. The naval architect then designs ships to carry the weapons and the men to use them to the correct part of the world. Thus, nations, like Britain and the USA with commitments the other side of the world, would be expected to expend more of the available space in their ships on facilities for getting the weapons and crew in a satisfactory state to a remote, perhaps hot, area than a nation which expects to make short harrying excursions from its home ports. It is important, therefore, to regard the ship as a complete weapon system and weapon and ship designers must work in the closest possible contact.

Nowhere, probably, was this more important than in the aircraft carrier. The type of aircraft carried so vitally affects an aircraft carrier that the ship was virtually designed around it; only by exceeding all the minimum demands of an aircraft and producing monster carriers, can any appreciable degree of flexibility be introduced. The guided missile destroyer results directly from the Defence Staff's assessment of likely enemy aircraft and guided weapons and their concept of how and where they would be used; submarine design is profoundly affected by diving depth and weapon systems which are determined by offensive and defensive considerations. The invention of the torpedo led to the motor torpedo boat which in turn led to the torpedo boat destroyer; the submarine, as an alternative carrier of the torpedo, led to the design of the anti-submarine frigate; the missile carrying nuclear submarine led to the hunter killer nuclear submarine. Thus, the particular demand of war, as is natural, produces a particular warship. Reference 6 gives a valuable historical survey while the host of specialized warships that arose from the second world war are described in Ref. 7.

Particular demands of the sea have resulted in many other interesting and important ships: the self-righting lifeboats, surface effect vessels, container ships, cargo drones, hydrofoil craft and a host of others. All are governed by the basic rules and tools of naval architecture which this book seeks to explore. Precision in the use of these tools must continue to be inspired by knowledge of sea disasters (Ref. 18), Liberty ships of the second world war (Ref. 8), the loss of the *Royal George*, the loss of HMS *Captain* (Ref. 9), and the loss of the *Vasa*:

> In 1628, the *Vasa* set out on a maiden voyage which lasted little more than two hours. She sank in good weather through capsizing while still in view of the people of Stockholm. (Ref. 10.)

Authorities

CLASSIFICATION SOCIETIES

The authorities with the most profound influence on shipbuilding, merchant ship design and ship safety are the classification societies. Among the most important are Lloyd's Register of Shipping, det Norske Veritas, the American Bureau of Shipping, Bureau Veritas, Registro Italiano, Germanische Lloyd and Nippon Kaiji Kyokai. These meet to discuss standards under the auspices of the International Association of Classification Societies (IACS).

It is odd that the two most influential bodies in the shipbuilding and shipping industries should both derive their names from the same owner of a coffee shop, Edward Lloyd, at the end of the seventeenth century. Yet the two organizations are entirely independent with quite separate histories. Lloyd's Insurance Corporation is concerned with mercantile and other insurance. Lloyd's Register of Shipping is concerned with the maintenance of proper technical standards in ship construction and the classification of ships, i.e. the record of all relevant technical details and the assurance that these meet the required standards. Vessels so registered with Lloyd's Register are said to be classed with the Society and may be awarded the classification ✠ 100 A1. The cross denotes that the ship has been built under the supervision of surveyors from Lloyd's Register while 100 A shows that the vessel is built in accordance with the recommended standards. The final 1 indicates that the safety equipment, anchors and cables are as required. Other provisos to the classification are often added.

The maintenance of these standards is an important function of Lloyd's Register who require surveys of a specified thoroughness at stated intervals by the Society's surveyors. Failure to conform may result in removal of the ship from class and a consequent reduction in its value. The total impartiality of the Society is its great strength. It is also empowered to allot load line (see Chapter 5) certificates to ships, to ensure that they are adhered to and, as agents for certain foreign governments, to assess tonnage measurement and to ensure compliance with safety regulations. Over 1000 surveyors, scattered all over the world, carry out the required surveys, reporting to headquarters in London or other national centres where the classification of the ships are considered.

The standards to which the ships must be built and maintained are laid down in the first of the two major publications of Lloyd's Register, *Rules and Regulations for the Classification of Ships*. This is issued annually and kept up to date to meet new demands. The other major publication is the *Register Book* in several volumes, which lists every known ship, whether classed with the Society or not, together with all of its important technical particulars. Separate books appear for the building and classification of yachts and there are many other publications to assist surveyors.

GOVERNMENT BODIES

The Department of Transport is the United Kingdom Government's authority for declaring the standards of safety for merchant ships, related to damage, collision, subdivision, life-saving equipment, loading, stability, fire protection, navigation, carriage of dangerous goods, load line standards and many allied subjects. This department is also the authority on tonnage measurement standards. It is responsible for seeing that the safety standards, many of which are governed by international agreements, are maintained (through the Marine Safety Agency) and for enquiring into sea disasters (through the Marine Accident Investigation Branch). Ship surveyors in the Marine Division and similar national authorities in other countries, like the US Coastguards, carry, thus, an enormous responsibility.

The Ministry of Defence (formerly the Admiralty) is responsible for the procurement of its own ships. This highly specialized task is performed by members of the Royal Corps of Naval Constructors, a civilian body formed in 1883 for these purposes. The need for such a body arises from the different design philosophy associated with merchant ships and warships. The owner of the merchant ship wants to make money from his capital investment and requires therefore a short building time, the minimum of maintenance, the maximum time at sea and simplicity of design. These factors render the construction of robust, simple ships possible by rule. The nation, however, gets no return for *its* capital investment save protection from potential enemies. While attempting to keep costs within reason, the constructor's prime task is to enable industry to produce an efficient maintainable ship-weapon system, neglecting no aspect which might put the warship at a disadvantage in the face of an enemy. This means that each warship has to be tailor made.

INTERNATIONAL BODIES

The International Maritime Organization (IMO), represents some 150 of the maritime nations of the world. The organization sponsors international action with a view to improving and standardizing questions relating to ship safety and measurement. It sponsors the International Conventions on Safety of Life at Sea which agree to the application of new standards of safety. The same organization sponsors, also, international conferences on the load line and standardizing action on tonnage measurement and many other maritime problems.

LEARNED SOCIETIES

The Institution of Naval Architects was formed in 1860 when interest in the subject in Britain quickened and it has contributed much to the development of naval architecture (Ref. 4). It became the Royal Institution of Naval Architects in 1960. Abroad, among the many societies worthy of mention are the Association Technique Maritime et Aèronautique in France, the Society of Naval Architects and Marine Engineers in the USA and the Society of Naval Architects of Japan.

References

1. Stoot, W. F. Some aspects of naval architecture in the eighteenth century, *TINA*, 1959.
2. Stoot, W. F. Ideas and personalities in the development of naval architecture, *TINA*, 1959.
3. Wheatley, H. B. (Ed.). *The diary of Samuel Pepys*, Bell, 1952.
4. Barnaby, K. C. *The Institution of Naval Architects—an historical survey*, Allen and Unwin, 1960. Also Newton, R. N. *Ibid*, 1960–80.
5. Murray, J. M. Merchant Ships 1860–1960, *TRINA*, 1960.
6. Sims, A. J. Warships 1860–1960, *TRÌNA*, 1960.
7. Various papers, *TINA*, 1947.
8. McCallum, J. The World Concord—A Case History, *TRINA*, 1981.
9. Hawkey, A. *HMS Captain*, Bell, 1963.
10. Bengt, Ohrelius. *Vasa, the King's ship*. Translated by M. Michael, Cassell, 1962.
11. Corlett, E. C. B. The Steamship Great Britain, *TRINA*, 1971.
12. Coates, J. F. The Naval Architecture of European Oared Ships, *TRINA*, 1993.
13. Brown, D. K. History as a Design Tool, *TRINA*, 1992.
14. Engineers and Risk Issues; Code of Professional Practice, Engineering Council 1993.
15. Rawson, K. J. Ethics and Fashion in Design, *TRINA*, 1989.
16. Safety Aspects of Ship Design and Technology, House of Lords' Report Paper No 30, 1992.
17. Buxton, I., The Development of the Merchant Ship 1880–1990, *Mariner's Mirror*, Feb. 1993.
18. Lloyd's Register of Shipping Statistical Tables. Annually.

2 Some tools

No occupation can be properly developed without tools, whether it be gardening or naval architecture or astro-navigation. Many tools needed for the study of naval architecture are already to hand, provided by mathematics, applied mechanics and physics and it will be necessary to assume as the book progresses that knowledge in all allied subjects has also progressed. Knowledge, for example, of elementary differential and integral calculus is assumed to be developing concurrently with this chapter. Moreover, the tools need to be sharp; definitions must be precise, while the devices adopted from mathematics must be pointed in such a way as to bear directly on ship shapes and problems. As a means of examining this science, these are the tools.

It is convenient too, to adopt a terminology or particular language and a shorthand for many of the devices to be used. In this chapter, which lays a firm foundation from which to build up the subject, some of the machines of use to the naval architect are examined. Finally, there are short notes on statistics and approximate formulae.

Basic geometric concepts

The main parts of a typical ship together with the terms applied to the principal parts are illustrated in Fig. 2.1. Because, at first, they are of little interest or influence, superstructures and deckhouses are ignored and the hull of the ship is considered as a hollow body curved in all directions, surmounted by a watertight deck. Most ships have only one plane of symmetry, called the *middle line plane* which becomes the principal plane of reference. The shape of the ship cut by this plane is known as the sheer plan or profile. The *design waterplane* is a plane perpendicular to the middle line plane, chosen as a plane of reference at

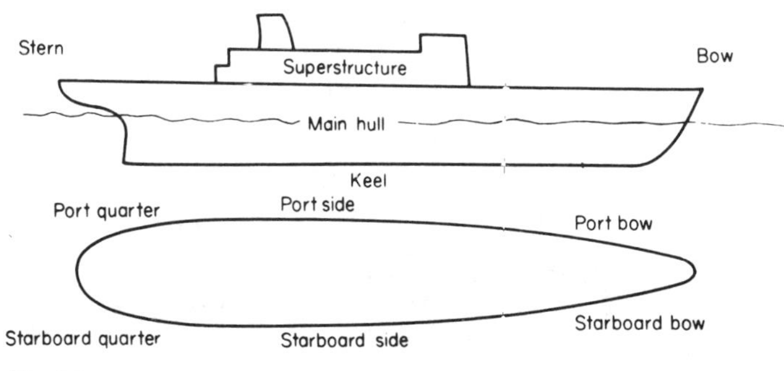

Fig. 2.1

or near the horizontal: it may or may not be parallel to the keel. Planes perpendicular to both the middle line plane and the design waterplane are called *transverse planes* and a transverse section of the ship does, normally, exhibit symmetry about the middle line. Planes at right angles to the middle line plane, and parallel to the design waterplane are called *waterplanes*, whether they are in the water or not, and they are usually symmetrical about the middle line. Waterplanes are not necessarily parallel to the keel. Thus, the curved shape of a ship is best conveyed to our minds by its sections cut by orthogonal planes. Figure 2.2 illustrates these planes.

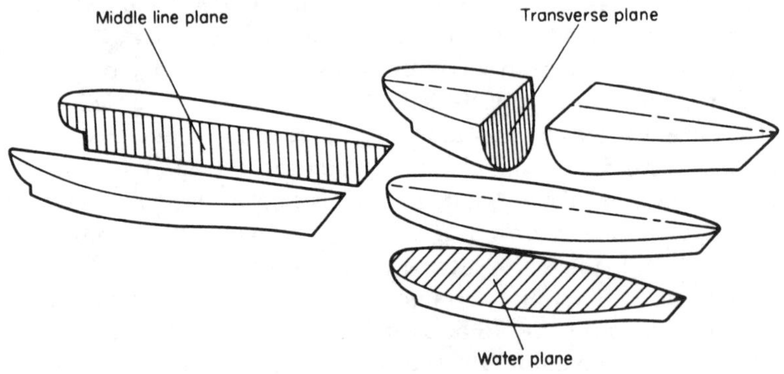

Fig. 2.2

Transverse sections laid one on top of the other form a *body plan* which, by convention, when the sections are symmetrical, shows only half sections, the forward half sections on the right-hand side of the middle line and the after half sections on the left. Half waterplanes placed one on top of the other form a *half breadth plan.* Waterplanes looked at edge on in the sheer or body plan are called *waterlines.* The sheer, the body plan and the half breadth collectively are called the *lines plan* or *sheer drawing* and the three constituents are clearly related. (See Fig. 2.3.)

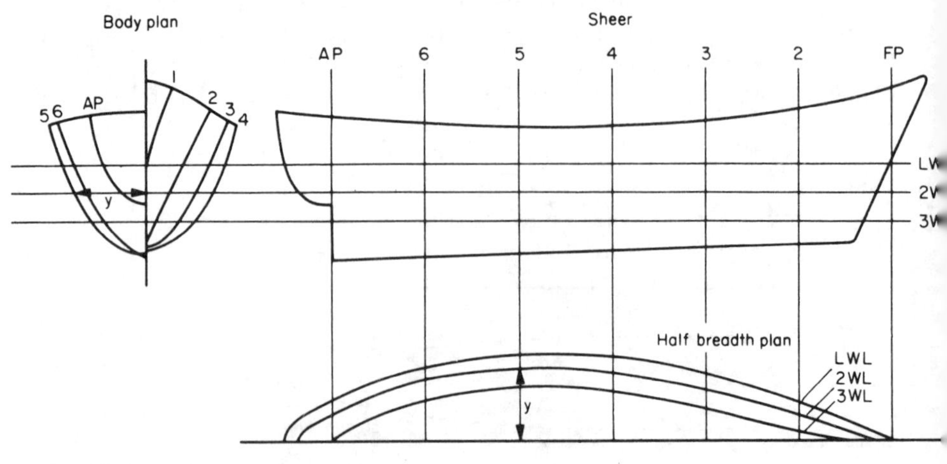

Fig. 2.3 Lines plan

It is convenient if the waterplanes and the transverse planes are equally spaced and datum points are needed to start from. That waterplane to which the ship is being designed is called the *load waterplane* (LWP) or *design waterplane* and additional waterplanes for examining the ship's shape are drawn above it and below it, equally spaced, usually leaving an uneven slice near the keel which is best examined separately.

A reference point at the fore end of the ship is provided by the intersection of the load waterline and the stem contour and the line perpendicular to the LWP through this point is called the *fore perpendicular* (FP). It does not matter where the perpendiculars are, provided that they are precise and fixed for the ship's life, that they embrace most of the underwater portion and that there are no serious discontinuities between them. The *after perpendicular* (AP) is frequently taken through the axis of the rudder stock or the intersection of the LWL and transom profile. If the point is sharp enough, it is sometimes better taken at the after cut up or at a place in the vicinity where there is a discontinuity in the ship's shape. The distance between these two convenient reference lines is called the *length between perpendiculars* (LBP or L_{PP}). Two other lengths which will be referred to and which need no further explanation are the *length overall* and the *length on the waterline*.

The distance between perpendiculars is divided into a convenient number of equal spaces, often twenty, to give, including the FP and the AP, twenty-one evenly spaced ordinates. These ordinates are, of course, the edges of transverse planes looked at in the sheer or half breadth and have the shapes half shown in the body plan. Ordinates can also define any set of evenly spaced reference lines drawn on an irregular shape. The distance from the middle line plane along an ordinate in the half breadth is called an *offset* and this distance appears again in the body plan where it is viewed from a different direction. All such distances for all waterplanes and all ordinates form a *table of offsets* which defines the shape of the hull and from which a lines plan can be drawn. A simple table of offsets is used in Fig. 3.30 to calculate the geometric particulars of the form.

A reference plane is needed about mid-length of the ship and, not unnaturally, the transverse plane midway between the perpendiculars is chosen. It is called *amidships* or *midships* and the section of the ship by this plane is the midship section. It may not be the largest section and it should have no significance other than its position halfway between the perpendiculars. Its position is usually defined by the symbol ⌽.

The shape, lines, offsets and dimensions of primary interest to the theory of naval architecture are those which are wetted by the sea and are called *displacement* lines, ordinates, offsets, etc. Unless otherwise stated, this book refers normally to displacement dimensions. Those which are of interest to the shipbuilder are the lines of the frames which differ from the displacement lines by the thickness of hull plating or more, according to how the ship is built. These are called *moulded* dimensions. Definitions of displacement dimensions are similar to those which follow but will differ by plating thicknesses.

The *moulded draught* is the perpendicular distance in a transverse plane from the top of the flat keel to the design waterline. If unspecified, it refers to

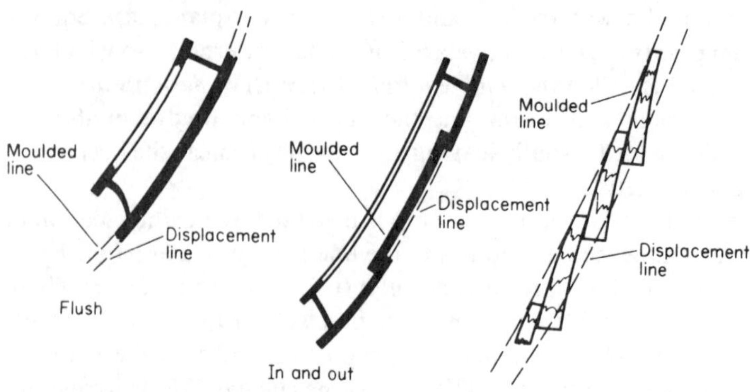

Fig. 2.4 Moulded and displacement lines

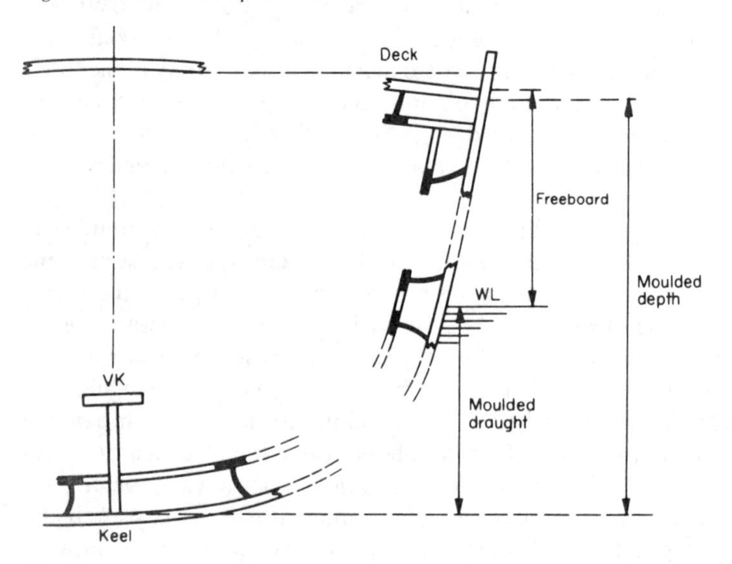

Fig. 2.5

amidships. The draught amidships is the mean draught unless the mean draught is referred directly to draught mark readings.

The *moulded depth* is the perpendicular distance in a transverse plane from the top of the flat keel to the underside of deck plating at the ship's side. If unspecified, it refers to this dimension amidships.

Freeboard is the difference between the depth at side and the draught. It is the perpendicular distance in a transverse plane from the waterline to the upperside of the deck plating at side.

The *moulded breadth extreme* is the maximum horizontal breadth of any frame section. The terms breadth and beam are synonymous.

Certain other geometric concepts of varying precision will be found useful in defining the shape of the hull. *Rise of floor* is the distance above the keel that a tangent to the bottom at or near the keel cuts the line of maximum beam amidships. See Fig. 2.6.

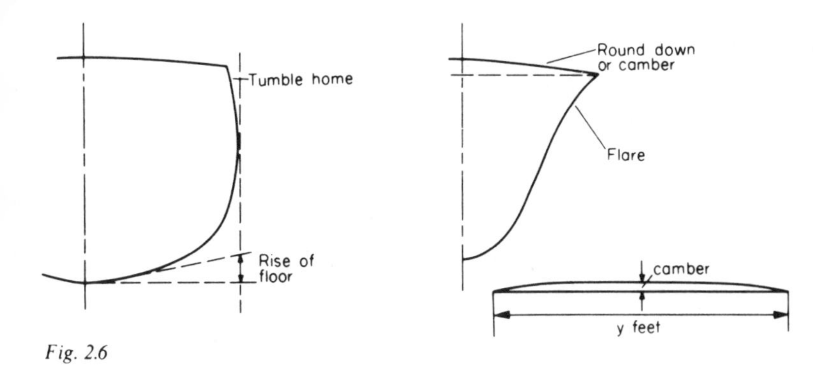

Fig. 2.6

Tumble home is the tendency of a section to fall in towards the middle line plane from the vertical as it approaches the deck edge. The opposite tendency is called *flare*. See Fig. 2.6.

Deck camber or *round down* is the curve applied to a deck transversely. It is normally concave downwards, a parabolic or circular curve, and measured as x inches in y feet.

Sheer is the tendency of a deck to rise above the horizontal in profile.

Rake is the departure from the vertical of any conspicuous line in profile such as a funnel, mast, stem contour, superstructure, etc. (Fig. 2.7).

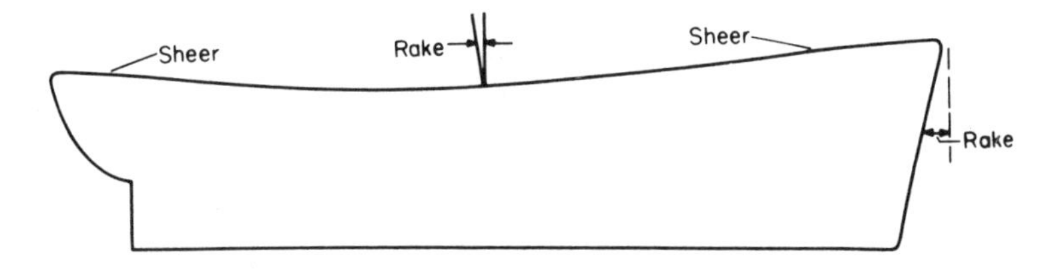

Fig. 2.7

There are special words applied to the angular movements of the whole ship from equilibrium conditions. Angular bodily movement from the vertical in a transverse plane is called *heel*. Angular bodily movement in the middle line plane is called *trim*. Angular disturbance from the mean course of a ship in the horizontal plane is called *yaw* or *drift*. Note that these are all angles and not rates, which are considered in later chapters.

There are two curves which can be derived from the offsets which define the shape of the hull by areas instead of distances which will later prove of great value. By erecting a height proportional to the area of each ordinate up to the LWP at each ordinate station on a horizontal axis, a curve is obtained known as the *curve of areas*. Figure 2.8 shows such a curve with number 4 ordinate, taken as an example. The height of the curve of areas at number 4 ordinate represents the area of number 4 ordinate section; the height at number 5 is proportional to the area of number 5 section and so on. A second type of area curve can be obtained by examining each ordinate section. Figure 2.8 again takes 4 ordinate

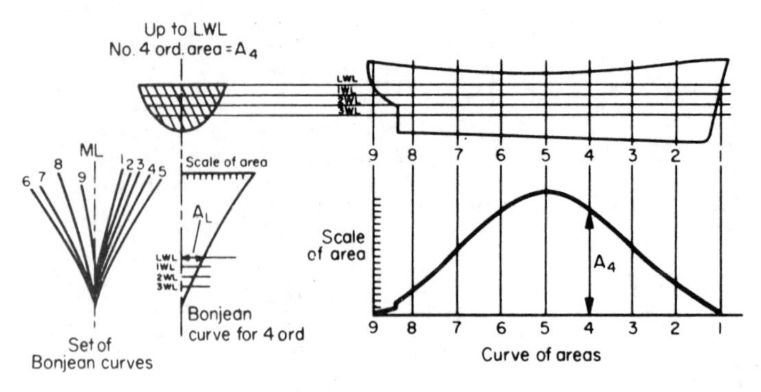

Fig. 2.8

section as an example. Plotting outwards from a vertical axis, distances corresponding to the areas of a section up to each waterline, a curve known as a *Bonjean curve* is obtained. Thus, the distance outwards at the LWL is proportional to the area of the section up to the LWL, the distance outwards at 1WL is proportional to the area of section up to 1WL and so on. Clearly, a Bonjean curve can be drawn for each section and a set produced.

The *volume of displacement*, ∇, is the total volume of fluid displaced by the ship. It is best conceived by imagining the fluid to be wax and the ship removed from it; it is then the volume of the impression left by the hull. For convenience of calculation, it is the addition of the volumes of the main body and appendages such as the slices at the keel, abaft the AP, rudder, bilge keels, propellers, etc., with subtractions for condensor inlets and other holes.

Finally, in the definition of hull geometry there are certain coefficients which will later prove of value as guides to the fatness or slimness of the hull.

The *coefficient of fineness of waterplane*, C_{WP}, is the ratio of the area of the waterplane to the area of its circumscribing rectangle. It varies from about 0·70 for ships with unusually fine ends to about 0·90 for ships with much parallel middle body.

$$C_{WP} = \frac{A_W}{L_{WL}B}$$

The *midship section coefficient*, C_M, is the ratio of the midship section area to the area of a rectangle whose sides are equal to the draught and the breadth extreme amidships. Its value usually exceeds 0·85 for ships other than yachts.

$$C_M = \frac{A_M}{BT}$$

The *block coefficient*, C_B, is the ratio of the volume of displacement to the volume of a rectangular block whose sides are equal to the breadth extreme, the mean draught and the length between perpendiculars.

$$C_B = \frac{\nabla}{BTL_{PP}}$$

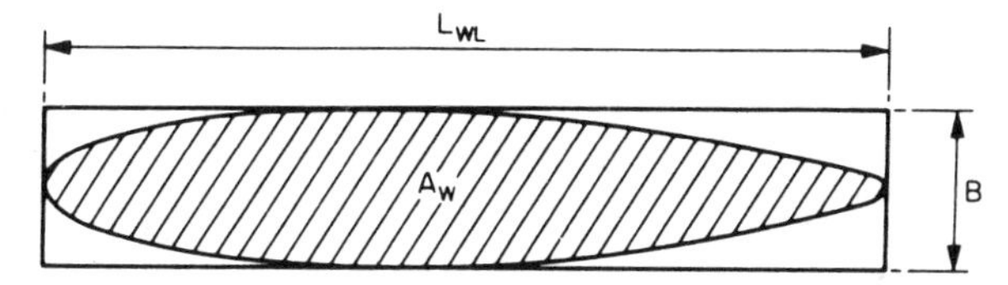

Fig. 2.9 Waterplane coefficient

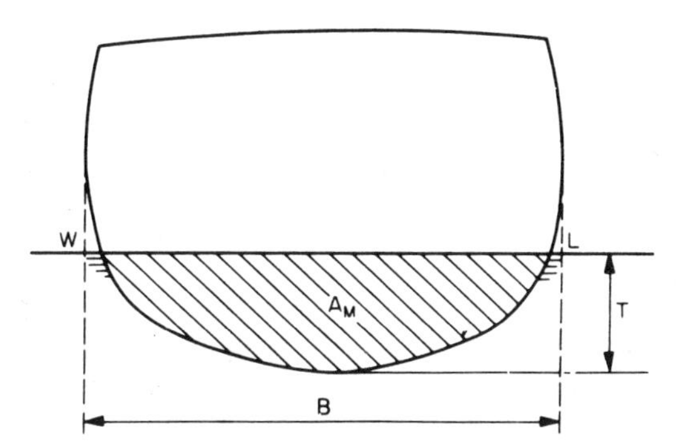

Fig. 2.10 Midship coefficient

Mean values of block coefficient might be 0·88 for a large oil tanker, 0·60 for an aircraft carrier and 0·50 for a yacht form.

The *longitudinal prismatic coefficient*, C_P, or simply *prismatic coefficient* is the ratio of the volume of displacement to the volume of a prism having a length equal to the length between perpendiculars and a cross-sectional area equal to the midship sectional area. Expected values generally exceed 0·55.

$$C_P = \frac{\nabla}{A_M L_{PP}}$$

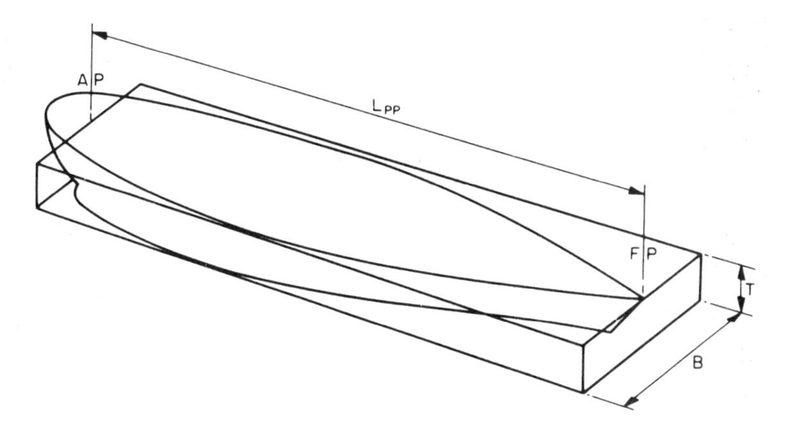

Fig. 2.11 Block coefficient

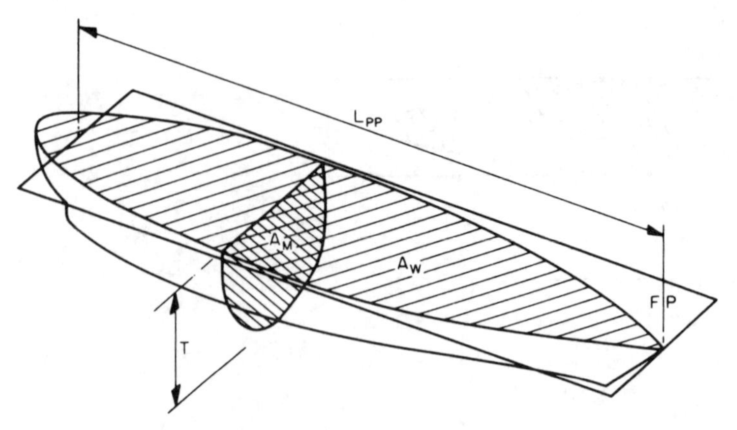

Fig. 2.12 Longitudinal prismatic coefficient

The *vertical prismatic coefficient*, C_{VP} is the ratio of the volume of displacement to the volume of a prism having a length equal to the draught and a cross-sectional area equal to the waterplane area.

$$C_{VP} = \frac{\nabla}{A_W T}$$

Before leaving these coefficients for the time being, it should be observed that the definitions above have used displacement and not moulded dimensions because it is generally in the very early stages of design that these are of interest. Practice in this respect varies a good deal. Where the difference is significant, as for example in the structural design of tankers by Lloyd's Rules, care should be taken to check the definition in use. It should also be noted that the values of the various coefficients depend on the positions adopted for the perpendiculars.

Properties of irregular shapes

Now that the geometry of the ship has been defined, it is necessary to anticipate what properties of these shapes are going to be useful and find out how to calculate them.

PLANE SHAPES

Waterplanes, transverse sections, flat decks, bulkheads, the curve of areas and expansions of curved surfaces are some of the plane shapes whose properties are of interest. The area of a surface in the plane of Oxy defined in Cartesian coordinates, is

$$A = \int y \, dx$$

in which all strips of length y and width δx are summed over the total extent of x. Because y is rarely, with ship shapes, a precise mathematical function of x the integration must be carried out by an approximate method which will presently be deduced.

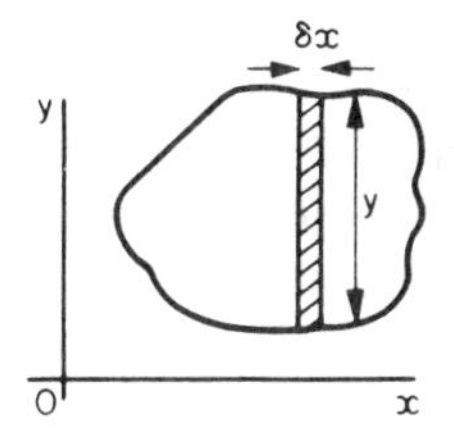

Fig. 2.13

There are first moments of area about each axis. (For the figures shown in Fig. 2.14, x_1 and y_1 are lengths and x and y are co-ordinates.)

$$M_{yy} = \int xy_1\,dx \quad \text{and} \quad M_{xx} = \int x_1 y\,dy$$

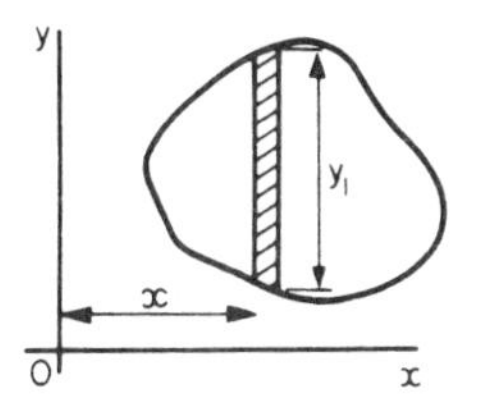

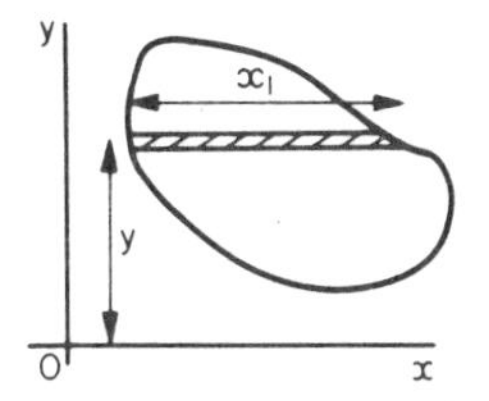

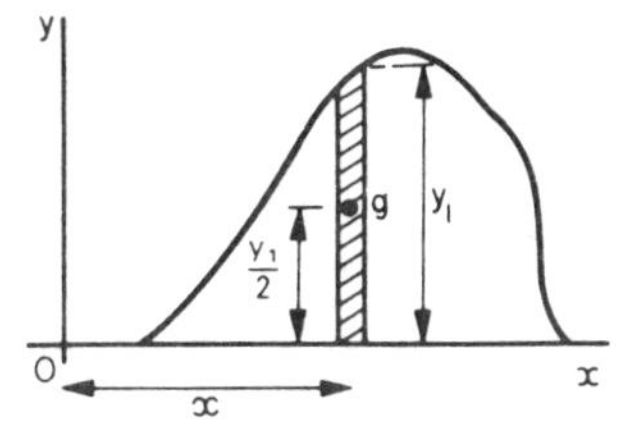

Fig. 2.14

Dividing each expression by the area gives the co-ordinates of the centre of area, $(\bar{x}, \bar{y})$:

$$\bar{x} = \frac{1}{A}\int xy_1\,dx \quad \text{and} \quad \bar{y} = \frac{1}{A}\int x_1 y\,dy$$

For the particular case of a figure bounded on one edge by the x-axis

$$M_y^* = \int \frac{1}{2}y^2\,dx \quad \text{and} \quad \bar{y} = \frac{1}{2A}\int y_1^2\,dx$$

For a plane figure placed symmetrically about the x-axis such as a waterplane, $M_{xx} = \int x_1 y\,dy = 0$ and the distance of the centre of area, called in the particular case of a waterplane, the *centre of flotation* (CF), from the y-axis is given by

$$\bar{x} = \frac{M_{yy}}{A} = \frac{\int xy_1\,dx}{\int y_1 dx}$$

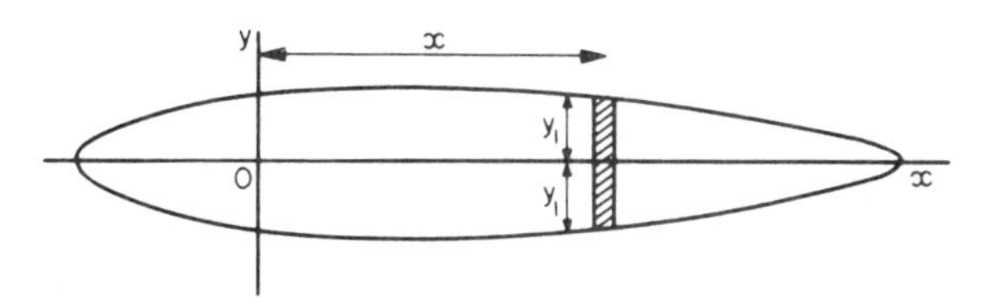

Fig. 2.15

* *Note that* $M_y \equiv M_{xx}$.

It is convenient to examine such a symmetrical figure in relation to the second moment of area, since it is normally possible to simplify work by choosing one symmetrical axis for ship shapes. The second moments of area or moments of inertia about the two axes for the waterplane shown in Fig. 2.15 are given by

$$I_T = \frac{1}{3}\int y_1^3 \, dx \quad \text{about Ox for each half}$$

$$I_{yy} = \int x^2 y_1 \, dx \quad \text{about Oy for each half}$$

The parallel axis theorem shows that the second moment of area of a plane figure about any axis, Q, of a set of parallel axes is least when that axis passes through the centre of area and that the second moment of area about any other axis, R, parallel to Q and at a distance h from it is given by (Fig. 2.16),

$$I_R = I_Q + Ah^2$$

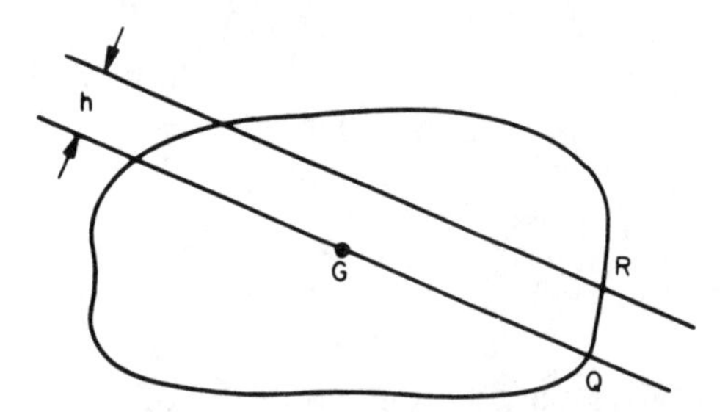

Fig. 2.16

From this, it follows that the least longitudinal second moment of area of a waterplane is that about an axis through the centre of flotation and given by (Fig. 2.17)

$$I_L = I_{yy} - A\bar{x}^2$$

i.e.

$$I_L = \int x^2 y_1 \, dx - A\bar{x}^2$$

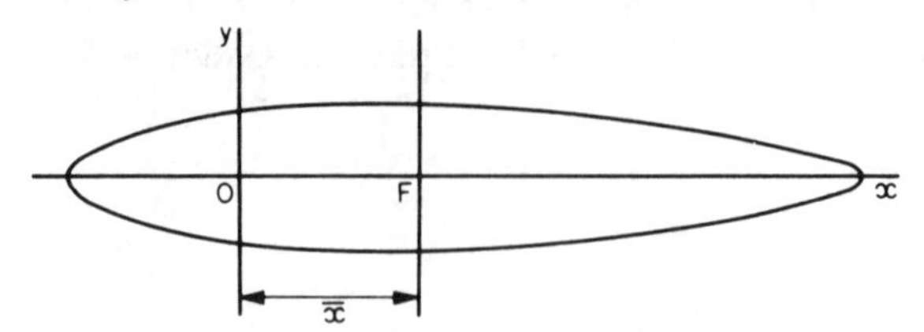

Fig. 2.17

THREE-DIMENSIONAL SHAPES

It has already been shown how to represent the three-dimensional shape of the ship by a plane shape, the curve of areas, by representing each section area by a length (Fig. 2.8). This is one convenient way to represent the three-dimensional shape of the main underwater form (less appendages). The volume of displacement is given by

$$\nabla = \int_{x_1}^{x_2} A \, dx$$

i.e. it is the sum of all such slices of cross-sectional area A over the total extent of x (Fig. 2.18).

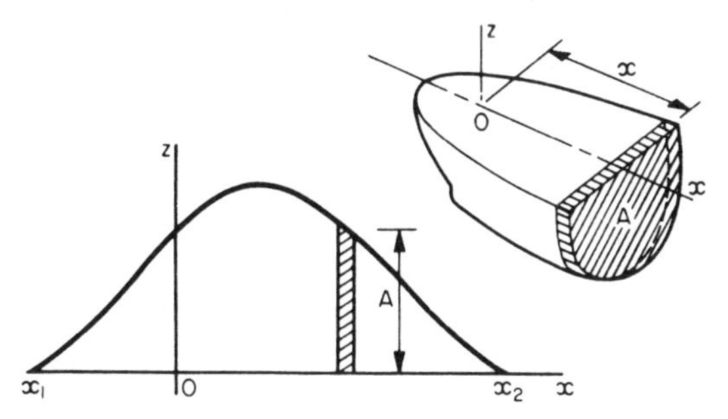

Fig. 2.18

The shape of the ship can equally be represented by a curve of waterplane areas on a vertical axis (Fig. 2.19), the breadth of the curve at any height, z, above the keel representing the area of the waterplane at that draught. The volume of displacement is again the sum of all such slices of cross-sectional area A_W, over the total extent of z from zero to draught T,

$$\nabla = \int_0^T A_W \, dz$$

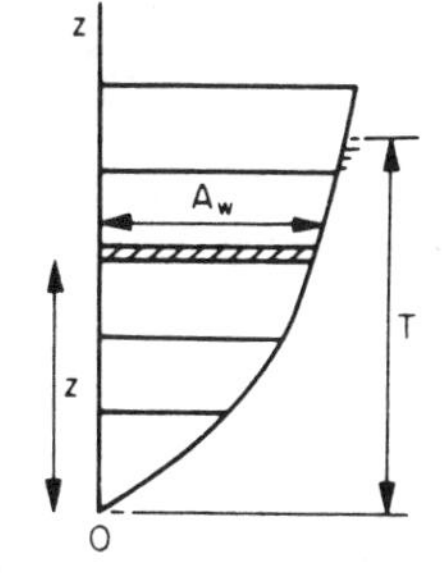

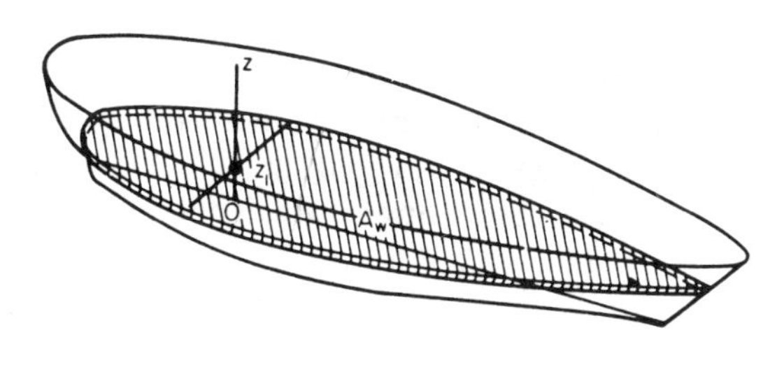

Fig. 2.19

The first moments of volume in the longitudinal direction about Oz and in the vertical direction about the keel are given by

$$M_L = \int Ax \, dx \quad \text{and} \quad M_V = \int_0^T A_W z \, dz$$

Dividing by the volume in each case gives the co-ordinates of the centre of volume. The centre of volume of fluid displaced by a ship is known as the *centre of buoyancy*; its projections in the plan and in section are known as the longitudinal centre of buoyancy (LCB) and the vertical centre of buoyancy (VCB)

$$\text{LCB from O}y = \frac{1}{\nabla}\int Ax\,\mathrm{d}x$$

$$\text{VCB above keel} = \frac{1}{\nabla}\int A_{\mathrm{W}}z\,\mathrm{d}z$$

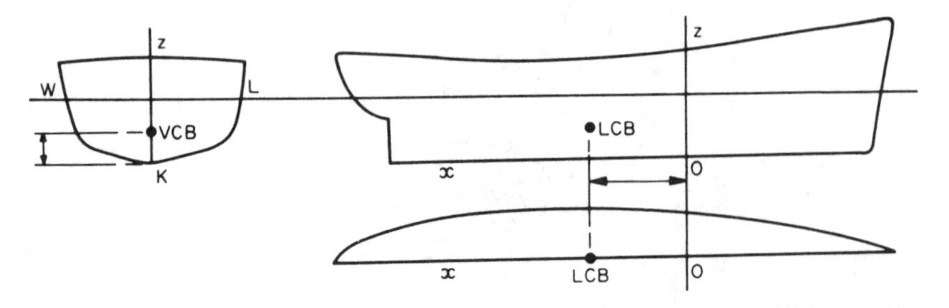

Fig. 2.20 Centre of buoyancy projections

Should the ship not be symmetrical below the waterline, the centre of buoyancy will not lie in the middle line plane. Its projection in plan may then be referred to as the transverse centre of buoyancy (TCB). Had z been taken as the distance below the waterline, the second expression would, of course, represent the position of the VCB below the waterline. Defining it formally, the *centre of buoyancy* of a floating body is the centre of volume of the displaced fluid in which the body is floating. The first moment of volume about the centre of volume is zero.

The *weight* of a body is the total of the weights of all of its constituent parts. First moments of the weights about particular axes divided by the total weight, define the co-ordinates of the centre of weight or centre of gravity (CG) relative to those axes. Projections of the centre of gravity of a ship in plan and in section are known as the longitudinal centre of gravity (LCG) and vertical centre of gravity (VCG) and transverse centre of gravity (TCG).

$$\text{LCG from O}y = \frac{1}{W}\int x\,\mathrm{d}W$$

$$\text{VCG above keel} = \frac{1}{W}\int z\,\mathrm{d}W$$

$$\text{TCG from middle line plane} = \frac{1}{W}\int y\,\mathrm{d}W$$

Defining it formally, the *centre of gravity* of a body is that point through which, for statical considerations, the whole weight of the body may be assumed to act. The first moment of weight about the centre of gravity is zero.

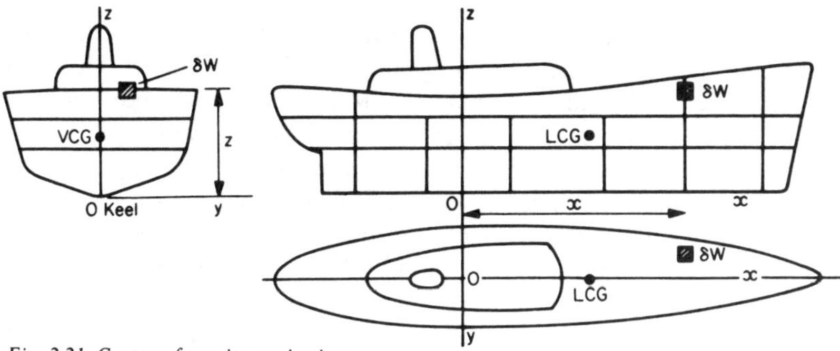

Fig. 2.21 Centre of gravity projections

METACENTRE

Consider any body floating upright and freely at waterline WL, whose centre of buoyancy is at B. Let the body now be rotated through a small angle in the plane of the paper without altering the volume of displacement (it is more convenient to draw if the body is assumed fixed and the waterline rotated to W_1L_1). The centre of buoyancy for this new immersed shape is at B_1. Lines through B and B_1 normal to their respective waterlines intersect at M which is known as the *metacentre* since it appears as if the body rotates about it for small angles of rotation. The *metacentre* is the point of intersection of the normal to a slightly

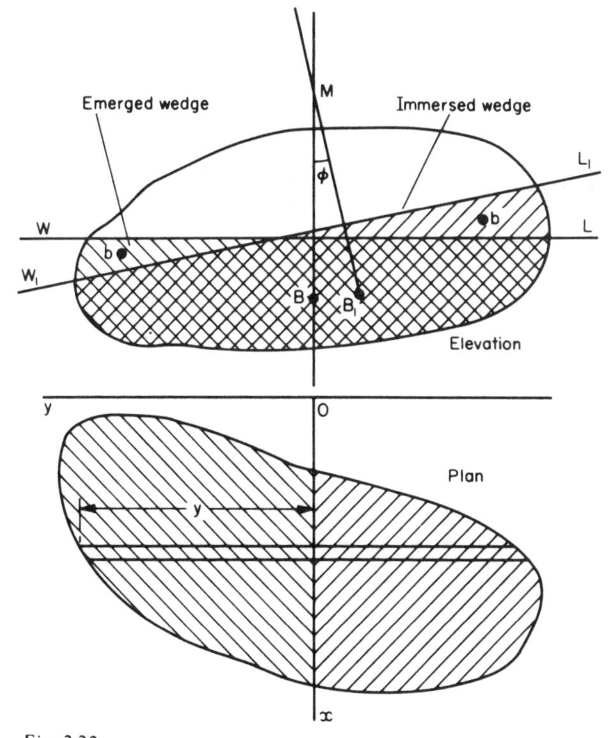

Fig. 2.22

inclined waterplane of a body, rotated without change of displacement, through the centre of buoyancy pertaining to that waterplane and the vertical plane through the centre of buoyancy pertaining to the upright condition. The term metacentre is reserved for small inclinations from an upright condition. The point of intersection of normals through the centres of buoyancy pertaining to successive waterplanes of a body rotated infinitesimally at any angle of inclination without change of displacement, is called the *pro-metacentre*.

If the body is rotated without change of displacement, the volume of the immersed wedge must be equal to the volume of the emerged wedge. Furthermore, the transfer of this volume from the emerged to the immersed side must be responsible for the movement of the centre of buoyancy of the whole body from B to B_1 ; from this we conclude:

(*a*) that the volumes of the two wedges must be equal
(*b*) that the first moments of the two wedges about their line of intersection must, for equilibrium, be equal and
(*c*) that the transfer of first moment of the wedges must equal the change in first moment of the whole body.

Writing down these observations in mathematical symbols,

$$\text{Volume of immersed wedge} = \int y \times \tfrac{1}{2} y\phi \, \mathrm{d}x$$
$$= \text{Volume of emerged wedge}$$
$$\text{1st moment of immersed wedge} = \int (\tfrac{1}{2} y^2 \phi) \times \tfrac{2}{3} y \, \mathrm{d}x$$
$$= \text{1st moment of emerged wedge}$$
$$\text{Transfer of 1st moment of wedges} = 2 \times \int \tfrac{1}{3} y^3 \phi \, \mathrm{d}x = \tfrac{2}{3}\phi \int y^3 \, \mathrm{d}x$$
$$\text{Transfer of 1st moment of whole body} = \nabla \times \overline{\mathrm{BB}'} = \nabla \, . \, \overline{\mathrm{BM}} \, . \, \phi$$
$$\therefore \quad \nabla \, . \, \overline{\mathrm{BM}} \, . \, \phi = \tfrac{2}{3}\phi \int y^3 \, \mathrm{d}x$$

But we have already seen that $I = \frac{2}{3} \int y^3 \, \mathrm{d}x$ about the axis of inclination for both half waterplanes

$$\therefore \quad \overline{\mathrm{BM}} = \frac{I}{\nabla}$$

This is an important geometric property of a floating body. If the floating body is a ship there are two $\overline{\mathrm{BM}}$s of particular interest, the transverse $\overline{\mathrm{BM}}$ for rotation about a fore-and-aft axis and the longitudinal $\overline{\mathrm{BM}}$ for rotation about a transverse axis, the two axes passing through the centre of flotation of the waterplane.

HOLLOW SHAPES

The hull of a ship is, of course, a hollow body enclosed by plating. It will be necessary to find the weight and positions of the centre of gravity of such shapes.

A pseudo-expansion of the shape is first obtained by a method described fully in a textbook on laying off. Briefly, the girths of section are plotted at each ordinate and increased in height by a factor to allow for the difference between projected and slant distances in plan. A mean value of this factor is found for each ordinate.

It is now necessary to apply to each ordinate a mean plating thickness which must be found by examining the plating thicknesses (or weights per unit area, sometimes called poundages) along the girth at each ordinate (Fig. 2.23). The variation is usually not great in girth and an arithmetic mean t' will be given by dividing the sum of each plate width × plate thickness by the girth. If the weight density of the material is w, the weight of the bottom plating is thus given by $W = w \int g't' \, dx$ and the position of the LCG is given by

$$\bar{x} = \frac{w}{W}\int g't'x \, dx$$

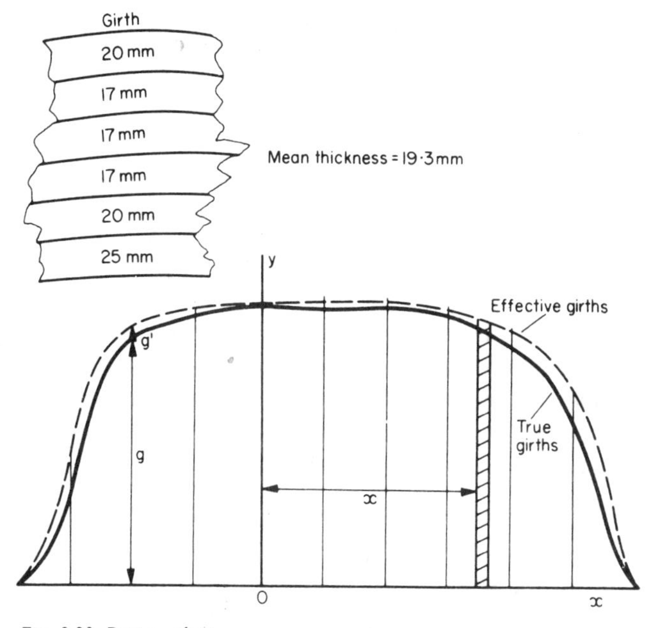

Fig. 2.23 Bottom plating

To find the position of the VCG, it is necessary to return to the sections and to find the position by drawing. Each section is divided by trial and error, into four lengths of equal weight. The mid-points of two adjacent sections are joined and the mid-points of these lines are joined. The mid-point of the resulting line is the required position of the c.g. and its height h above the keel measured. For the whole body, then, the position of the VCG above the keel is given by

$$\bar{h} = \frac{w}{W}\int g't'h \, dx$$

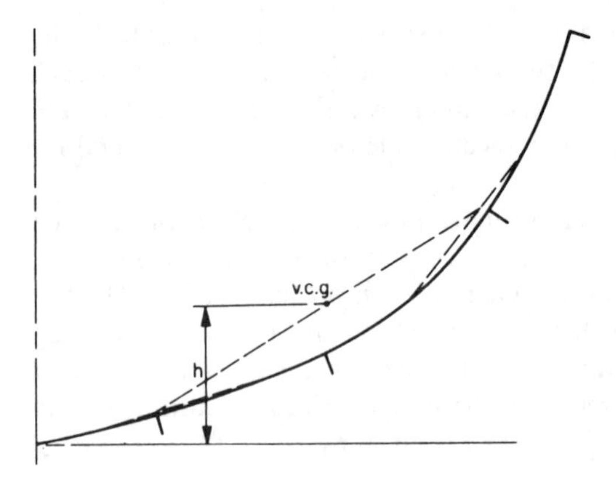

Fig. 2.24 VCG of bottom plating

Various factors can be applied to the weight density figure to account for different methods of construction. An allowance for the additional weight of laps, if the plating is raised and sunken or clinker, can be made; an addition can be made for rivet heads. It is unwise to apply any general rule and these factors must be calculated for each case.

SYMBOLS AND CONVENTIONS

It would be simpler if everyone used the same symbols for the same things. Various international bodies attempt to promote this and the symbols used in this book, listed at the beginning, follow the general agreements. As explained in the Introduction, industry, following a declaration of intent by the Government, is in the process of changing from the units given in British Standards 1991 to SI metric units. Force, for example, which by BS 1991 might be expressed in tonf will, by SI, be expressed in newtons or meganewtons. Industry will take time to adjust and this book uses both sets of units with a view to encouraging a facility in both. The symbols and units associated with hydrodynamics are those agreed by the International Towing Tank Conference.

Approximate integration

A number of different properties of particular interest to the naval architect have been expressed as simple integrals because this is a convenient form of shorthand. It is not necessary to be familiar with the integral calculus, however, beyond understanding that the elongated S sign, $\int$, means the sum of all such typical parts that follow the sign over the extent of whatever follows d. Some textbooks at this stage would use the symbol Σ which is simply the Greek letter S. It is now necessary to adopt various rules for calculating these integrals. The obvious way to calculate $A = \int y\,\mathrm{d}x$ is to plot the curve on squared paper and then count up all the small squares. This could be extended to calculate the first moment, $M = \int xy\,\mathrm{d}x$, in which the number of squares in a column, y, is multiplied by the number of squares from the origin, x, and this added for all of the columns that go to make up the shape. Clearly, this soon becomes laborious

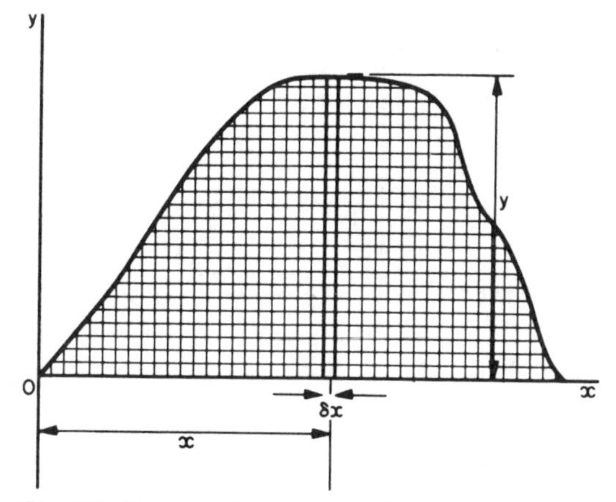

Fig. 2.25 The squared paper approach

and other means of determining the value of an integral must be found. The integral calculus will be used to deduce some of the rules but those who are not yet sufficiently familiar with that subject—and indeed, by those who are—they should be regarded merely as tools for calculating the various expressions shown above in mathematical shorthand.

TRAPEZOIDAL RULE

A trapezoid is a plane four-sided figure having two sides parallel. If the lengths of these sides are y_1 and y_2 and they are h apart, the area of the trapezoid is given by

$$A = \tfrac{1}{2}h(y_1 + y_2)$$

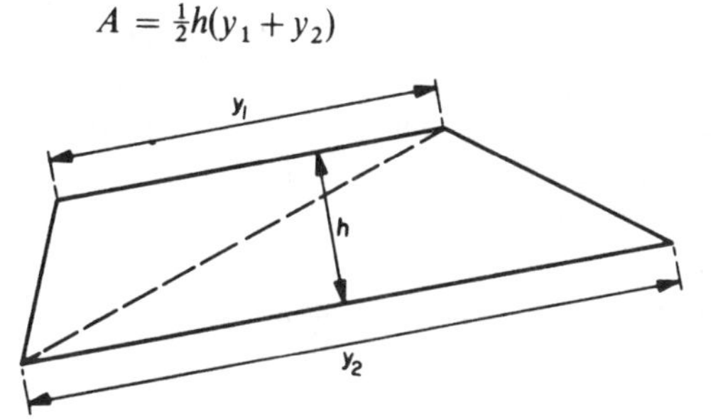

Fig. 2.26 A trapezoid

A curvilinear figure can be divided into a number of approximate trapezoids by covering it with n equally spaced ordinates, h apart, the breadths at the ordinates in order being $y_1, y_2, y_3, \ldots, y_n$.

Commencing with the left-hand trapezoid, the areas of each trapezoid are given by

$$\tfrac{1}{2}h(y_1 + y_2)$$

$$\tfrac{1}{2}h(y_2 + y_3)$$

$$\tfrac{1}{2}h(y_3 + y_4) \ldots$$

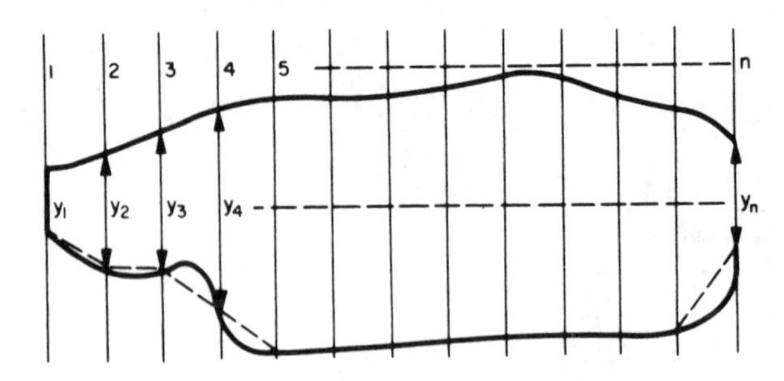

Fig. 2.27 Curvilinear figure represented by trapezoids

By addition, the total area A of the figure is given by

$$A = \tfrac{1}{2}h(y_1 + 2y_2 + 2y_3 + \ldots + y_n)$$

$$= h(\tfrac{1}{2}y_1 + y_2 + y_3 + \ldots + \tfrac{1}{2}y_n)$$

This is termed the *Trapezoidal Rule*. Clearly, the more numerous the ordinates, the more accurate will be the answer. Thus, to evaluate the expression $A = \int y\,dx$ the shape is divided into evenly spaced sections h apart, the ordinates measured and substituted in the rule given above. If the ordinates represent cross-sectional areas of a solid, then the integration gives the volume of that solid, $\nabla = \int A\,dx$.

Expressions can be deduced for moments, but these are not as convenient to use as those that follow.

SIMPSON'S RULES

Generally known as Simpson's rules, these rules for approximate integration were, in fact, deduced by other mathematicians many years previously. They are a special case of the Newton–Cotes' rules. Let us deduce a rule for integrating a curve y over the extent of x. It will be convenient to choose the origin to be in the middle of the base $2h$ long, having ordinates y_1, y_2 and y_3. The choice of origin in no way affects the results as the student should verify for himself.

Assume that the curve can be represented by an equation of the third order,

$$y = a_0 + a_1x + a_2x^2 + a_3x^3$$

The area under the curve is given by

$$A = \int_{-h}^{h} y\,dx = \int_{-h}^{h} (a_0 + a_1x + a_2x^2 + a_3x^3)\,dx$$

$$= \left[a_0x + a_1\frac{x^2}{2} + a_2\frac{x^3}{3} + a_3\frac{x^4}{4}\right]_{-h}^{+h} = 2a_0h + \tfrac{2}{3}a_2h^3 \qquad \text{Equation (i)}$$

Assume now that the area can be given by the expression

$$A = Ly_1 + My_2 + Ny_3 \qquad \text{Equation (ii)}$$

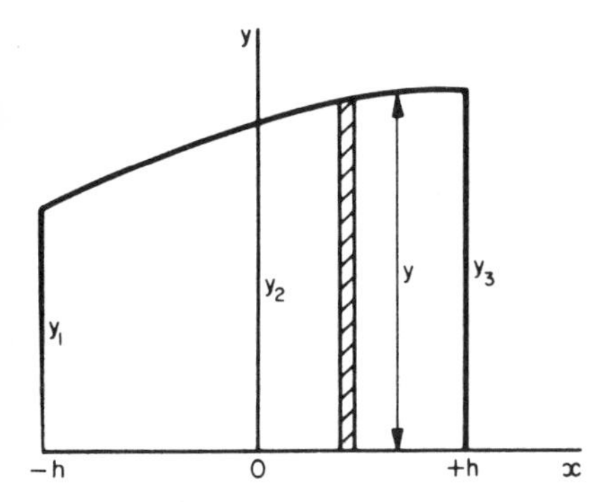

Fig. 2.28

Now

$$y_1 = a_0 - a_1 h + a_2 h^2 - a_3 h^3$$

$$y_2 = a_0$$

$$y_3 = a_0 + a_1 h + a_2 h^2 + a_3 h^3$$

Substituting in Equation (ii)

$$A = (L+M+N)a_0 - (L-N)a_1 h + (L+N)a_2 h^2 - (L-N)a_3 h^3$$

Equation (iii)

Equating the coefficients of a in equations (i) and (iii)

$$\left.\begin{aligned} L+M+N &= 2h \\ L-N &= 0 \\ L+N &= \tfrac{2}{3}h \end{aligned}\right\} \therefore \quad L = N = \tfrac{1}{3}h \quad \text{and} \quad M = \tfrac{4}{3}h$$

The area *can* be represented by Equation (ii), therefore, provided that the coefficients are those deduced and

$$A = \tfrac{1}{3}hy_1 + \tfrac{4}{3}hy_2 + \tfrac{1}{3}hy_3 = \tfrac{1}{3}h(y_1 + 4y_2 + y_3)$$

This is known as *Simpson's First Rule* or *3 Ordinate Rule*. A curved figure can be divided by any uneven number of equally spaced ordinates h apart. The area within ordinates numbers 1 and 3 is

$$A_1 = \tfrac{1}{3}h(y_1 + 4y_2 + y_3)$$

within 3 and 5 ordinates

$$A_2 = \tfrac{1}{3}h(y_3 + 4y_4 + y_5)$$

within 5 and 7 ordinates

$$A_3 = \tfrac{1}{3}h(y_5 + 4y_6 + y_7)$$

and so on.

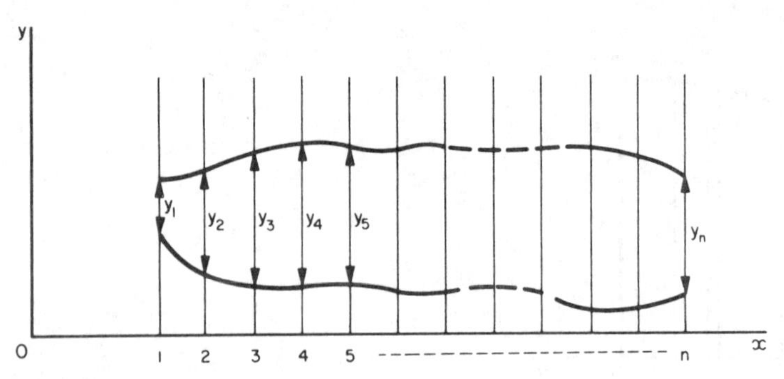

Fig. 2.29

The total area is therefore given by

$$A = \tfrac{1}{3}h(y_1 + 4y_2 + 2y_3 + 4y_4 + 2y_5 + 4y_6 + 2y_7 + \ldots + y_n)$$

i.e.

$$A = \tfrac{2}{3}h(\tfrac{1}{2}y_1 + 2y_2 + y_3 + 2y_4 + y_5 + 2y_6 + y_7 + \ldots + \tfrac{1}{2}y_n)$$

This is the generalized form of the first rule applied to areas. The common multiplier is $\frac{1}{3} \times$ the common interval h and the individual multipliers are 1, 4, 2, 4, 2, 4, ..., 2, 4, 1.

The rule is one for evaluating an integral. While it has been deduced using area as an example, it is equally applicable to any integration using Cartesian or polar co-ordinates. To evaluate the integral $M_x = \int xy\,dx$, for example, instead of multiplying the value of y at each ordinate by the appropriate Simpson multiplier, the value of xy is so treated. Similarly, to evaluate $I_x = \int x^2 y\,dx$, the value of $x^2 y$ at each ordinate is multiplied by the appropriate Simpson multiplier. All of these operations are best performed in a methodical fashion as shown in the following example and more fully in worked examples later in the chapter. Students should develop a facility in the use of Simpson's rules by practice.

EXAMPLE 1. Calculate the value of the integral $\int P\,dv$ between the values of v equal to 7 and 15. The values of P at equal intervals of v are as follows:

v	7	9	11	13	15
P	9	27	36	39	37

Solution: The common interval, i.e. the distance between successive values of v, is 2. Setting out Simpson's rule in tabular form,

v	P	Simpson's multipliers	Functions of $\int P\,dv$
7	9	1	9
9	27	4	108
11	36	2	72
13	39	4	156
15	37	1	37
			382

$$\int P\,dv = \tfrac{1}{3} \times 2 \times 382 = 254{\cdot}7$$

The units are, of course, those appropriate to $P \times v$.

To apply the trapezoidal rule to a curvilinear shape, we had to assume that the relationship between successive ordinates was linear. In applying Simpson's first rule, we have assumed the relationship to be bounded by an expression of the third order. Much greater accuracy is therefore to be expected and for most functions in naval architecture the accuracy so obtained is quite sufficient. Where there is known to be a rapid change of form, it is wise to put in an intermediate ordinate and the rule can be adapted to do this. Suppose that the rapid change is known to be between ordinates 3 and 4 (Fig. 2.30)

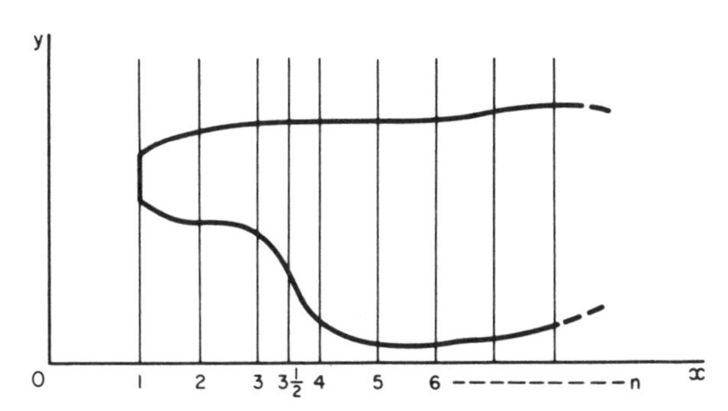

Fig. 2.30

$$\text{Area between 1 and 3 ords.} = \tfrac{1}{3}h(y_1 + 4y_2 + y_3)$$

$$\text{Area between 3 and 4 ords.} = \tfrac{1}{3}\frac{h}{2}(y_3 + 4y_{3\frac{1}{2}} + y_4) = \tfrac{1}{3}h(\tfrac{1}{2}y_3 + 2y_{3\frac{1}{2}} + \tfrac{1}{2}y_4)$$

$$\text{Area between 4 and 6 ords.} = \tfrac{1}{3}h(y_4 + 4y_5 + y_6)$$

$$\text{Total area} = \tfrac{1}{3}h(y_1 + 4y_2 + 1\tfrac{1}{2}y_3 + 2y_{3\frac{1}{2}} + 1\tfrac{1}{2}y_4 + 4y_5 + 2y_6 + \ldots + y_n)$$

Note that, unless a second half ordinate is inserted, n must now be even.

Rules can be deduced, in a similar manner, for figures bounded by unevenly spaced ordinates or different numbers of evenly spaced ordinates. For four evenly spaced ordinates the rule becomes

$$A = \tfrac{3}{8}h(y_1 + 3y_2 + 3y_3 + y_4)$$

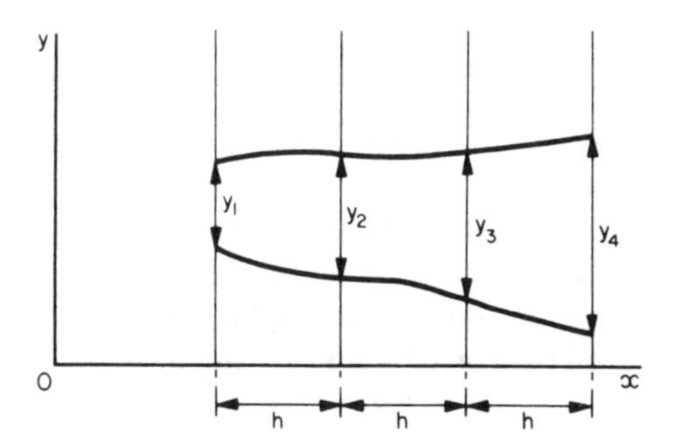

Fig. 2.31

This is known as *Simpson's Second Rule.* Extended for a large number of ordinates, it becomes

$$A = \tfrac{3}{8}h(y_1 + 3y_2 + 3y_3 + 2y_4 + 3y_5 + 3y_6 + 2y_7 + \ldots + y_n)$$

Thus, the common multiplier in this case is $\frac{3}{8}$ times the common interval and the individual multipliers, 1, 3, 3, 2, 3, 3, 2, 3, 3, 2, . . . , 3, 3, 1. It is suitable for 4, 7, 10, 13, 16, etc., ordinates. It can be proved in a manner exactly similar to that employed for the first rule, assuming a third order curve, and it can be used like the first rule to integrate any continuous function.

Another particular Simpson's rule which will be useful is that which gives the area between two ordinates when three are known. The area between ordinates 1 and 2 is given by

$$A_1 = \tfrac{1}{12}h(5y_1 + 8y_2 - y_3)$$

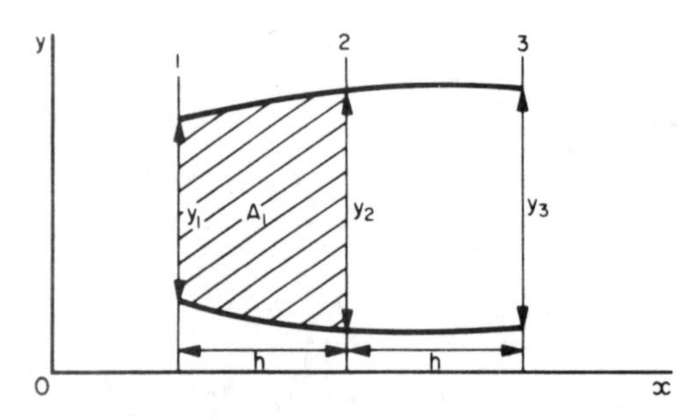

Fig. 2.32

This rule cannot be used for moments. The first moment of area of that portion between ordinates 1 and 2 about number 1 ordinate is

$$M_x = \tfrac{1}{24}h^2(3y_1 + 10y_2 - y_3)$$

These two rules are known loosely as the 5,8 *minus one Rule* and the 3,10 *minus one Rule.* They are somewhat less accurate than the first two rules. Incidentally, applying the 5,8 minus one rule backwards, the unshaded area of Fig. 2.32 is

$$A_2 = \tfrac{1}{12}h(-y_1 + 8y_2 + 5y_3)$$

and adding this to the expression for A_1, the total area is

$$A = \tfrac{1}{3}h(y_1 + 4y_2 + y_3)$$

If the common multiplier has been forgotten, the student can quickly deduce it by applying the particular rule to a rectangle.

Rules can be combined one with another just as the unit for each rule is combined in series to deal with many ordinates. It is important that any discontinuity in a curve falls at the end of a unit, e.g. on the 2 multiplier for the first and second rules; if this is so, the rules can be used on curves with discontinuities. In general, because of differences in the common interval necessary each side of a discontinuity, it will be convenient to deal with the two parts separately.

The first rule deals with functions bounded by 3, 5, 7, 9, 11, 13, etc., ordinates and the second rule with 4, 7, 10, 13, etc., ordinates. Let us deduce a rule for six ordinates

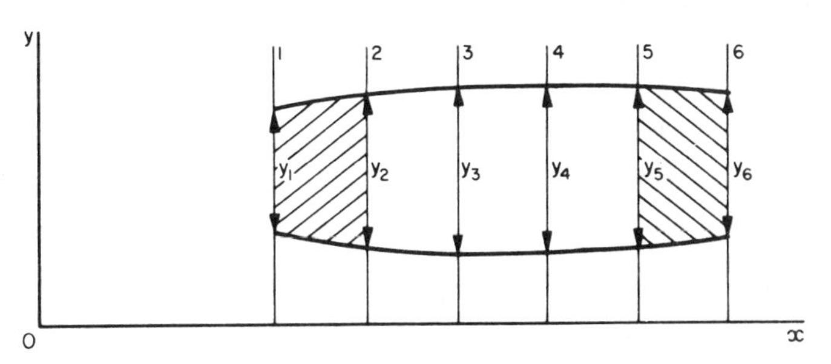

Fig. 2.33

$$\text{Area between 1 and 2 ords.} = \tfrac{1}{12}h(5y_1+8y_2-y_3) = h(\tfrac{5}{12}y_1+\tfrac{8}{12}y_2-\tfrac{1}{12}y_3)$$

$$\text{Area between 2 and 5 ords.} = \tfrac{3}{8}h(y_2+3y_3+3y_4+y_5)$$
$$= h(\tfrac{3}{8}y_2+\tfrac{9}{8}y_3+\tfrac{9}{8}y_4+\tfrac{3}{8}y_5)$$

$$\text{Area between 5 and 6 ords.} = \tfrac{1}{12}h(-y_4+8y_5+5y_6)$$
$$= h(-\tfrac{1}{12}y_4+\tfrac{8}{12}y_5+\tfrac{5}{12}y_6)$$

Total area

$$A = h(\tfrac{5}{12}y_1+\tfrac{25}{24}y_2+\tfrac{25}{24}y_3+\tfrac{25}{24}y_4+\tfrac{25}{24}y_5+\tfrac{5}{12}y_6)$$
$$= \tfrac{25}{24}h(0{\cdot}4y_1+y_2+y_3+y_4+y_5+0{\cdot}4y_6)$$

These few Simpson's rules, applied in a repetitive manner, have been found satisfactory for hand computation for many years. The digital computer makes somewhat different demands and the more generalized Newton–Cotes' rules, summarized in Table 2.1, may be found more suitable for some purposes.

Table 2.1
Newton–Cotes' rules

Number of ordinates	Multipliers for ordinates numbers 1	2	3	4	5	
2	$\frac{1}{2}$	$\frac{1}{2}$				
3	$\frac{1}{6}$	$\frac{4}{6}$	$\frac{1}{6}$			
4	$\frac{1}{8}$	$\frac{3}{8}$	$\frac{3}{8}$	$\frac{1}{8}$		
5	$\frac{7}{90}$	$\frac{32}{90}$	$\frac{12}{90}$	$\frac{32}{90}$	$\frac{7}{90}$	
6	$\frac{19}{288}$	$\frac{75}{288}$	$\frac{50}{288}$	$\frac{50}{288}$	$\frac{75}{288}$	...

Table 2.1 (continued)
Newton-Cotes' rules

Number of ordinates	Multipliers for ordinates numbers 1	2	3	4	5
7	$\frac{41}{840}$	$\frac{216}{840}$	$\frac{27}{840}$	$\frac{272}{840}$	$\frac{27}{840}$ ...
8	$\frac{751}{17280}$	$\frac{3577}{17280}$	$\frac{1323}{17280}$	$\frac{2989}{17280}$	$\frac{2989}{17280}$ ...
9	$\frac{989}{28350}$	$\frac{5888}{28350}$	$-\frac{928}{28350}$	$\frac{10496}{28350}$	$-\frac{4540}{28350}$ ...

Area $= L \times \Sigma$ (Multiplier $\times$ ordinate)

Ordinates are equally spaced with end ordinates coinciding with ends of curve. Multipliers are always symmetrical and are indicated by dots.

TCHEBYCHEFF'S RULES

Returning to Equation (ii) on p. 24, the rule required was forced to take the form of the sum of equally spaced ordinates, each multiplied by a coefficient. The rule could have been forced to take many forms, most of them inconvenient.

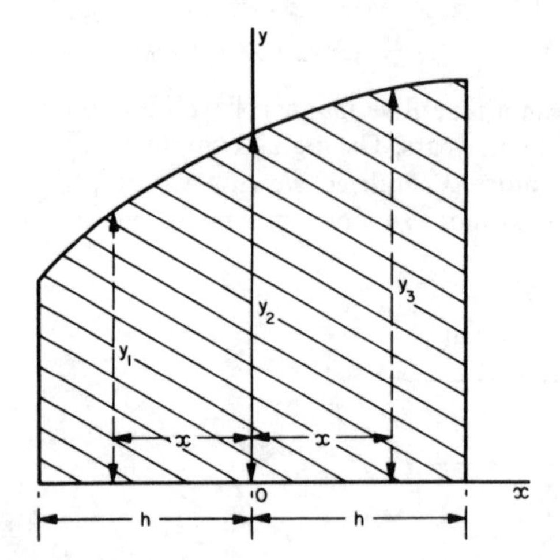

Fig. 2.34

One form which does yield a convenient rule results from assuming that the area can be represented by the sum of ordinates placed a special distance x (which may be zero) from the origin, all multiplied by the same coefficient, i.e. instead of assuming the form as before, assume that the area can be represented by

$$A = p(y_1 + y_2 + y_3), \quad y_2 \text{ being at the origin}$$

Now

$$y_1 = a_0 - a_1 x + a_2 x^2 - a_3 x^3$$

$$y_2 = a_0$$

$$y_3 = a_0 + a_1 x + a_2 x^2 + a_3 x^3$$

adding:

$$A = p(3a_0 + 2a_2 x^2)$$

equating coefficients of a above with those of Equation (i) on p. 24

$$3p = 2h \qquad \therefore \quad p = \tfrac{2}{3}h$$

and

$$2px^2 = \tfrac{2}{3}h^3 = ph^2$$

$$\therefore \quad x = \frac{1}{\sqrt{2}}h = 0{\cdot}7071h$$

The total shaded area in Fig. 2.34 can therefore be calculated by erecting ordinates equal to 0·7071 of the half length from the mid-point, measuring their heights and that of the mid-ordinate, adding the three heights together and multiplying the total by two-thirds of the half length.

$$A = \tfrac{2}{3}h(y_1 + y_2 + y_3)$$

where y_1 and y_3 are 0·7071 of the half length h, from the mid-point.

This is Tchebycheff's rule for three ordinates. Similar rules can be deduced for 2, 3, 4, 5, 6, 7 and 9 ordinates when spacings become:

Table 2.2
Tchebycheff's rule spacings

Number of ordinates, n	Spacing each side of mid-ordinate as a factor of the half length h					Degree of curve
2	0·57735					3
3	0	0·70711				3
4	0·18759	0·79465				5
5	0	0·37454	0·83250			5
6	0·26664	0·42252	0·86625			7
7	0	0·32391	0·52966	0·88386		7
8	0·10268	0·40620	0·59380	0·89733		5
9	0	0·16791	0·52876	0·60102	0·91159	9
10	0·08375	0·31273	0·50000	0·68727	0·91625	5

The eight and ten ordinate rule spacings have been deduced by applying the four and five ordinate rules each side of the mid-point of the half length. The common multiplier for all rules is the whole length $2h$ divided by n, the number of ordinates used, $2h/n$.

Tchebycheff's rules are used not infrequently, particularly the ten ordinate rule, for calculating displacement from a 'Tchebycheff body plan', i.e. a body plan drawn with ordinate positions to correspond to the Tchebycheff spacings. Areas of the sections are calculated by Simpson's rules or by other convenient means, merely added together and multiplied by $2h/n$ to give volume of displacement. Lines are, in fact, often faired on a Tchebycheff body plan to avoid the more prolonged calculation by Simpson's rules with each iteration. Since fairing is basically to a curve or areas, this assumes the use of Tchebycheff ordinates to define the body plan.

GAUSS RULES

It has been seen that the Simpson rules and Newton–Cotes' rules employ equally spaced ordinates with unequal multipliers and the Tchebycheff rules use constant multipliers with unequal ordinate spacing. A third set of rules, the Gauss rules, uses unequal spacing of ordinates and unequal multipliers as shown in Table 2.3.

Table 2.3
Gauss' rules

Number of ordinates	Spacing each side of mid-ordinate as a factor of the half length. Multiplier			
2	Spacing	0·57735		
	Multiplier	0·50000		
3	Position	0	0·77460	
	Multiplier	0·44444	0·27778	
4	Position	0·33998	0·86114	
	Multiplier	0·32607	0·17393	
5	Position	0	0·53847	0·90618
	Multiplier	0·28445	0·23931	0·11846
	Integral = Sum of products × whole base			

The Gauss rules have the merit of being more accurate than either the Simpson or Tchebycheff rules, but their application involves more tedious calculation when manual methods are used. For this reason, they have been largely neglected by naval architects in the past. When an electronic computer is used, it is as easy and quick to manipulate the more complicated multipliers and the basic calculations are similar for all three rules. By using Gauss rules, the naval architect can either obtain greater accuracy by using the same number of ordinates or he can obtain the same accuracy with fewer ordinates and in less time.

It can be shown that:

(*a*) a Simpson rule with an even number of ordinates is only marginally more accurate than the next lower odd ordinate rule; odd ordinate Simpson rules are therefore to be preferred, (*b*) a Tchebycheff rule with an even number of ordinates gives the same accuracy as the next highest odd ordinate rule. Even ordinate Tchebycheff rules are therefore to be preferred,

(*c*) a Tchebycheff rule with an even number of ordinates gives an accuracy rather better than the next highest odd ordinate Simpson rule, i.e. the two-ordinate Tchebycheff rule is more accurate than the three-ordinate Simpson rule,

(*d*) The five-ordinate Gauss rule gives an accuracy comparable with that achieved with nine-ordinate Simpson or Tchebycheff rules.

EXAMPLE 2. Integrate $y = \tan x$ from $x = 0$ to $x = \pi/3$ by the five ordinate rules of (*a*) Simpson, (*b*) Newton–Cotes, (*c*) Tchebycheff, (*d*) Gauss.

Solution: The precise solution is

$$\int_0^{\pi/3} \tan x = [-\log\cos x]_0^{\pi/3}$$

$$= 0{\cdot}69315$$

$$= 0{\cdot}22064\pi$$

(*a*) Simpson

x	y	SM	$f(A)$
0	0	$\frac{1}{2}$	0
$\pi/12$	0·26795	2	0·53590
$\pi/6$	0·57735	1	0·57735
$\pi/4$	1·00000	2	2·00000
$\pi/3$	1·73205	$\frac{1}{2}$	0·86603
			3·97928

$$\text{Area} = \frac{2}{3} \times \frac{\pi}{12} \times 3{\cdot}97928 = 0{\cdot}22107\pi$$

(*b*) Newton–Cotes

x	y	M	$f(A)$
0	0	7	0
$\pi/12$	0·26795	32	8·57440
$\pi/6$	0·57735	12	6·92820
$\pi/4$	1·00000	32	32·00000
$\pi/3$	1·73205	7	12·12435
			59·62695

$$\text{Area} = \frac{1}{90} \times \frac{\pi}{3} \times 59{\cdot}62695 = 0{\cdot}22084\pi$$

(*c*) Tchebycheff

x		y
$\pi/6(1-0{\cdot}83250)$	$=\ 5{\cdot}025°$	0·08793
$\pi/6(1-0{\cdot}37454)$	$=\ 18{\cdot}764°$	0·33972
$\pi/6$	$=\ 30°$	0·57735
$\pi/6(1+0{\cdot}37454)$	$=\ 41{\cdot}236°$	0·87655
$\pi/6(1+0{\cdot}83250)$	$=\ 54{\cdot}975°$	1·42674
		3·30829

$$\text{Area} = \frac{\pi}{15} \times 3{\cdot}30829 = 0{\cdot}22055\pi$$

(*d*) Gauss

x		y	M	$f(A)$
$\pi/6(1-0{\cdot}90618)$	$=\ 2{\cdot}815°$	0·04918	0·11846	0·00583
$\pi/6(1-0{\cdot}53847)$	$=\ 13{\cdot}846°$	0·24647	0·23931	0·05898
$\pi/6$	$=\ 30°$	0·57735	0·28445	0·16423
$\pi/6(1+0{\cdot}53847)$	$=\ 46{\cdot}154°$	1·04114	0·23931	0·24916
$\pi/6(1+0{\cdot}90618)$	$=\ 57{\cdot}185°$	1·55090	0·11846	0·18372
				0·66192

$$\text{Area} = \frac{\pi}{3} \times 0{\cdot}66192 = 0{\cdot}22064\pi$$

ANALOGUE COMPUTERS AND SIMULATORS

The physical laws connecting pressure, volume and absolute temperature of a gas and connecting electrical potential, resistance and current of a simple circuit are respectively:

$$\frac{P \times V}{T} = \text{Constant} \qquad \text{and} \qquad \frac{I \times R}{E} = \text{Constant}$$

Such is the similarity of the laws that the physical law for the gas could be represented by a simple electrical circuit; with potential kept constant a variation applied to R and current I measured by an ammeter could represent the relative changes to pressure and volume of a gas at constant temperature. The variations in this case are, of course, obvious but more complicated circuits subject to the basic laws of Ohm and Kirchhoff and with the parameters varied with time can be built up to represent and measure more difficult physical relationships.

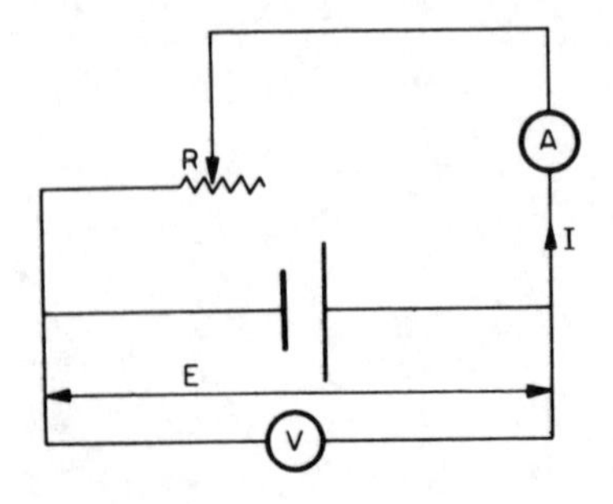

Fig. 2.35

This is the basis of the *analogue computer*. Let us anticipate some of the work of Chapter 13 to demonstrate a particular application of the electrical analogue. The motion of a ship in response to applied forces can be represented by a mathematical equation. The applied forces may arise from the deliberate action of those on board in moving a control surface such as a rudder (or hydroplane in the case of a submarine) or from external agencies such as waves striking the ship. In its simplest form, the equation may be a linear differential equation involving one degree of freedom but, in order to achieve greater accuracy, may include non-linear and cross-coupling terms.

The same form of equation can be represented electrically by a suitably contrived circuit. Consider now the case of a ship turning under the action of the rudder and let the parameter to be studied be the heading of the ship. If the components of the electrical circuit are correctly chosen then, by varying an input signal in conformity with the rudder movements, the voltage across two points of the circuit can be measured to represent the ship's heading. By extending the circuitry, more variables can be measured such as the angle of heel during turn, drift angle, etc.

Clearly, a difficulty arises in determining the correct values of the electrical components to suit a particular ship. These may have to be computed by theoretical means, or measured by model or full scale trials. Having determined them, however, the circuit will represent faithfully the ship's behaviour. That is to say, it will 'simulate' the ship's behaviour. It can be made to do this in real time, e.g. the ship's head can be made to vary at the same rate as in the full scale ship in real life.

The realism can be heightened by mounting the computer on a platform which can turn, heel and pitch in response to signals from the computer and the input can be derived from a steering wheel so that the operator gains the impression of actually being on board a ship. The complete system is referred to as a *simulator* and such devices are used extensively for training personnel particularly in the operation of vehicles (aircraft, ships) where hazardous situations can arise which would be too dangerous, or impossible, to reproduce in reality. The simulators for pilotage in crowded and restricted waters are an example. The degree of realism can be varied to suit the need, the most comprehensive involving virtual reality techniques.

It will be realized from the above that the analogue computer is a specialized tool. It can be made reasonably flexible by permitting adjustment of the electrical constants over a wide range. In this way, the same computer can be adjusted to represent different ships or to study another problem provided that the problem can be represented by the same basic equations. Compared, however, with the digital computer, described below, it is much less versatile.

One big advantage of being able to work in real time is that it is often possible to introduce into the electrical circuit actual components of the real system. If the behaviour of a particular hydraulic control system is difficult to represent mathematically, (it may be non-linear in a complex manner, for example) it can itself be built into the simulator. This would be impossible with a digital computer.

DIGITAL COMPUTERS

In essence, a digital computer is an electronic device capable of holding large amounts of data and of performing simple arithmetical computations at high speed. An operator need not know in detail how it works but should understand its basic characteristics, strengths and weaknesses.

Any computer system includes input units which accept information in a suitably coded form (tape or disk readers, teletypes, document readers or light pens); storage or memory units for holding instructions (typically disks, tapes or drums); a calculation unit by which data is manipulated; output units for presenting results (visual display units (VDUs), printers, or graph plotters); and a power unit.

The immediate output may be a magnetic tape or disk which can be decoded later on separate print-out devices. Sometimes the output tape is used directly to control a machine tool or automatic draughting equipment. Input and output units may be remote from the computer itself, providing a number of outstations with access to a large central computer.

As with any other form of communication, that between the designer and the computer must be conducted in a language understood by both. There are many such languages (e.g. Basic, Pascal, Cobol) suitable for scientific, engineering and commercial work. Modern computer languages resemble stilted English. The computer itself translates the words into the more complex machine language through a compiler.

Teletypes and VDUs associated with design systems are usually interactive, enabling a designer to engage in a dialogue with the computer or, more accurately, with the software in the computer. Software is of two main types; that which controls the general activities within the computer (e.g. how data is stored and accessed) and that which directs how a particular problem is to be tackled. The computer may prompt the operator by asking for more data or for a decision between possible options it presents to him. The software can include decision aids, i.e. it can recommend a particular option as the best in the circumstances—and give its reasons if requested. If the operator makes, or appears to make, a mistake the machine can challenge the input.

Displays can be in colour and present data in graphical form. Colour can be useful in differentiating between different elements of the total display. Red can be used to highlight hazardous situations because humans associate red with danger. However, for some applications monochrome is superior. Shades of one colour can more readily indicate the relative magnitude of a single parameter, e.g. shades of blue indicating water depth. Graphical displays are often more meaningful to humans than long tabulations of figures. Thus a plot of points which should lie on smooth curves will quickly highlight a rogue reading. This facility is used as an input check in large finite element calculations (hydrodynamic or structural). The computer can cause the display to rotate so that, in the case of a ship's hull for instance, a complex shape can be viewed from a number of directions. Taking this concept one stage further the computer can generate pictures of a ship's interior and enable the customer to in effect 'walk through' the vessel. Realism and usefulness are enhanced by the representation of colours, surface textures and lighting (Ref. 1).

It does not follow that because a computer can be used to provide a service it should be so used. It can be expensive of money and time. Thus in the example above, the computer may be cheaper than a full scale mock-up but would simple artist's sketches do? A computer is likely to be most cost effective as part of a comprehensive system. Thus if used for store withdrawals it can note when stocks of a component reach a predetermined level and prompt the storeman by printing out a replacement order.

For what work then is a computer likely to be suitable? To answer this its special abilities must be considered. It can carry out repetitive calculations accurately and rapidly (e.g. solve multiple differential equations); store large quantities of data, updating and retrieving it rapidly on demand; process stored data, sorting it and causing it to be printed in a variety of formats.

The ship designer can therefore use it to provide a large library of accurate and up to date facts (e.g. data on performances, size, weight, cost, reliability). This data may be accessed through direct interrogation by an on-line terminal or by means of catalogues printed out regularly to some agreed format. He or she can carry out calculations as discrete actions or as part of an integrated suite of design calculations. Data may be fed in separately for each calculation or be called up by the computer from its own memory. Or a designer can analyse data on achievement of planned dates, cashflow or manpower usage, monitoring the achievement against the plan. Applications of especial interest are considered in the following paragraphs.

Computer-aided design. Systems by which the designer can use a terminal to assist in the design process by giving immediate access to the data stored and to a complete suite of design programs have been developed for many industries. Several systems have been developed for ship design, some concentrating on the initial design phase and others on the detailed design process and its interaction with production. The aim is to obtain the right balance between man and machine using the strengths of one to complement the other (Ref. 2).

Automated draughting equipment. This equipment has been available for some years. Typically, a reader and keyboard are used by an operator to record from

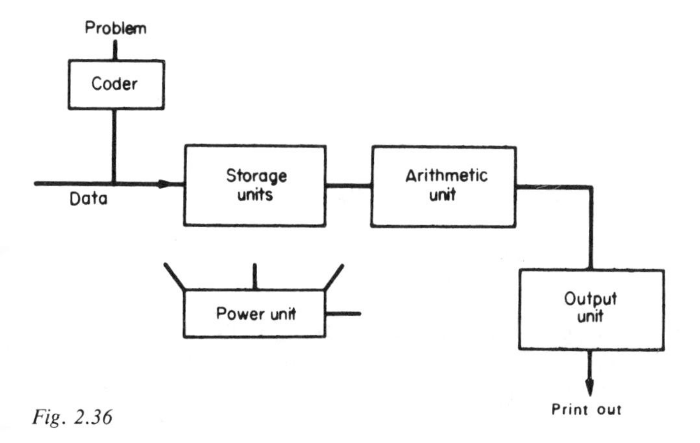

Fig. 2.36

a general arrangement drawing of an equipment all relevant information needed to define its components. By feeding the tape so generated into a plotter the detail drawings of all components can be generated automatically. In association with a VDU the equipment can be used to generate a new drawing or to develop a rough sketch into a full working drawing. Where a symbol or component is repeated many times over the operator can generate it once, store it in the computer and recall it as an entity as many times as required. Very successful applications of this have been achieved in the design of printed circuit boards and in the structural design of supertankers.

Simulation modelling. Provided that the factors governing a real life situation are understood, it may be possible to represent it by a set of mathematical relationships. In other words it can be modelled and the model used to study the effects of changing some of the factors much more quickly, cheaply and safely than could be achieved with full scale experimentation. Consider tankers arriving at a terminal. Factors influencing the smooth operation are numbers of ships, arrival intervals, ship size, discharge rate and storage capacities. A simulation model could be produced to study the problem, reduce the queueing time and to see the effects of additional berths and different discharge rates and ship size. The economics of such procedure is also conducive to this type of modelling.

Approximate formulae and rules

Approximate formulae and rules grew up with the craftsman approach to naval architecture and were encouraged by the secrecy that surrounded it. Many were bad and most have now been discarded. There remains a need, however, for coarse approximations during the early, iterative processes of ship design. Usually, the need is met by referring to a similar, previous design and correcting the required figure according to the dimensions of new and old, e.g. supposing an estimate for the transverse $\overline{\mathrm{BM}} = I_T/\nabla$ is required, the transverse second moment of area I_T is proportional to $L \times B^3$ and ∇ is proportional to $L \times B \times T$

$$\therefore \quad \overline{\mathrm{BM}} \propto \frac{LB^3}{LBT} = \frac{B^2}{T}$$

$$\therefore \quad \overline{\mathrm{BM}} = \frac{B^2}{KT}$$

where K is a constant for geometrically similar ships.

$$\text{or} \qquad \frac{\overline{\mathrm{BM}}_1}{\overline{\mathrm{BM}}_2} = \frac{B_1^2}{B_2^2} \times \frac{T_2}{T_1}$$

This presumes that the forms are similar. This formula is one that may suffice for a check; K varies between 10 and 15 for ship shapes. More important, it shows how ship dimensions affect this geometric property, viz. that the beam contributes as its square, the draught inversely and the length not at all. The effects of changes to dimensions can, therefore, be assessed.

An approximate formula for the longitudinal $\overline{\mathrm{BM}}$ is

$$\overline{\mathrm{BM}}_{\mathrm{L}} = \frac{3A_{\mathrm{W}}^2 L}{40B\,\nabla}$$

NORMAND'S FORMULA

This is known also as Morrish's rule. It gives the distance of the centre of buoyancy of a ship-like form below the waterline.

$$\text{Distance of VCB below waterline} = \frac{1}{3}\left(\frac{T}{2} + \frac{\nabla}{A_{\mathrm{W}}}\right)$$

The formula is remarkably accurate for conventional ship forms and is used frequently in the early design stages.

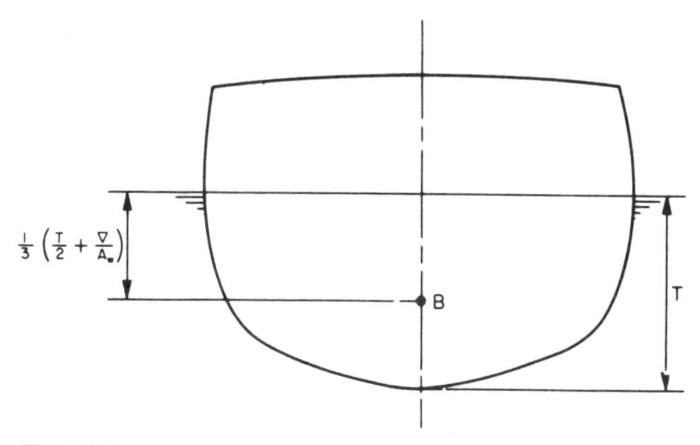

Fig. 2.37

WEIGHT CONVENTIONS

Plating thickness in ships is usually specified in decimals of an inch or in millimetres. In warship practice, thickness used to be specified as a weight per square foot of surface. Since steel weighs about 480 $\mathrm{lbf/ft^3}$, plating one inch thick weighs about 40 $\mathrm{lbf/ft^2}$.

Sections used in ship construction are also specified in different ways. The sections commonly used in merchant ships have varying web thicknesses and are specified in the dimensions of depth × flange × flange × web thickness, e.g. $10 \times 3\frac{1}{2} \times 3\frac{1}{2} \times 0{\cdot}50$ channel has a depth of 10 in., flanges $3\frac{1}{2}$ in. wide and a web 0·50 in. thick. Reference to standard tables would show that its flange thickness is 0·56 in. Nowadays, sections are defined almost exclusively in millimetres, web × table × thickness with a mass per metre in kg/m.

Statistics

A feature of present day naval architecture, as in other engineering disciplines, is the increasing use made of statistics by the practising naval architect. This is not because the subject itself has changed but rather that the necessary mathematical methods have been developed to the stage where they can be applied to the subject.

It will be concluded, for example in Chapter 6, that the hull girder stress level accepted from the standard calculation should reflect the naval architect's opinion as to the probability of exceeding the standard assumed loading during the life of the ship. Again, in the study of ship motions the extreme amplitudes of motion used in calculations must be associated with the probability of their occurrence and probabilities of exceeding lesser amplitudes are also of considerable importance.

It is not appropriate in a book of this nature to develop in detail the statistical approach to the various aspects of naval architecture. Students should refer to a textbook on statistics for detailed study. However, use is made in several chapters of certain general concepts of which the following are important.

PROBABILITY

Consider an aggregate of n experimental results (e.g. amplitudes of pitch from a ship motion trial) of which m have the result R and $(n-m)$ do not have this result. Then, the probability of obtaining the result R is $p = m/n$. The probability that R will not occur is $1-p$. If an event is impossible its probability is zero. If an event is a certainty its probability is unity.

PROBABILITY CURVE

When a large amount of information is available, it can be presented graphically by a curve. The information is plotted in such a way that the area under the curve is unity and the probability of the experimental result lying between say R and $R+\delta R$ is represented by the area under the curve between these values of the abscissa. There are a number of features about this probability curve which may best be defined and understood by an example.

Consider the following example of experimental data.

EXAMPLE 3. Successive amplitudes of pitch to the nearest half degree recorded during a trial are:

$4, 2, 3\frac{1}{2}, 2\frac{1}{2}, 3, 2, 3\frac{1}{2}, 1\frac{1}{2}, 3, 1, 3\frac{1}{2}, \frac{1}{2}, 2, 1, 1\frac{1}{2}, 1, 2, 1\frac{1}{2}, 1\frac{1}{2}, 4, 2\frac{1}{2}, 3\frac{1}{2}, 3, 2\frac{1}{2}, 2, 2\frac{1}{2}, 2\frac{1}{2}, 3, 2, 1\frac{1}{2}.$

Solution: As a string of figures these values have little significance to the casual reader. Using the concepts given above however, the occurrence of specific pitch amplitudes is given in columns (1) and (2) below:

(1) Pitch amplitude	(2) Number of occurrences	(3) $(1)\times(2)$	(4) $(1)-\mu$	(5) $(4)^2$	(6) $(5)\times(2)$
$\frac{1}{2}$	1	$\frac{1}{2}$	−1·82	3·312	3·31
1	3	3	−1·32	1·741	5·23
$1\frac{1}{2}$	5	$7\frac{1}{2}$	−0·82	0·672	3·36
2	6	12	−0·32	0·102	0·61
$2\frac{1}{2}$	5	$12\frac{1}{2}$	+0·18	0·032	0·16
3	4	12	0·68	0·462	1·85
$3\frac{1}{2}$	4	14	1·18	1·392	5·57
4	2	8	1·68	2·822	5·64
	30	$69\frac{1}{2}$			25·73

Selecting the information from this table,

ARITHMETIC MEAN $= \mu = \dfrac{69\frac{1}{2}}{30} = 2{\cdot}32$ deg

MEDIAN is the mid measurement of the 30 which occurs midway between 2 and $2\frac{1}{2} = 2{\cdot}25$ deg. The median bisects the area

MODE = 2 deg. The mode is the most common value

RANGE $= 4 - \frac{1}{2} = 3\frac{1}{2}$ deg. The range is the difference of extreme values of variate

STANDARD DEVIATION $= \sigma = \sqrt{\left(\dfrac{25{\cdot}73}{30}\right)} = 0{\cdot}926$ deg

VARIANCE $= \sigma^2 = 0{\cdot}858\ (\text{deg})^2$

These values provide a much clearer physical picture of the motion being measured than does the original test data. The significance of these various features of the probability curve will be clear to the student with some knowledge of statistics.

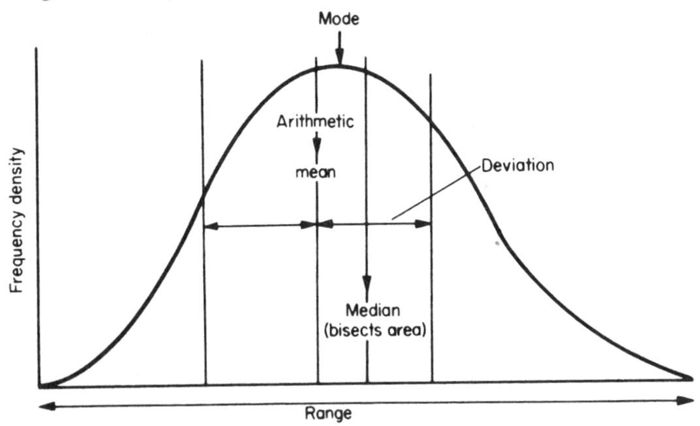

Fig. 2.38

Worked examples

EXAMPLE 4. Calculate the area, position of the centre of flotation and the second moments of area about the two principal axes of the waterplane defined by the following ordinates, numbered from forward. It is 220 m long.

Ordinate number	1	2	3	4	5	6	7	8	9	10	11
$\frac{1}{2}$ Ord. (m)	0·2	2·4	4·6	6·7	8·1	9·0	9·4	9·2	8·6	6·3	0·0

Solution:

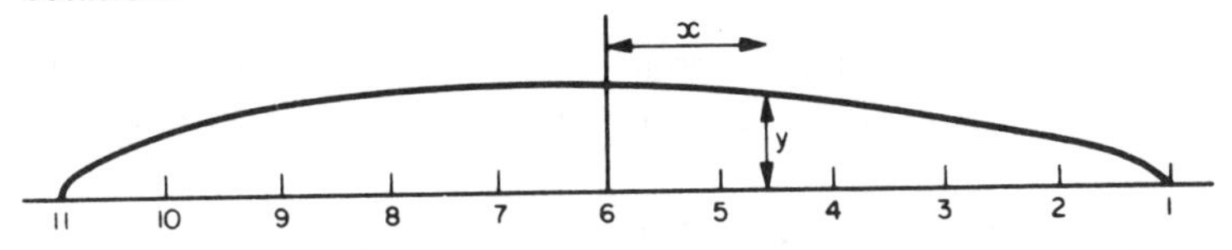

Fig. 2.39

Ord. no.	½ Ord. y	S.M.	Func. of y	Lever x	Func. of xy	Lever x	Func. of x^2y	½ Ord.³ y^3	Func. of y^3
1	0·2	½	0·1	5	0·5	5	2·5	0·0	0·0
2	2·4	2	4·8	4	19·2	4	76·8	13·8	27·6
3	4·6	1	4·6	3	13·8	3	41·4	97·3	97·3
4	6·7	2	13·4	2	26·8	2	53·6	300·8	601·6
5	8·1	1	8·1	1	8·1	1	8·1	531·4	531·4
6	9·0	2	18·0	0	68·4	0	0·0	729·0	1458·0
7	9·4	1	9·4	−1	−9·4	−1	9·4	830·6	830·6
8	9·2	2	18·4	−2	−36·8	−2	73·6	778·7	1557·4
9	8·6	1	8·6	−3	−25·8	−3	77·4	636·1	636·1
10	6·3	2	12·6	−4	−50·4	−4	201·6	250·0	500·0
11	0·0	½	0·0	−5	0·0	−5	0·0	0·0	0·0
			98·0		−122·4 68·4 −54·0		544·4		6240·0

$$\text{Area} = \int y\,dx \text{ for each half}$$

There are eleven ordinates and therefore ten spaces, $\frac{220}{10} = 22$ m apart. The total of the $f(y)$ column must be multiplied by $\frac{2}{3}$ times 22, the common interval, to complete the integration and by 2, for both sides of the waterplane.

$$\text{Total area} = 2 \times \tfrac{2}{3} \times 22 \times 98{\cdot}0 = 2{,}874{\cdot}7 \text{ m}^2$$

$$\text{1st moment} = \int xy\,dx \text{ for each half}$$

Instead of multiplying each ordinate by its actual distance from Oy, we have made the levers the number of ordinate distances to simplify the arithmetic so that the total must be multiplied by a lever factor 22. We have also chosen Oy as number 6, the mid-ordinate as being somewhere near the centre of area, to ease the arithmetic; it may well transpire that number 7 ordinate would have been closer. The moments each side are in opposition and must be subtracted to find the out-of-balance moment. First moment about number 6 ordinate, $M_x = 2 \times \frac{2}{3} \times 22 \times 22 \times 54{\cdot}0$.

Now the distance of the centre of area from O$y = M_x/A$; three of the multipliers are common to both and cancel out, leaving,

$$\text{CF abaft 6 ord. } \bar{x} = 22 \times \frac{f(xy)}{f(y)} = \frac{22 \times 54}{98} = 12{\cdot}1 \text{ m}$$

$$\text{2nd Moment about O}y = \int x^2 y\,dx \text{ for each half}$$

We have twice multiplied by the number of ordinate spacings instead of the actual distances so that the lever factor this time is 22 × 22. The second moments all act together (x is squared and therefore always positive) and must be added

$$\therefore \quad I \text{ about 6 ord.} = 2 \times \tfrac{2}{3} \times 22 \times 22 \times 22 \times 544{\cdot}4 = 7{,}729{,}000 \text{ m}^4$$

Now this is not the least I and is not of much interest; the least is always that about an axis through the centre of area and is found from the parallel axis theorem

$$I_L = I - A\bar{x}^2 = 7{,}729{,}000 - 2874{\cdot}7 \times (12{\cdot}1)^2 = 7{,}309{,}000 \text{ m}^4$$

$$\text{2nd Moment about } Ox = \tfrac{1}{3}\int y^3\,dx \text{ for each half}$$

To integrate $y^3, f(y)^3$ must be multiplied by $\frac{2}{3}$ times the common interval.

$$I_T = 2\times\tfrac{1}{3}\times\tfrac{2}{3}\times 22\times 6240{\cdot}0 = 61{,}000\ \text{m}^4$$

This time, the axis already passes through the centre of area.

EXAMPLE 5. A ship has a main body defined by the waterplane areas given below. The waterlines are 1 ft 3 in. apart. In addition, there is an appendage having a volume of displacement of 330 ft^3 with a centre of volume $3\frac{1}{2}$ in. below No. 4 WL.

What are the volume of displacement and the position of the VCB?

Waterline	1	2	3	4
Area (ft^2)	1230	1100	870	480

Solution:

WL	Area A	S.M.	$f(A)$	Lever, y	$f(Ay)$
1	1230	1	1230	0	0
2	1100	3	3300	1	3300
3	870	3	2610	2	5220
4	480	1	480	3	1440
			7620		9960

$$\text{Volume of displacement for main body} = \tfrac{3}{8}\times 1{\cdot}25\times 7620 = 3572\ \text{ft}^3$$

$$\text{VCB below 1 WL for main body} = \frac{9960}{7620}\times 1{\cdot}25 = 1{\cdot}63\ \text{ft}$$

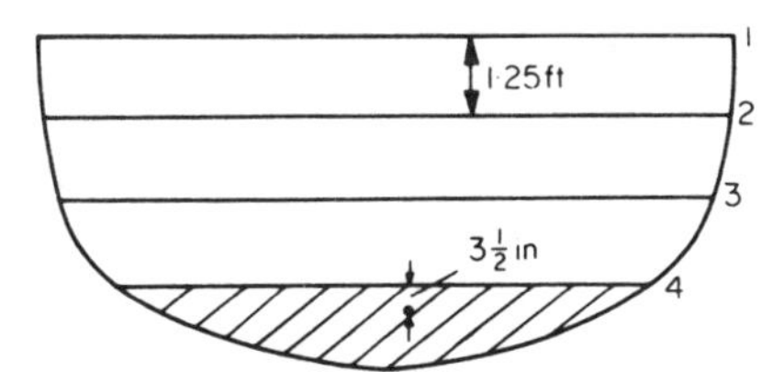

Fig. 2.40

Adding volumes and moments for main body and appendage:

Main body volume = 3572 ft^3

Appendage volume = 330 ft^3

Total volume = 3902 ft^3

Main body moment below 1 WL = $3572\times 1{\cdot}63$ = 5830 ft^4

Appendage moment below 1 WL = $330\,(3{\cdot}75+0{\cdot}29)$ = 1333 ft^4

7163 ft^4

$$\text{Whole body VCB below 1 WL} = \frac{7163}{3902} = 1{\cdot}84\ \text{ft}$$

The value of a sketch, however simple, cannot be over-emphasized in working through examples.

All of these examples are worked out by slide rule. The number of digits worked to in any number should be pruned to be compatible with this level of accuracy.

EXAMPLE 6. Calculate by the trapezoidal rule the area of a transverse half section of a tanker bounded by the following waterline offsets. Waterlines are 3 m apart:

WL	1	2	3	4	5	6
Offset (m)	24·4	24·2	23·7	22·3	19·1	3·0

Compare this with the areas given by Simpson's rules,

(*a*) with six ordinates
(*b*) with five and a half ordinate equal to 16·0 m

Calculate for (*b*) also the vertical position of the centre of area.

Solution:

WL	Offset, y	Trap. mult.	$f_1(y)$	Simp. mult.	$f_2(y)$
1	24·4	$\frac{1}{2}$	12·2	0·4	9·8
2	24·2	1	24·2	1	24·2
3	23·7	1	23·7	1	23·7
4	22·3	1	22·3	1	22·3
5	19·1	1	19·1	1	19·1
6	3·0	$\frac{1}{2}$	1·5	0·4	1·2
			103·0		100·3

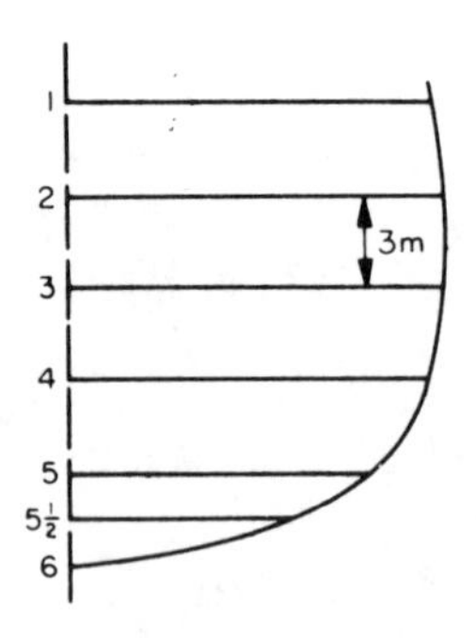

Fig. 2.41

$$\text{Area by trapezoidal rule} = 3 \times 103{\cdot}0 = 309{\cdot}0 \text{ m}^2$$

$$\text{Area by Simpson's rule} = \tfrac{25}{24} \times 3 \times 100{\cdot}3 = 313{\cdot}4 \text{ m}^2$$

WL	Offset, y	S.M.	$f(y)$	Lever, x	$f(xy)$
1	24·4	1	24·4	0	0
2	24·2	4	96·8	1	96·8
3	23·7	2	47·4	2	94·8
4	22·3	4	89·2	3	267·6
5	19·1	$1\frac{1}{2}$	28·65	4	114·6
$5\frac{1}{2}$	16·0	2	32·0	$4\frac{1}{2}$	144·0
6	3·0	$\frac{1}{2}$	1·5	5	7·5
			319·95		725·3

$$\text{Area} = \tfrac{1}{3} \times 3 \times 319{\cdot}95 = 319{\cdot}95 \text{ m}^2$$

$$\text{Centre of area below 1 WL} = \frac{725{\cdot}3}{319{\cdot}95} \times 3{\cdot}0 = 6{\cdot}80 \text{ m}$$

EXAMPLE 7. An appendage to the curve of areas abaft the after perpendicular, ordinate 21, is fair with the main body curve and is defined as follows:

Ordinate	20	21	22
Area (ft^2)	114·6	110·0	15·0

The ordinates are equally spaced 18 ft apart. Calculate the volume of the appendage and the longitudinal position of its centre of buoyancy.

Solution:

Ord. no.	Area, A	S.M.	$f(A)$	S.M.	$f_M(A)$
20	114·6	−1	−114·6	−1	−114·6
21	110·0	8	880·0	10	1100·0
22	15·0	5	75·0	3	45·0
			840·4		1030·4

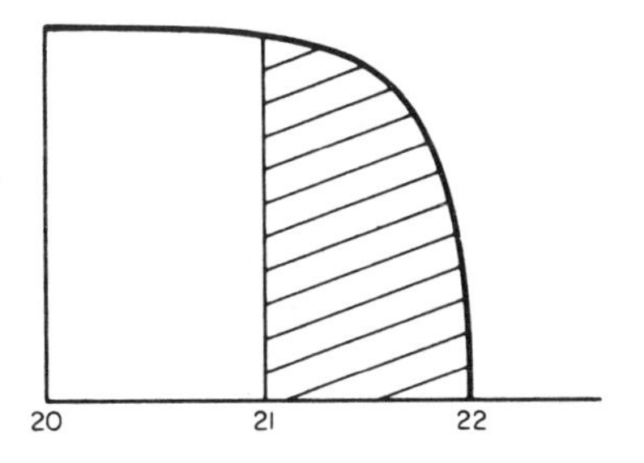

Fig. 2.42

$$\text{Volume of appendage} = \tfrac{1}{12} \times 18 \times 840{\cdot}4 = 1260{\cdot}6 \text{ ft}^3$$

$$\text{c.b. from 22 ordinate} = \tfrac{1}{24} \times 18^2 \times 1030{\cdot}4 \times \frac{12}{18 \times 840{\cdot}8}$$

$$= \frac{18 \times 1030{\cdot}4}{2 \times 840{\cdot}8} = 11{\cdot}03 \text{ ft}$$

$$\text{c.b. abaft the AP} = 18{\cdot}0 - 11{\cdot}03 = 6{\cdot}97 \text{ ft}$$

EXAMPLE 8. It is necessary to calculate the volume of a wedge of fluid immersed by the rotation of a vessel through 20 degrees. The areas and distances of the centres of areas from the axis of rotation of the immersed half waterplanes have been calculated at 5 degree intervals as follows:

Angle of inclination (deg)	0	5	10	15	20
Area (m^2)	650	710	920	1030	1810
Centre of area from axis (m)	3·1	3·2	3·8	4·8	6·0

Calculate the volume of the immersed wedge.

Solution: The Theorem of Pappus Guldinus states that the volume of a solid of revolution is given by the area of the plane of revolution multiplied by the distance moved by its centre of area. If r is the distance of the centre of area of a typical plane from the axis, for a rotation $\delta\theta$, it moves a distance $r\,\delta\theta$ and the volume traced out by the area A is

$$\delta V = Ar\,\delta\theta \quad (\theta \text{ in radians})$$

The total volume is therefore given by

$$V = \int Ar\,\mathrm{d}\theta$$

This time, therefore, an approximate integration has to be performed on Ar over the range of θ.

Ord.	A	r	Ar	S.M.	$f(Ar)$
0	650	3·1	2015	1	2015
5	710	3·2	2272	4	9088
10	920	3·8	3496	2	6992
15	1030	4·8	4944	4	19776
20	1810	6·0	10860	1	10860
					48731

$$\text{Volume} = \tfrac{1}{3} \times 5 \times \frac{\pi}{180} \times 48731 = 1420 \text{ m}^3$$

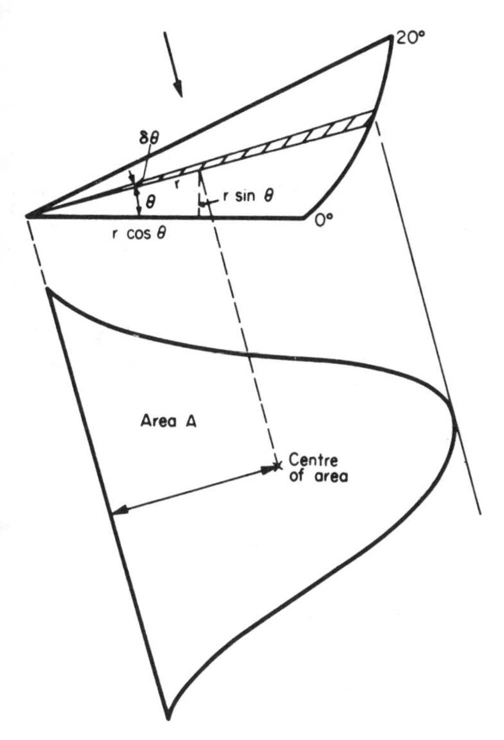

Fig. 2.43

Problems

1. A ship, 200 m between perpendiculars, has a beam of 22 m and a draught of 7 m. If the prismatic coefficient is 0·75 the area of the waterplane 3500 m^2 and mass displacement in salt water is 23,000 tonnes, estimate

(*a*) block coefficient
(*b*) waterplane coefficient
(*c*) midship section coefficient and
(*d*) the distance of the centre of buoyancy above the keel.

2. The length, beam and mean draught of a ship are respectively 115, 15·65 and 7·15 m. Its midship section coefficient is 0·921 and its block coefficient is 0·665. Find

(*a*) displacement in tonnef and newtons in salt water
(*b*) area of immersed midship section
(*c*) prismatic coefficient of displacement.

3. Two similar right circular cones are joined at their bases. Each cone has a height equal to the diameter of its base. The composite body floats so that both

apexes are in the water surface. Calculate

(*a*) the midship section coefficient
(*b*) the prismatic coefficient
(*c*) the waterplane coefficient.

4. A curve has the following ordinates, spaced 1·68 m apart: 10·86, 13·53, 14·58, 15·05, 15·24, 15·28, 15·22 m. Calculate the area by Simpson's first rule and compare it with the area given by the trapezoidal rule. What is the ratio of the two solutions?

5. The half ordinates of the load waterplane of a vessel are 1·2, 4·6, 8·4, 11·0, 12·0, 11·7, 10·3, 7·5 and 3·0 ft respectively and the overall length is 120 ft. What is its area?

6. A curvilinear figure has the following ordinates at equidistant intervals: 12·4, 27·6, 43·8, 52·8, 44·7, 29·4 and 14·7. Calculate the percentage difference from the area found by Simpson's first rule when finding the area by (*a*) the trapezoidal rule, (*b*) Simpson's second rule.

7. The effective girths of the outer bottom plating of a ship, 27·5 m between perpendiculars, are given below, together with the mean thickness of plating at each ordinate. Calculate the volume of the plating. If the plating is of steel of mass density 7700 kg/m^3, calculate the weight in meganewtons.

Ord. No.	AP	10	9	8	7	6	5	4	3	2	FP
Girth (m)	14·4	22·8	29·4	34·2	37·0	37·4	36·8	28·6	24·2	22·6	23·2
Thickness (mm)	10·2	10·4	10·6	11·4	13·6	13·6	12·8	10·4	10·1	10·1	14·2

8. The half ordinates of a vessel, 144 m between perpendiculars, are given below.

Ord. No.	AP	8	7	6	5	4	3	2	FP
½ Ord. (m)	17·0	20·8	22·4	22·6	21·6	18·6	12·8	5·6	0·0

In addition, there is an appendage, 21·6 m long, abaft the AP, whose half ordinates are:

Ord. No.	12	11	10	AP
½ Ord. (m)	0·0	9·6	14·0	17·0

Find the area and position of the centre of area of the complete waterplane.

9. The loads per foot due to flooding, at equally spaced positions on a transverse bulkhead are given below. The bulkhead is 45 ft deep. Calculate the total load on the bulkhead and the position of the centre of pressure

Ord. No.	1	2	3	4	5	6	7	8	9	10
Load (tonf/ft)	0	15	30	44	54	63	69	73	75	74

10. A tank is 8 m deep throughout its length and 20 m long and its top is flat and horizontal. The sections forward, in the middle and at the after end are all triangular, apex down and the widths of the triangles at the tank top are respectively 15, 12 and 8 m.

Draw the calibration curve for the tank in tonnes of fuel against depth and state the capacity when the depth of oil is 5·50 m. SG of oil fuel = 0·90. Only five points on the curve need be obtained.

11. Areas of waterplanes, 2·5 m apart, of a tanker are given below.

Calculate the volume of displacement and the position of the VCB. Compare the latter with the figure obtained from Normand's (or Morrish's) rule.

Waterplane	1	2	3	4	5	$5\frac{1}{2}$	6
Area (m^2)	4010	4000	3800	3100	1700	700	200

12. The waterline of a ship is 70 m long. Its half ordinates, which are equally spaced, are given below. Calculate the least second moment of area about each of the two principal axes in the waterplane.

Ord. No.	1	2	3	4	5	6	7	$7\frac{1}{2}$	8
$\frac{1}{2}$ Ord. (m)	0·0	3·1	6·0	8·4	10·0	10·1	8·6	6·4	0·0

13. The half ordinates of the waterplane of a ship, 440 ft between perpendiculars, are given below. There is, in addition, an appendage abaft the AP with a half area of 90 ft^2 whose centre of area is 8 ft from the AP; the moment of inertia of the appendage about its own centre of area is negligible.

Calculate the least longitudinal moment of inertia of the waterplane.

Section	AP	10	9	8	7	6	5	4	3	2	FP
$\frac{1}{2}$ Ord. (ft)	6·2	16·2	22·5	26·0	27·5	27·4	23·7	19·2	14·5	8·0	0·0

14. The half breadths of the 16 ft waterline of a ship which displaces 18,930 tonf in salt water are given below. In addition, there is an appendage abaft the AP, 30 ft long, approximately rectangular with a half breadth of 35·0 ft. The length BP is 660 ft.

Calculate the transverse $\overline{BM}$ and the approximate value of $\overline{KM}$.

Station	FP	2	3	4	5	6	7	8	9	10	AP
$\frac{1}{2}$ Breadth (ft)	0·0	21·0	32·0	37·0	40·6	43·0	43·8	43·6	43·0	40·0	37·0

15. The shape of a flat, between bulkheads, is defined by the ordinates, spaced 4 ft apart, given below. If the plating weighs 7 lbf per square foot of surface area, calculate the weight of the plating and the distance of the c.g. from No. 1 ordinate.

Ord. No.	1	2	3	4	5	6
Breadth (ft)	53·0	50·0	45·0	38·0	30·0	14·0

16. Each of the two hulls of a catamaran has the following dimensions.

Ord. No.	1	2	3	4	5	$5\frac{1}{2}$	6
$\frac{1}{2}$ Ord. (in.)	0·0	4·0	6·2	7·2	6·4	4·9	0·0

The length and volume of displacement of each hull are respectively 18 ft and 5·3 ft^3. The hull centre lines are 6 ft apart. Calculate the transverse $\overline{BM}$ of the boat.

17. Compare the areas given by Simpson's rules and the trapezoidal rule for the portions of the curve defined below:

(*a*) between ordinates 1 and 4
(*b*) between ordinates 1 and 2

Ord. No.	1	2	3	4
Ord. (m)	39·0	19·0	12·6	10·0

The distance between ordinates is 10 m.

18. Apply Normand's (or Morrish's) rule to a right circular cylinder floating with a diameter in the waterplane. Express the error from the true position of the VCB as a percentage of the draught.

19. Deduce a trapezoidal rule for calculating longitudinal moments of area.

20. Deduce the five ordinate rules of (*a*) Newton–Cotes, (*b*) Tchebycheff.

21. Compare with the correct solution to five decimal places $\int_0^\pi \sin x\, dx$ by the three ordinate rules of (*a*) Simpson, (*b*) Tchebycheff, (*c*) Gauss.

22. A quadrant of 16 m radius is divided by means of ordinates parallel to one radius and at the following distances: 4, 8, 10, 12, 13, 14 and 15 m. The lengths of these ordinates are respectively: 15·49, 13·86, 12·49, 10·58, 9·33, 7·75 and 5·57 m. Find:

(*a*) the area to two decimal places by trigonometry
(*b*) the area using only ordinates 4 m apart by Simpson's rule
(*c*) the area using also the half ordinates
(*d*) the area using all the ordinates given
(*e*) the area using all the ordinates except 12·49.

23. Calculate, using five figure tables, the area of a semicircle of 10 m radius by the four ordinate rules of (*a*) Simpson, (*b*) Tchebycheff, (*c*) Gauss and compare them with the correct solution.

24. Show that $\overline{KB}$ is approximately $T/6(5-2C_{VP})$.

25. From strains recorded in a ship during a passage, the following table was deduced for the occurrence of stress maxima due to ship motion. Calculate for this data (*a*) the mean value, (*b*) the standard deviation.

Max. stress (MN/m^2)	10	20	30	40	50	60
Occurrences	852	1335	772	331	140	42

26. Construct a probability curve from the following data of maximum roll angle from the vertical which occurred in a ship crossing the Atlantic. What are (*a*) the mean value, (*b*) the variance, (*c*) the probability of exceeding a roll of 11 degrees

Max. roll angle, deg.	1	3	5	7	9	11	13	15	17
Occurrences	13,400	20,550	16,600	9720	4420	1690	510	106	8

References

1. Thornton, A. T. Design Visualisation of Yacht Interiors. *TRINA*, 1992.
2. Holmes, S. J. The Application and Development of Computer Systems for Warship Design. *NA*, July 1981.

3 Flotation and trim

A ship, like any other three-dimensional body, has six degrees of freedom. That is to say, any movement can be resolved into movements related to three orthogonal axes, three translations and three rotations. With a knowledge of each of these six movements, any combination movement of the ship can be assessed. The three principal axes have already been defined in Chapter 2. This chapter will be confined to an examination of two movements

(*a*) behaviour along a vertical axis, O*z* in the plane O*xz*

(*b*) rotation in the plane O*xz* about a transverse horizontal axis, O*y*.

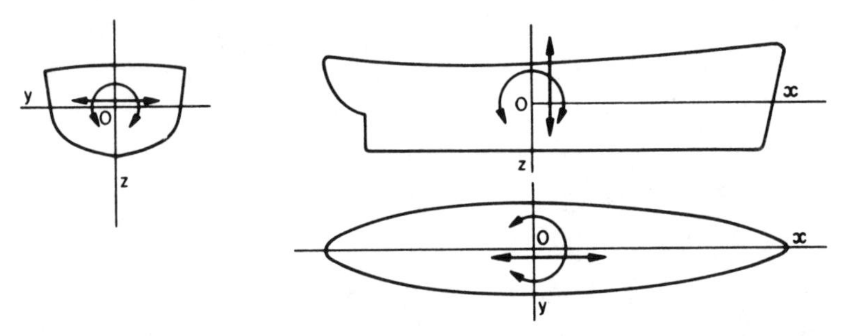

Fig. 3.1

Attention will be confined to static behaviour, i.e. conditions applying when the ship is still. Generally, it is the change from one static condition to another that will be of interest and so it is convenient to imagine any movement occurring very slowly. Dynamic behaviour, involving time, motion and momentum will be dealt with in later chapters.

Flotation

PROPERTIES OF FLUIDS

The *mass density* of a fluid ρ, is the mass of the fluid per unit volume. The *weight density* w, of a fluid is the weight of the fluid per unit volume. In SI units, $w = \rho g$ so that, if ρ is in kg/m^3, w is in newtons/m^3. In British units, $w = \rho(g/g)$ so that, if ρ is in lb/ft^3, w has the same numerical value in lbf/ft^3. Similarly, in old metric units, w has the same numerical value in tonnef/m^3 as ρ has in tonne/m^3. Since they vary with pressure and temperature, the values must be related to a standard condition of pressure and temperature. The former is normally taken to be one atmosphere, 14·7 lbf/in^2 (10^5 Pa = 1 Bar) and the latter sometimes 15°C (59°F), and for water sometimes 4°C (39°F) when its density is a maximum.

The ratio of the density of a solid or a liquid to the density of pure water is the *specific gravity*, γ. Since it is the basic reference for all such materials, the weight properties of pure distilled water are reproduced in Fig. 9.1.

The inverse of the weight density is called the *reciprocal weight density u*, or *specific volume*. The standard value assumed in British units for salt water is 35 ft³/tonf and it is for this value that hydrostatic data for the ship are normally compiled; the SI equivalent to 35 ft³/tonf is 0·975 m³/tonnef or 99·5 m³/MN. Corrections are applied for variations of reciprocal weight density from this value. Table 3.1 gives values of mass density for common fluids and for steel, air and mahogany.

Table 3.1
Properties of some common materials

Material	Mass density, ρ		Reciprocal mass density		Specific gravity, γ
	(lb/ft³)	(kg/m³)	(ft³/ton)	(m³/Mg)	
Fresh water (standard)	62·4	1000	35·9*	1·00	1·00
Fresh water (British preferred value)	62·2	996*	36·0	1·00	1·00
Salt water	64·0	1025	35·0	0·975	1·03
Furnace fuel oil	59·1	947	37·8	1·05	0·95
Diesel oil	52·5	841	42·6	1·19	0·84
Petrol	43·5	697	51·5	1·44	0·70
Steel	480	7689	4·7	0·13	7·70
Mahogany	53	849	42·3	1·18	0·85
Air	0·0807	1·293	27,800	774·775	—

* Not used; these are for comparison only.

The reciprocal weight density is found merely by inverting the weight density and adjusting the units; for example, the weight density w lbf/ft³ results in a reciprocal weight density u ft³/tonf as follows:

$$u\ \text{ft}^3/\text{tonf} = \frac{1}{w}\frac{\text{ft}^3}{\text{lbf}} \times \frac{2240\ \text{lbf}}{1\ \text{tonf}}$$

Thus, 35 ft³/tonf corresponds to 64 lbf/ft³ and 62·2 lbf/ft³ to 36 ft³/tonf. With SI units, the reciprocal weight density u must be expressed in m³/MN which involves g. Hence, weight density reciprocal

$$u = \frac{1}{\rho}\frac{\text{m}^3}{\text{kg}}\ \frac{\text{s}^2}{9{\cdot}807\ \text{m}}\ \frac{\text{kg m}}{\text{s}^2\ \text{newton}}\ \frac{10^6\text{N}}{\text{MN}}$$

$$= \frac{10^6}{9{\cdot}807\rho}\ \text{m}^3/\text{MN} \quad \text{with } \rho \text{ in kg/m}^3$$

for steel, for example,

$$u = \frac{10^6}{9{\cdot}807 \times 7689} = 13{\cdot}26\ \text{m}^3/\text{MN}$$

The student is advised always to write in the units in his calculation to ensure that they cancel to the required dimensions.

ARCHIMEDES' PRINCIPLE

This states that when a solid is immersed in a liquid, it experiences an upthrust equal to the weight of the fluid displaced.

Thus, the tension in a piece of string by which a body is suspended, is reduced when the body is immersed in fluid by an amount equal to the volume of the body times the weight density of the fluid; a diver finds an article heavier to lift out of water than under it, by an amount equal to its volume times the weight density of water. This upthrust is called the buoyancy of the object. If, by chance, the body has the same weight density as the fluid, the upthrust when it was totally immersed would be equal to its weight; the string would just go limp and the diver would find the object to be apparently weightless. If the body were to have a smaller weight density than the fluid, only sufficient of the body to cause an upthrust equal to its weight could be immersed without force; if the body is pushed further down the buoyancy exceeds the weight and it bobs up, like a beach ball released from below its natural position in the sea.

This leads to a corollary of Archimedes' principle known as the *Law of Flotation*. When a body is floating freely in a fluid, the weight of the body equals the buoyancy, which is the weight of the fluid displaced.

The *buoyancy* of a body immersed in a fluid is the vertical upthrust it experiences due to displacement of the fluid (Fig. 3.2). The body, in fact, experiences all of the hydrostatic pressures which obtained before it displaced the fluid.

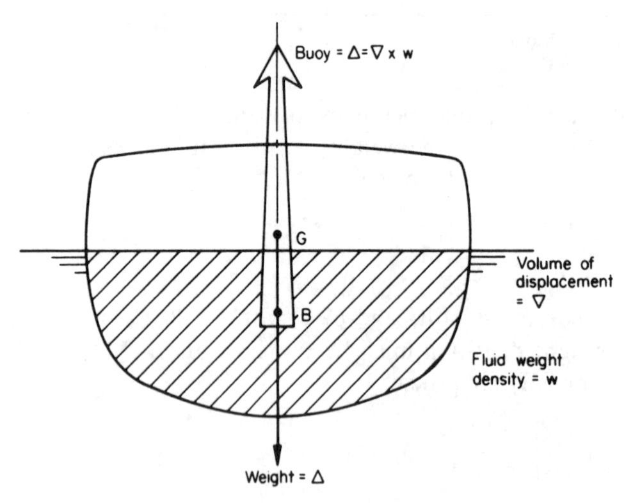

Fig. 3.2

The buoyancy is the resultant of all of the forces due to hydrostatic pressure on elements of the underwater portion (Fig. 3.3). Now, the hydrostatic pressure at a point in a fluid is equal to the depth of the point times the weight density of the fluid, i.e. it is the weight of a column of the fluid having unit cross-section and

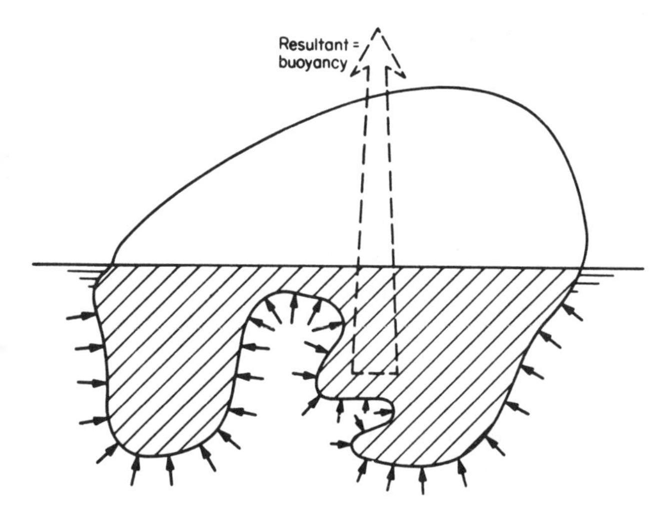

Fig. 3.3

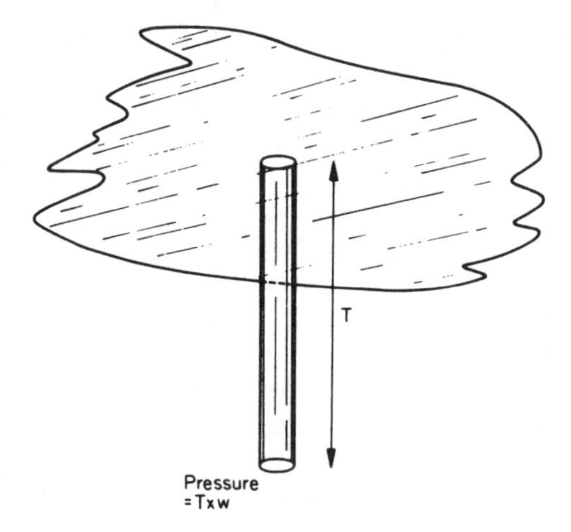

Fig. 3.4 Hydrostatic pressure at a point

length equal to the depth of immersion, T

$$p = Tw$$

Let us examine the pressure distribution around a rectangular block $a \times b \times c$ floating squarely in a fluid at a draught T. The pressures on the vertical faces of the block all cancel out and contribute nothing to the vertical resultant; the hydrostatic pressure at the bottom face is Tw and so the total vertical upthrust is this pressure multiplied by the area:

$$\text{upthrust} = (Tw)ab$$

But this is the displaced volume, abT, times the weight density of the fluid, w, which is in accordance with the law of flotation. Worked example 3 is relevant.

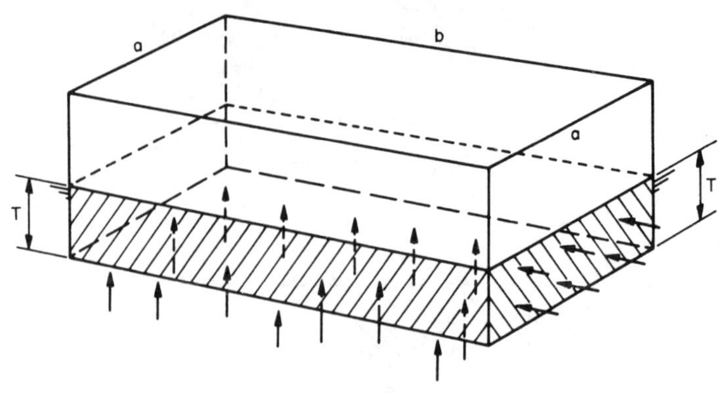

Fig. 3.5

In the general case, imagine a body floating freely in a fluid as shown in Fig. 3.3. The body is supported by the summation of all the pressure forces acting on small elements of the surface area of the body. Now imagine the body removed and replaced by a thin film of material with the same surface shape, and further imagine that the interior volume created by this film is filled with the same fluid as that in which the body is floating. If the film has negligible weight, a state of equilibrium will be produced when the level of fluid reaches the outside level. Thus, the forces acting on the film from the outside just support those acting on the film from the inside, i.e. the weight of fluid displaced by the body.

Buoyancy is plainly a force. It is the upthrust caused by displacement of fluid. For very many years naval architects have used the terms *buoyancy* and *displacement* interchangeably and the latter has become an alternative word for the upthrust of the fluid upon a floating vessel. Both weight and buoyancy vary in the same way with gravity so that anywhere in the world, the ship will float at exactly the same waterline in water of the same density. For static considerations we are interested wholly in forces and weights and so the concept of a *force displacement* Δ identical to buoyancy is satisfactory and convenient and is retained in this book.

The same is not quite true for the study of dynamic behaviour of a vessel which depends upon mass rather than weight. It is necessary to introduce the concept of *mass displacement* Σ. However it is force which causes change and normally the single word displacement refers to a force and is defined by the symbol Δ. Thus:

$$\Delta = g\,\Sigma = \varrho\, g\, \nabla = w\nabla$$

Retention of the metric tonneforce (tonnef) seems likely for many years yet in rule-of-thumb practice, although students should find no difficulty in dealing with displacement in terms of newtons. The tonnef is the force due to gravity acting on a mass of one tonne and has the merit of being numerically identical to the tonne and quite close to the old British tonf:

1 tonf = 1·016 tonnef = 9964 newtons (1 tonnef = 9807 newtons)

It is convenient thus to work in terms of the mass of a ship $\Sigma = n$ tonnes which, under gravity leads to a force displacement, $\Delta = \Sigma g = n \times g = n$ tonnef

With an understanding of the relationship of buoyancy and weight, the wonder at why steel objects can float diminishes. It is natural to expect a laden cargo ship to wallow deeply and a light ballasted ship to tower high, and we now know that the difference in buoyancy is exactly equal to the difference in loading. An important and interesting example is the submarine; on the surface, it has a buoyancy equal to its weight like any other floating body. When submerged, sufficient water must have been admitted to make weight and buoyancy roughly equal, any small out-of-balance force being counteracted by the control surfaces. A surface effect machine, or cushion craft obeys Archimedes' principle when it hovers over the sea; the indentation of the water beneath the machine has a volume which, multiplied by the weight density of the sea, equals the weight of the craft (ignoring air momentum effects).

The watertight volume of a ship above the water line is called the *reserve of buoyancy*. It is clearly one measure of the ship's ability to withstand the effects of flooding following damage and is usually expressed as a percentage of the load displacement.

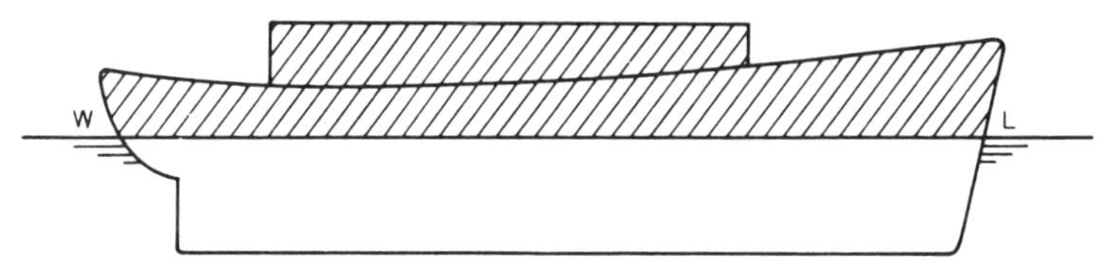

Fig. 3.6 Reserve of buoyancy

VERTICAL MOVEMENT

Figure 3.2 shows the forces acting upon a floating body which are

(*a*) the weight, vertically downwards, which may be taken for static considerations as acting as if it were all concentrated at the centre of gravity, as for any rigid body;
(*b*) The buoyancy, vertically upwards, which may be assumed concentrated at the centre of buoyancy, which is the centre of volume of the underwater shape.

It must be made clear that when the ship is still, the weight and buoyancy forces must act in the same straight line BG, otherwise a couple would act upon the ship, causing it to change its attitude. What happens when a small weight is placed on the vertical line through BG? (Strictly speaking, over the centre of flotation, as will become clear later.) Clearly the vessel sinks—not completely but by an amount so that the additional buoyancy equals the additional weight and vertical equilibrium is again restored. The ship undergoes a parallel sinkage having a buoyancy W and the centre of buoyancy B moves towards the addition by an amount $\overline{\mathrm{BB}'}$. Taking moments about B

$$W\overline{\mathrm{Bb}} = (\Delta + W)\overline{\mathrm{BB}'}$$

$$\therefore \quad \overline{\mathrm{BB}'} = \frac{W\overline{\mathrm{Bb}}}{\Delta + W} \qquad \text{Equation (i)}$$

Thus, the new buoyancy ($\Delta + W$) acts now through the new centre of buoyancy B′, whose position has just been found. It has been assumed that there is no trim; when there is, Fig. 3.7 shows the projections on to a transverse plane and the same result holds.

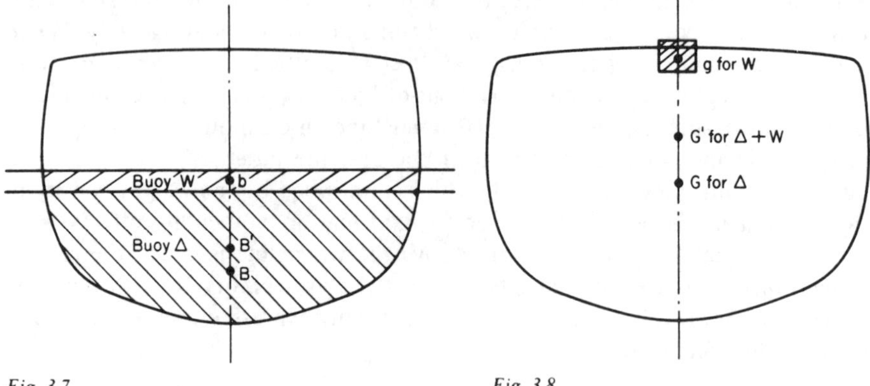

Fig. 3.7

Fig. 3.8

In the same way, the ship has a new centre of gravity. Taking moments about G (see Fig. 3.8)

$$W\overline{\text{Gg}} = (\Delta + W)\overline{\text{GG}'}$$

$$\therefore \quad \overline{\text{GG}'} = \frac{W\overline{\text{Gg}}}{\Delta + W} \qquad \text{Equation (ii)}$$

Thus, the new weight ($\Delta + W$) acts now through the new centre of gravity G′, whose position has just been found. Note the similarity of the expressions.

Equation (i) involves a knowledge of the position of b, the centre of buoyancy of the added layer of buoyancy. If the layer is thick, it must be determined from the ship geometry—in fact, it is probably as quick to determine the position of B′ directly from the shape of the whole ship. Frequently, however, additions are small in comparison with the total displacement and b can be taken half way up a slice assumed to have parallel sides. It is convenient to find the approximate thickness of this slice by using a device called the *tonf per inch immersion* (TPI) or *tonnef per centimetre immersion* (TPC). The general abbreviation is TPI.*

The tonf per inch immersion of a waterplane is the weight required to effect a parallel sinkage of the ship at that waterline of one inch. If the waterplane area is A_W ft^2,

$$\text{weight required} = \frac{A_W \times \frac{1}{12}\,\text{ft}^2}{u} \times \text{ft} \times \frac{\text{tonf}}{\text{ft}^3}$$

i.e.

$$\text{TPI} = \frac{A_W}{12u}\,\text{tonf}$$

$$\text{TPI} = \frac{A_W}{420} \quad \text{for salt water}$$

* Tonf (or tonnef) parallel immersion

For an additional weight W, therefore, the parallel sinkage in inches is given approximately by W/TPI; for wallsided vessels this expression is, of course, precise. This is an extremely useful expression and figures for TPI are calculated for all waterplanes for a ship, forming an important part of the hydrostatic data. Values vary from 6·0 for an ocean tug to 300 for a supertanker.

In old metric units, the weight required to effect a parallel sinkage of one metre is

$$A_W \, \mathrm{m}^2 \; \rho \frac{\mathrm{kg}}{\mathrm{m}^3} \frac{\mathrm{Mg}}{1000\,\mathrm{kg}} \frac{\mathrm{g}}{\mathrm{g}}$$

i.e. TPC $= \rho A_w 10^{-5}$ tonnef per metre with A_w in m² and ρ in kg/m³. For salt water, $\rho = 1025$ kg/m³, whence TPC $= 0{\cdot}01025\,A_w$.

In SI units, meganewtons per metre immersion

$$= A_W \, \mathrm{m}^2 \; \rho \frac{\mathrm{kg}}{\mathrm{m}^3} \frac{9{\cdot}807\,\mathrm{m}}{\mathrm{s}^2} \frac{\mathrm{newton\,s}^2}{\mathrm{kg\,m}} \frac{\mathrm{MN}}{10^6\mathrm{N}}$$

$$= 9{\cdot}807 \rho A_W 10^{-6}$$

which for salt water becomes $0{\cdot}01005\,A_w$ MN/m

tonnef per centimetre $= 0{\cdot}4 \times$ tonf per inch

MN per m immersion $= 0{\cdot}392 \times$ tonf per inch

Of course, arguments are completely reversed if weights are removed and there is a parallel rise. What happens if the weight of the ship remains the same and the density of the water in which it is floating is changed? Let us examine, at first, a body of displacement Δ floating in water of weight density w_1 which passes into water of lower weight density w_2. It will sink deeper because the water is less buoyant. The weight and buoyancy have not changed because nothing has been added to the body or taken away

$$\therefore \quad \Delta = \nabla_1 w_1 = \nabla_2 w_2$$

$$\frac{\nabla_1}{\nabla_2} = \frac{w_2}{w_1}$$

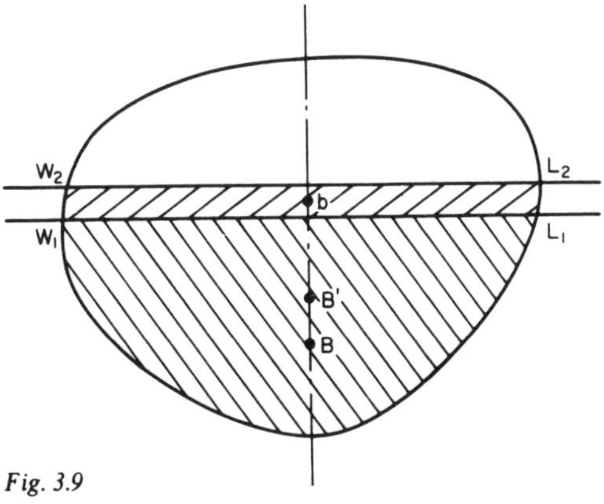

W2
W1
L2
L1
b
B'
B

Fig. 3.9

$$\text{The volume of the layer} = \nabla_2 - \nabla_1 = \nabla_2\left(1-\frac{w_2}{w_1}\right)$$

Now, this layer is made up of water of weight density w_2

$$\therefore \quad \text{weight of layer} = \nabla_2 w_2\left(1-\frac{w_2}{w_1}\right) = \Delta\left(1-\frac{w_2}{w_1}\right)$$

The approximate thickness of this layer in inches is given by this expression divided by the TPI in w_2 water; if the TPI is known in w_1 water,

$$\text{TPI}_2 = \text{TPI}_1\frac{w_2}{w_1} = \text{TPI}_1\frac{u_1}{u_2}$$

The approximate thickness of the layer is also, of course, given by its volume above, divided by the waterplane area

$$A_W = 12\,\text{TPI}_1 u_1 = \frac{12\,\text{TPI}_1}{w_1}$$

Taking first moments of the volume about B the rise of the centre of buoyancy,

$$\overline{\text{BB}^1} = \frac{(\nabla_2-\nabla_1)\overline{\text{Bb}}}{\nabla_2} = \left(1-\frac{w_2}{w_1}\right)\overline{\text{Bb}}$$

It has been assumed that the increases in draught are small and that TPI is sensibly constant over the change. There are occasions when a greater accuracy is needed. What is the relationship between the change in displacement and TPI? It will be remembered from Chapter 2 (Fig. 2.19) that the change in volume of displacement,

$$\text{change in } \nabla = \int_{\text{WL}_1}^{\text{WL}_2} A_W\,\text{d}T$$

$$\therefore \quad \text{change in } \Delta = \int_{\text{WL}_1}^{\text{WL}_2} \frac{A_W}{u}\,\text{d}T = 12\int_{\text{WL}_1}^{\text{WL}_2} (\text{TPI})\,\text{d}T \quad \text{if } A_W \text{ is in ft}^2 \text{ and } T \text{ in ft}$$

Writing this another way,

$$\frac{1}{12}\frac{\text{d}\Delta}{\text{d}T} = \text{TPI}$$

or for one metre trim, $\dfrac{\Delta\overline{\text{GM}_\text{L}}}{L}$

With an adjustment for units, the slope of the displacement curve with respect to draught, gives the TPI; displacement is the integral curve of TPI, i.e. displacement is represented by the area under the curve of TPI plotted against draught.

Displacement of a ship is calculated with some accuracy during its early design to a series of equally spaced waterlines. Although the curve of displacement is plotted, it cannot be read to the required accuracy at intermediate waterlines. To find the displacement at an intermediate waterline, therefore, to the nearest tabulated displacement is added (or subtracted) the integral of the TPI curve between the tabulated waterline and the required waterline. For example, suppose Fig. 3.10 shows an extract of the curves of displacement and TPI at the 19 ft and 21 ft waterlines. The displacement at say 19 ft $10\frac{1}{2}$ in. cannot accurately be read from the displacement curve; TPI can be read with sufficient accuracy at say, 19 ft, 19 ft $5\frac{1}{4}$ in. and 19 ft $10\frac{1}{2}$ in., integrated by Simpson's first rule and the displacement of the slice added to 4921 to give 5471, to an accuracy of 1 tonf, if such accuracy is needed. Worked example 4, illustrates this.

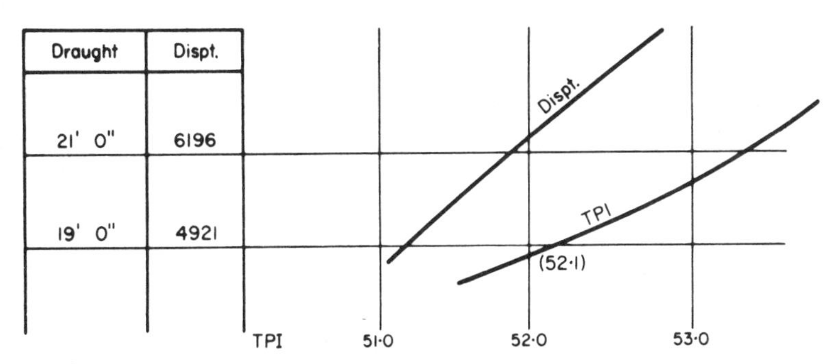

Fig. 3.10

Trim

Trim is the difference in draughts forward and aft. An excess draught aft is called trim *by the stern*, while an excess forward is called trim *by the bow*. It is important to know the places at which the draughts are measured and trim, unless it is obvious, is usually referred to *between perpendiculars* or *between marks*. Used as a verb, trim refers to the act of angular rotation about the Oy-axis, from one angular position to another. There is also another important use of the word trim which must be excluded for the time being; in relation to submarines, trim is also the relationship between weight and buoyancy. So far as surface ships are concerned,

$$\text{trim} = T_A - T_F$$

and the angle of trim $\theta = \dfrac{T_A - T_F}{L}$

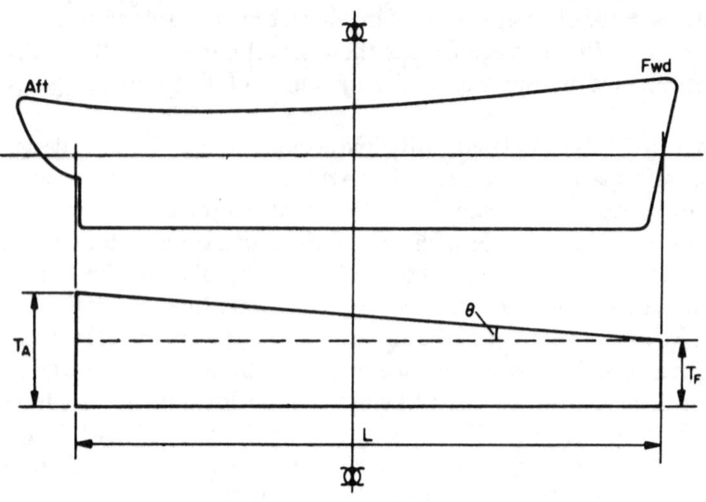

Fig. 3.11 Trim

where L is the horizontal distance between the points at which T_A and T_F are measured.

If a ship is trimmed without change of displacement, it must rotate about the centre of flotation. This can be shown by writing down the condition that there shall be no change in displacement, i.e. that the volumes of emerged and immersed wedges are equal (Fig. 3.12)

$$2\int y_F(x_F\theta)\,dx = 2\int y_A(x_A\theta)\,dx$$

i.e. $\int xy\,dx$ forward $= \int xy\,dx$ aft, which is the condition for the centre of area of the waterplane.

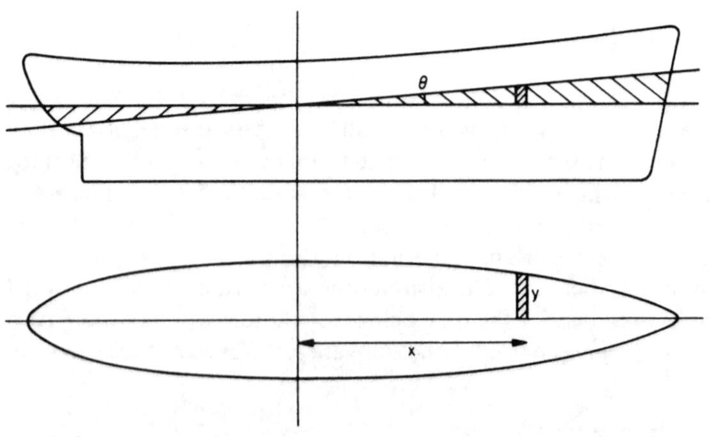

Fig. 3.12

A ship trims, therefore, about the centre of flotation of the waterplane. If a small weight is added to the ship, to avoid trim it must be added over the centre of

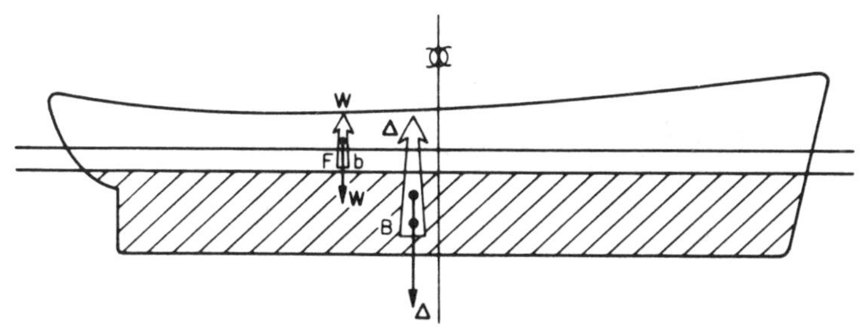

Fig. 3.13 Addition of weight at the centre of flotation

flotation; the centre of buoyancy of the additional layer will also be at the centre of flotation and there is no out-of-balance moment (Fig. 3.13).

CHANGES OF DRAUGHT WITH TRIM

Because a ship trims about the centre of flotation, the draught at that position does not alter with trim. It does alter everywhere else, including amidships where the mean draught between perpendiculars occurs. Now, the displacement of a ship is recorded for different mean draughts but at a specified design trim. Should the ship be floating at a different trim, it is necessary to imagine the ship being trimmed about the centre of flotation (which does not alter the displacement) until it is brought to the designed trim; the revised mean draught is calculated and so the displacement pertinent to that draught can be found. Suppose that a ship is floating t out of designed trim at waterline WL and needs to be brought to the designed trim W_1L_1 (Fig. 3.14). The change of trim $t = W_1W + LL_1 = NL_1$. By the principle of similar triangles, the correction to the mean draught,

$$d = \frac{a}{L}t$$

where a is the distance of the CF from amidships. From this revised mean draught, the displacement can be found from the ship's hydrostatic data. Had the displacement been found first from the hydrostatic data for the ship floating at the actual mean draught, it would have been necessary to correct the displacement for the amount out of designed trim; this is clearly the displacement of a layer of thickness at/L which is

$$\text{trim correction} = \frac{at}{L}\,.\,\text{TPI}$$

Whether this is positive or negative depends upon the position of the CF and whether the excess trim is by bow or stern. Instead of attempting to remember rules to govern all cases, the student is advised to make a sketch similar to Fig. 3.14 on every occasion. For the case drawn, when the CF is abaft amidships and there is an excess trim by the bow, the trim correction is negative, i.e. the displacement given by the hydrostatic data for the actual mean draught to waterline WL is too great.

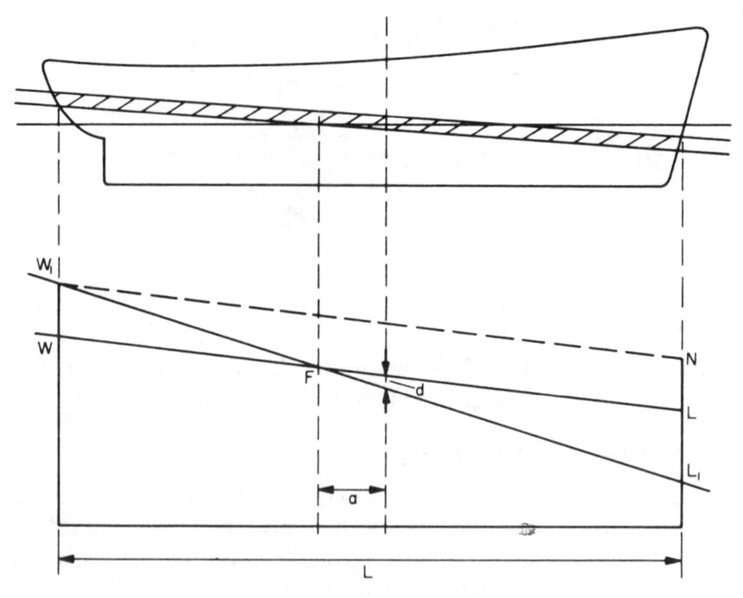

Fig. 3.14

The principle of similar triangles is especially useful in calculating draughts at various positions along the ship, e.g. at the draught marks which are rarely at the perpendiculars. To calculate the effect of changes of trim on draughts at the marks, a simple sketch should be made to illustrate what the change at the marks must be; in Fig. 3.15, a change of trim t between perpendiculars causes a change of draught at the forward marks of bt/L and at the after marks of ct/L. Remember, all distances are measured from the CF!

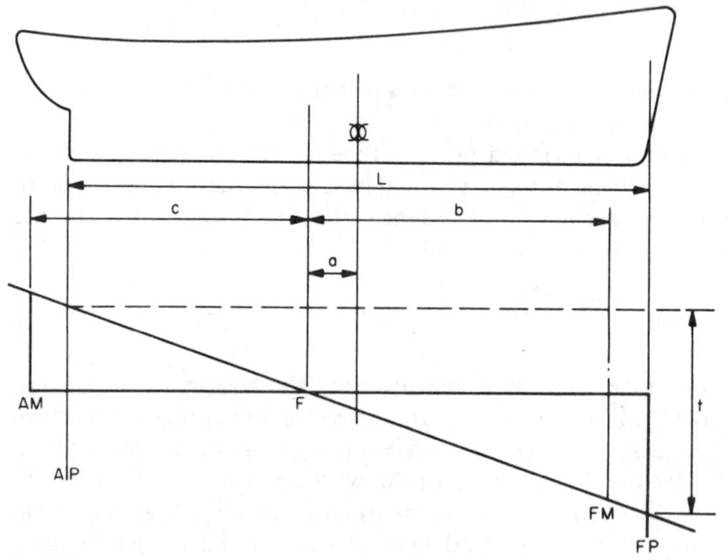

Fig. 3.15

MOMENT CAUSING TRIM

The trim rotation θ is caused by a moment M. What is the relationship between M and θ? In Chapter 2, we examined the rotation of any floating body, and discovered that it appeared to rotate about a point called the metacentre which was at a height I/∇ above the centre of buoyancy. Examine now the particular case of rotation in the longitudinal vertical plane. Let us apply a moment M at the CF causing a rotation θ, immersing a wedge of water aft and causing a wedge to emerge forward. The change in underwater shape due to the movement of the wedges, causes B to move aft to, say, B_1. The buoyancy Δ acts vertically upwards, at right angles to the new waterline; the weight continues to act vertically downwards through the centre of gravity G which has not moved. The application of a moment M therefore has caused a couple $\Delta\overline{GZ}$, $\overline{GZ}$ being the perpendicular distance between the lines of buoyancy and weight, i.e.

$$M = \Delta\overline{GZ}$$

For small angles, sin θ is approximately equal to θ

$$\therefore \quad M = \Delta\overline{GM}_L\theta$$

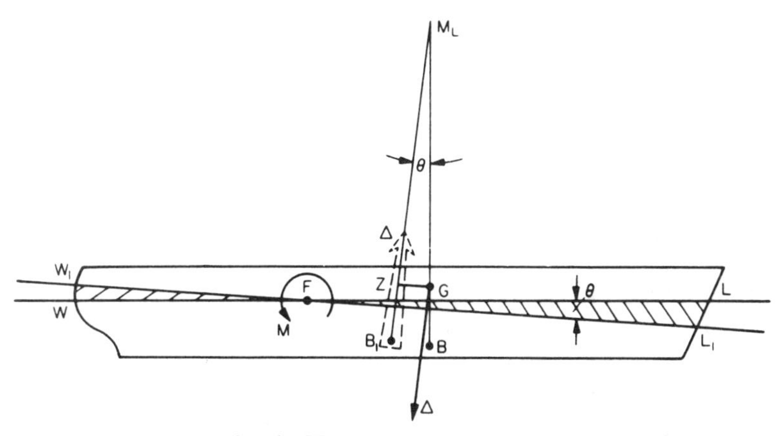

Fig. 3.16 Moment applied at the CF

If the angle of trim is such as to cause one inch of trim, $\theta = 1/12L$ and the moment to cause one inch of trim is

$$\Delta\overline{GM}_L\frac{1}{12L} \quad \text{or for one metre trim,} \quad \frac{\Delta\overline{GM}_L}{L}$$

This is an important new tool for calculating changes of trim and it is tabulated for each waterline in the hydrostatic data,

$$\textit{Moment to change trim one inch}, \text{ MCT1}'' = \frac{\Delta\overline{GM}_L}{12L}$$

Thus, the change of trim in inches, caused by the application of a moment M is given by $M/\text{MCT1}''$. Just as the length over which the trim is measured requires specifying, so does the MCT1″ require the length to be stated and, frequently, this standard is MCT1″ BP (between perpendiculars).

In practice, $\overline{GM}_L$ and $\overline{BM}_L$ are both large numbers and close together. ($\overline{BG}$ may be one per cent of $\overline{BM}_L$.) An approximation to MCT1″ is given by

$$\text{MCT1}'' \text{ approx.} = \frac{\Delta\overline{BM}_L}{12L}$$

$$= \frac{\Delta}{12L}\frac{I_L}{\nabla} = \frac{I_L}{420L} \text{ for salt water} \qquad \text{Equation (iii)}$$

In metric units, the moment to change trim one metre or the *one metre trim moment* is $\Delta\overline{GM}_L/L$ tonnef m with Δ in tonnef or MN m with Δ in MN. For salt water, approximately,

$$\left.\begin{aligned}\text{One metre trim moment} = \frac{\Delta\overline{BM}_L}{L} &= 1{\cdot}025\frac{I}{L} \text{ tonnef m}\\ &= 0{\cdot}01005\frac{I}{L} \text{ MN m}\end{aligned}\right\} \text{Equation (iii)}$$

with I and L in metre units.

$$\text{One metre trim moment} = 12{\cdot}19 \times \text{MCT1}'' \text{ tonnef m}$$

$$= 0{\cdot}120 \times \text{MCT1}'' \text{ MN m}$$

MCT can be used for the general case, typical values being:

	tonf ft	tonnef m	MN m
60 ft (18 m) fishing vessel	2·5	30	0·3
300 ft (90 m) submarine, surfaced	150	1800	18
350 ft (110 m) frigate	450	5500	54
500 ft (150 m) dry cargo vessel	1750	20000	210
800 ft (240 m) aircraft carrier	5500	70000	650
1000 ft (300 m) supertanker	16000	200000	1900

ADDITION OF WEIGHT

It is a principle of applied mechanics that, for statical considerations, the effects of a force P acting on a body are exactly reproduced by a parallel force P at a distance h from the original line plus a moment Ph. In other words, the body shown in Fig. 3.17 behaves in precisely the same way under the action of the force and moment shown dotted as it does under the action of the force shown full. This is a useful principle to apply to the addition of a weight to a ship which causes it to sink by a shape awkward to assess. If the weight be assumed replaced by a force W and a moment Wh at the centre of flotation, the separate effects of each can be readily calculated and added:

(*a*) the force W at the centre of flotation causes a parallel sinkage W/TPI,
(*b*) the moment Wh causes a change of trim Wh/MCT

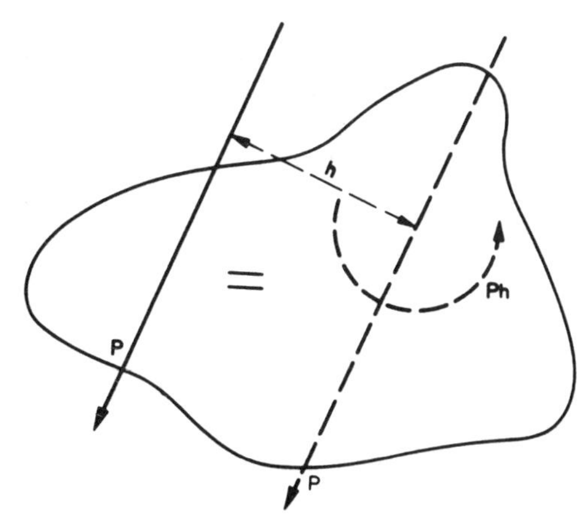

Fig. 3.17

The changes of draught at any position of the ship's length can be calculated for each movement and added, algebraically, taking care to note the sign of each change. See worked example 5.

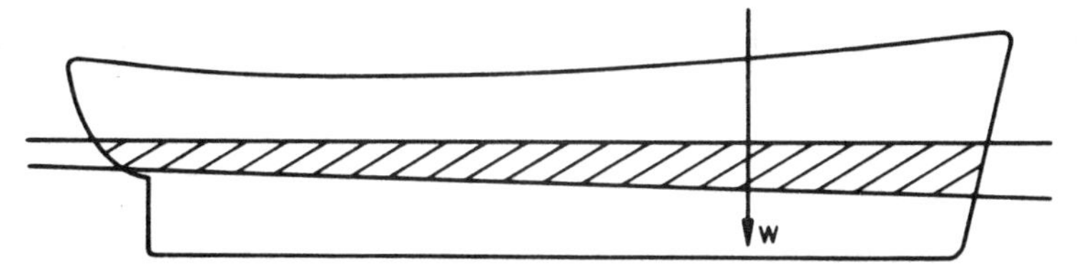

Fig. 3.18

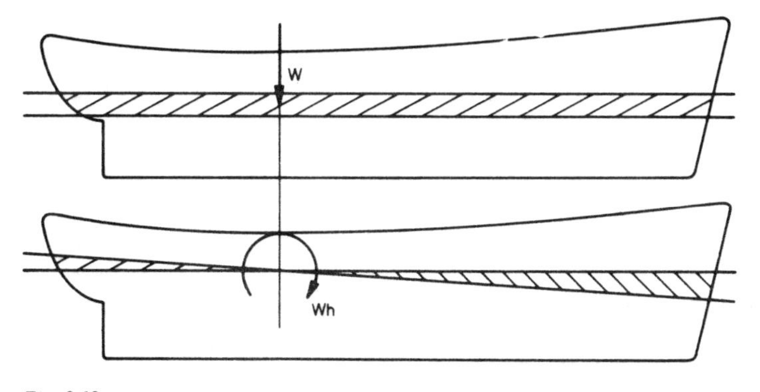

Fig. 3.19

One particular case of added weight (a negative addition, in fact) is worthy of special mention. When a ship grounds or docks, there is an upwards force at the point of contact. This is assumed replaced bv a vertical force at the CF plus a moment which cause, respectively, a parallel rise and a trim. What causes the force is of little interest in this context. A fall of water level, equal to the change of draught at the point of contact is calculated from the combined effects of a parallel rise and a trim. Any attempt to treat the problem in one step is likely to meet with failure and the approach of Fig. 3.20 is essential. Worked example 6 illustrates the method.

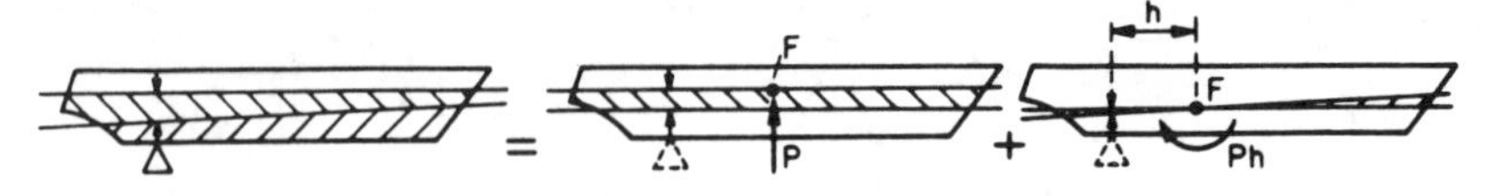

Fig. 3.20

While we are now in a position to assess the effects of a small added weight, let us return for a moment to the exact behaviour of the ship so that we have a real understanding of what is happening. Referring to Fig. 3.21, before the addition of weight W, B, G and M were in a vertical line. The addition of the weight alters the position of all three but they must, of course, finally be in a straight line, a line perpendicular to the new waterline. What are the movements of B, G and M? G must move along a straight line joining G and g and the distance it moves could be calculated by an expression similar to Equation (ii) on p. 58. B moves upwards along the straight line Bb_1 due to the parallel sinkage to B_1 and then moves to B_2 such that B_1B_2 is parallel to bb. $\overline{BB_1'}$ can be calculated from Equation (i) on p. 57. M moves up by virtue of B moving up to B_1 and, assuming I constant, down because ∇ has increased; $\overline{BM} = I/\nabla = wI/\Delta$ and $\overline{B_1M_1} = wI/(\Delta + W)$ so that

$$\overline{BM} - \overline{B_1M_1} = wI\left(\frac{1}{\Delta} - \frac{1}{\Delta + W}\right)$$

$$= \frac{wIW}{\Delta(\Delta + W)}$$

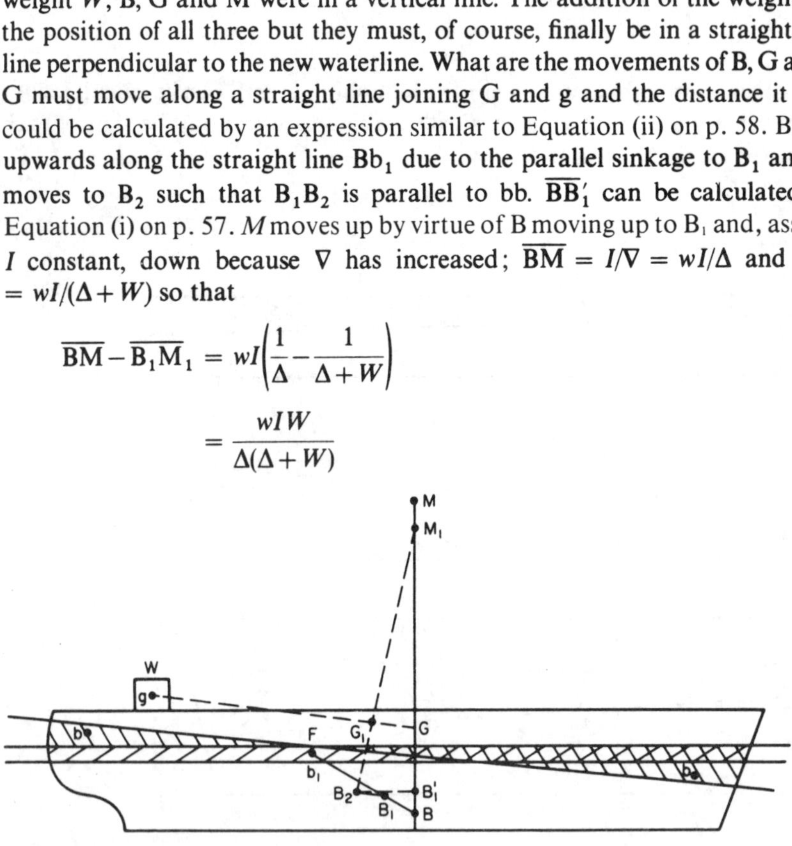

Fig. 3.21

We have assumed that I has not appreciably changed and that the vertical component of $\overline{B_1B_2}$ is negligible. Thus, from the three movements calculated, a drawing could be made and the angle of trim measured. Clearly, the use of MCT is quicker.

LARGE WEIGHT ADDITIONS

While the use of MCT and TP1 is permissible for moderate additions of weight, the assumptions implicit in their use break down if the addition becomes large or the resulting trim is large. No longer is it possible to assume that I does not change, that TPI is constant or the longitudinal position of the CF unchanged. An approach similar to Fig. 3.21 is essential, viz.:

(*a*) calculate new position of G as already described or by considering the two components,

$$\overline{GG'} = \frac{Wc}{\Delta + W} \quad \text{and} \quad \overline{G'G_1} = \frac{Wd}{\Delta + W}$$

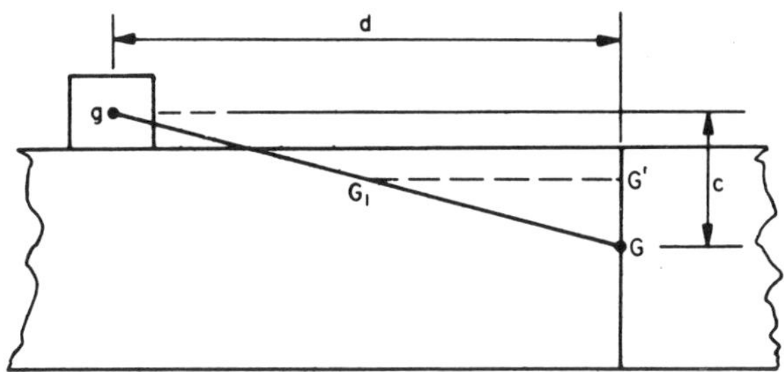

Fig. 3.22 Movement of G

(*b*) calculate new position of B in two steps

(i) the change in B due to the addition of a parallel slab; the displacement of the slab must, of course, be equal to the additional weight and its centre of buoyancy b_1 may be assumed on a line joining the centres of flotation of the two waterplanes at a height determined by the areas of the two waterplanes. Then

$$\overline{BB_1'} = \frac{We}{\Delta + W} \quad \text{and} \quad \overline{B_1'B_1} = \frac{Wf}{\Delta + W}$$

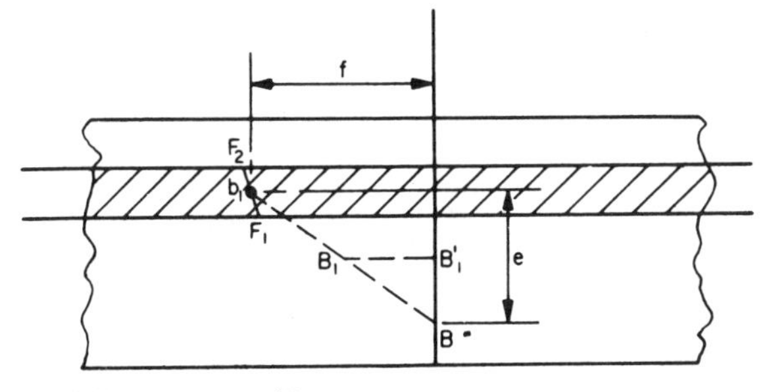

Fig. 3.23 Movement of B

(ii) the change in position of B_1 due to the trimming of the waterplane (Fig. 3.24). The geometry of the wedges must be examined and the positions of b found by integrator or rule. At this stage however, we encounter a

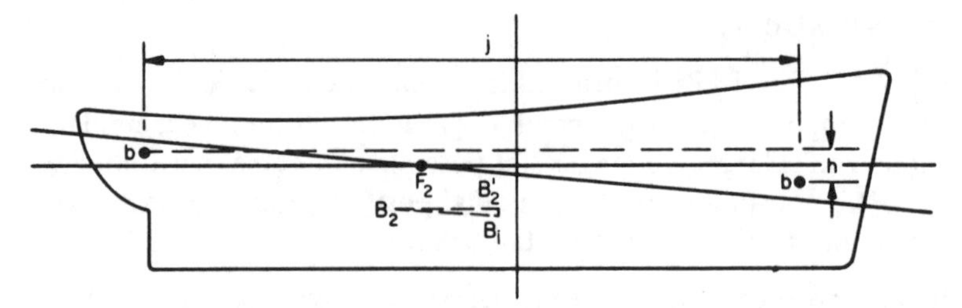

Fig. 3.24 Trim effects on $\mathbf{B}_1$

difficulty—we do not know what θ is. The calculation is therefore made for two or more trial values of θ, giving positions of B_2 applicable to each. As before

$$\overline{B_1 B_2'} = \frac{Wh}{\Delta + W} \quad \text{and} \quad \overline{B_2' B_2} = \frac{Wj}{\Delta + W}$$

(*c*) examine the trigonometry of B and G, knowing that they must be in vertical lines (Fig. 3.25). Having plotted the trial values of B_2, the choice of correct position of B_2 to correspond to the correct θ is best carried out graphically by trial and error. Having made such a judgment, the position is re-checked.

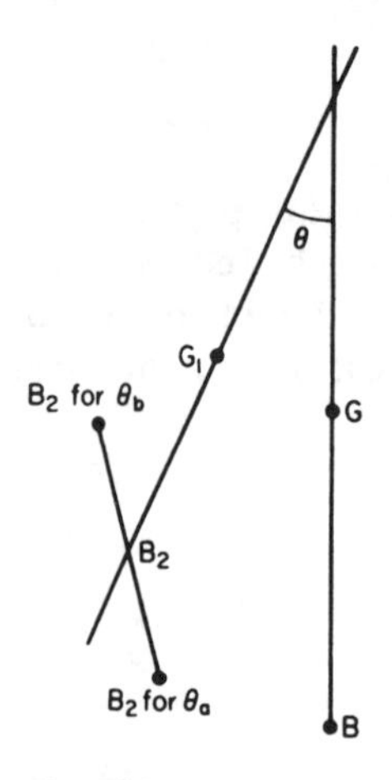

Fig. 3.25

This is a laborious process. A preferable approach is to regard the ship as an entirely new one with a revised position of G and a trim found from the Bonjean curves to give B vertically beneath G and the correct displacement. This, in fact, is how the design trim of the ship is initially determined.

DETERMINATION OF DESIGN TRIM

As described later, the early stages of ship design involve a trial and error process of matching the longitudinal positions of B and G. For a given displacement and position of G, it will at some stage be necessary to find the position of the LCB and the trim resulting. Similar requirements occur in bilging and launching calculations, which are discussed in later chapters.

To find the draught and trim corresponding to a specified displacement and LCG position, it is necessary first to estimate the approximate likely waterline. Accuracy in the estimate is not important. The trial waterline is placed on a profile on which the Bonjean curve has been drawn for each ordinate so that the areas up to the waterline can be read off (Fig. 3.26). These areas and their longitudinal moments are then integrated in the usual fashion to find the displacement and LCB position. The waterplane offsets are then found, by projecting the waterline on to an offset body plan and the properties of the waterplane, in particular LCF, TPI and MCT (using the approximation of Equation (iii) on p. 66) are calculated. The ship may then be subjected to the calculations for parallel sinkage to alter the trial waterline to a new one giving the required displacement; it may then be trimmed to give the required position of LCB. If the parallel sinkage or trim are large to achieve this, a second trial waterline will be necessary and a similar adjustment made.

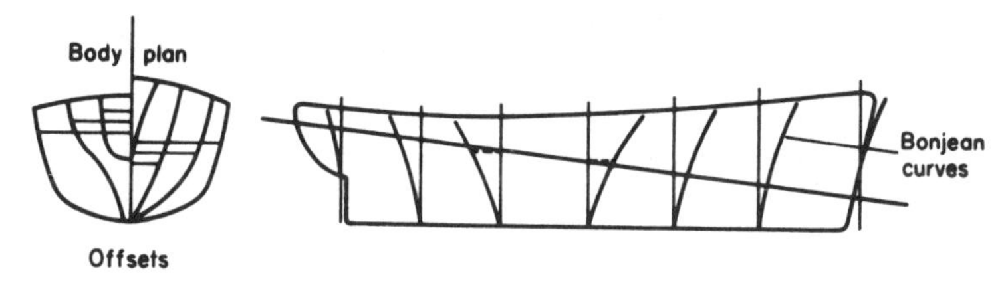

Fig. 3.26

CHANGE OF WATER DENSITY

A ship moving from sea water to river will be subjected to a parallel sinkage and a trim. Let us examine why.

It is necessary, to study the problem clearly, again to separate parallel sinkage and trim. Because the river water has a smaller weight density and is therefore less buoyant, the ship will sink deeper. We have already found on p. 60 that

$$\text{the volume of layer} = \nabla_2\left(1-\frac{w_2}{w_1}\right)$$

and

$$\text{weight of layer} = \nabla_2 w_2\left(1-\frac{w_2}{w_1}\right)$$

$$= \Delta\left(1-\frac{w_2}{w_1}\right)$$

w being weight density, not the reciprocal.

Now the volume that previously provided sufficient buoyancy had its centre of buoyancy at B, in line with G. The addition of a layer, which has its centre of buoyancy at F, the centre of flotation, causes an imbalance. As drawn (Fig. 3.27), a trim by the bow is caused and the wedges of added and lost buoyancy introduced by the trim restore equilibrium (Fig. 3.28).

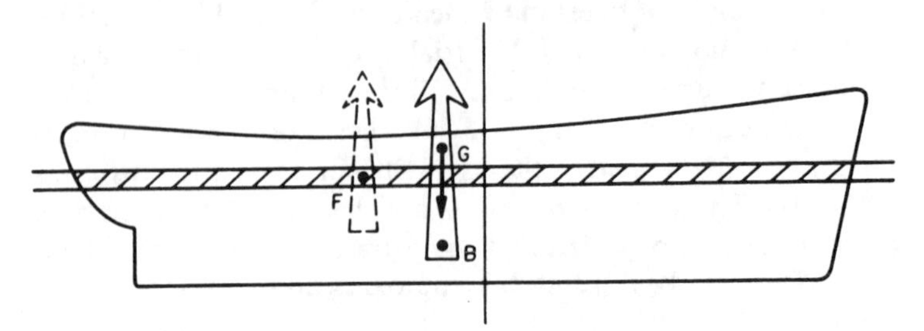

Fig. 3.27

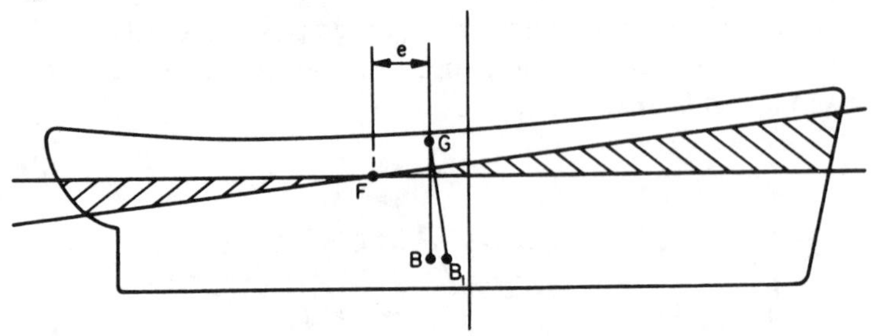

Fig. 3.28

If e is the horizontal distance between LCF and LCB, the moment causing trim is

$$\Delta e\left(1-\frac{w_2}{w_1}\right)$$

The trim is then

$$\frac{\Delta e}{\text{MCT}}=\Delta e\left(1-\frac{w_2}{w_1}\right)\times\frac{\text{L}}{\Delta\text{GM}_\text{L}}=\text{approx }\Delta e\left(1-\frac{w_2}{w_1}\right)\times\frac{L}{Iw_2}$$

It must be remembered that the trim occurs in w_2 water and MCT must be related to that; if the value of MCT is given for w water, approximately, since MCT is proportional to wI/L,

$$\text{MCT}_2=\text{MCT}\times\frac{w_2}{w}$$

It is not difficult to produce formulae for changes in draught for particular changes of condition and to invent rules which give the direction of trim. The student is advised to commit none to memory. Each problem should be tackled with a small diagram and an understanding of what actually happens. Summarizing, the process is

(*a*) using Archimedes' principle, calculate the volume and thence the weight of the layer,

(*b*) calculate the parallel sinkage using TPI pertinent to the new water,
(*c*) calculate the trim due to the movement of the layer buoyancy from B to F, using MCT pertinent to the new water. Worked example 7 is an illustration.

Hydrostatic data

Throughout its life, a ship changes its weight and disposition of cargo, its draught and trim, and freeboard; the density of the water in which it floats varies. Its stability, discussed later, also changes. If its condition in any stated set of circumstances is to be estimated, its condition in a precise state must be known so that the effect of the changes from that state can be calculated. This precise state is known as the design condition. For ocean going ships, it is calculated for sea water of reciprocal weight density 35 ft^3/tonf; (occasionally, for ships confined to fresh water, 36 ft^3/tonf is used). In metric units, a specific volume of 99·5 m^3/MN or 0·975 m^3/tonnef is employed for salt water. For this water the tools necessary to calculate changes from the design or load waterline are calculated for a complete range of waterlines. Also calculated are the geometrical properties of the underwater form for the range of waterlines.

Collectively, this information is known as the hydrostatic data. It is presented either in tabular form when intermediate values are interpolated or as a set of curves, which are called the *hydrostatic curves.*

HYDROSTATIC CURVES

The following properties are plotted against draught to form the hydrostatic curves

(i) Centre of bouyancy above keel, $\overline{KB}$
(ii) Transverse metacentre above keel, $\overline{KM}$
(iii) CB aft of amidships
(iv) CF aft of amidships
(v) Displacement
(vi) Tonf per inch immersion (tonnef per centimetre immersion)
(vii) Change of displacement for one inch (one metre) change of trim
(viii) Moment to change trim one inch (one metre trim moment).

Occasionally, certain other properties are also plotted, such as longitudinal $\overline{KM}$, area of wetted surface and some of the geometric coefficients.

The first four items are all items of ship geometry and are affected neither by the density of the water nor by the ship's weight. Items (v), (vi) and (vii) are all related to the weight density of the water. Item (viii) is also related to the weight density of the water and alone is affected by the vertical position of the ship's centre of gravity, albeit only slightly. Accepting the approximation of Equation (iii) on p. 69, all of these last four properties are corrected in the same way when their values are required in water of different densities; if the property is known in water of reciprocal weight density u_1 and it is required in water of reciprocal density u_2,

$$\text{property in } u_2 \text{ water} = \text{property in } u_1 \text{ water} \times \frac{u_1}{u_2}$$

u_1 will often be equal to 35 ft³/tonf or 0·975 m³/tonnef.

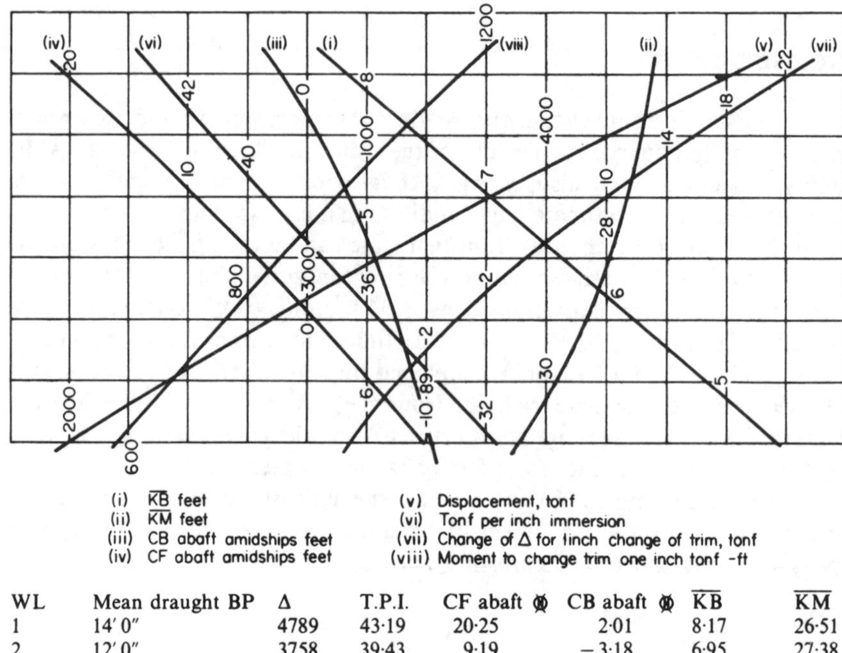

WL	Mean draught BP	Δ	T.P.I.	CF abaft ⊗	CB abaft ⊗	$\overline{KB}$	$\overline{KM}$
1	14′ 0″	4789	43·19	20·25	2·01	8·17	26·51
2	12′ 0″	3758	39·43	9·19	−3·18	6·95	27·38
3	10′ 0″	2851	35·58	−0·39	−7·17	5·76	28·79
4	8′ 0″	1998	31·69	−9·94	−10·31	4·57	31·01

HYDROSTATIC CURVES SHIP No. 7329
Length BP = 507′ 2″, Breadth extreme = 53′ 1″, Draught to 1WL = 14′ 0″, Trim BP = zero.

Fig. 3.29 Typical hydrostatic curves

CALCULATION OF HYDROSTATIC DATA

The traditional manner of calculating the hydrostatic data is so methodical that it lends itself very readily to programming for a computer. Each of the properties listed above is calculated for each of the waterlines of the ship by the methods already described in this and the previous chapters. The offsets are set down in tabular form as shown in Fig. 3.30 which has been compiled for illustration purposes for a simple ship shape—in practice 21 ordinates and several more waterlines are usual. Offsets are set down in tabular fashion to correspond to the orthogonal sections explained by Fig. 2.3. The table is known as the *displacement table* and the complete sheet of calculations as a *displacement sheet*.

The waterline offsets are set down in vertical columns and each is operated upon by the vertical column of Simpson's multipliers. The sums of these columns are functions (i.e. values proportional to) of waterplane areas which, operated upon by the horizontal multipliers and summed give a function of the displacement. Levers applied to these functions result in a function of VCB. This whole process applied, first, to the ordinates by the horizontal multipliers and, secondly, to the functions of ordinate areas by the vertical multipliers and levers results in a function of displacement and a function of LCB. Clearly, the two functions of displacement should be the same and this comprises the traditional corner check.

Ordinate number	S.M.	WATERLINES 5		4		3		2		1		Vertical sections FNS of area	Mults of area	Lever	FNS for momts
		1		4		2		4		1					
1	1	**1**	1	**1**	1	**1**	1	**1**	1	**1**	1				
		1		4		2		4		1		12	12	3	36
2	4	**5**	20	**6**	24	**7**	28	**8**	32	**9**	36				
		5		24		14		32		9		84	336	2	672
3	2	**7**	14	**8**	16	**9**	18	**10**	20	**11**	22				
		7		32		18		40		11		108	216	1	216
4	4	**9**	36	**10**	40	**11**	44	**12**	48	**13**	52				
		9		40		22		48		13		132	528	0	924
5	2	**6**	12	**7**	14	**8**	16	**9**	18	**10**	20				
		6		28		16		36		10		96	192	1	192
6	4	**4**	16	**5**	20	**6**	24	**7**	28	**8**	32				
		4		20		12		28		8		72	288	2	576
7	1	**1**	1	**1**	1	**1**	1	**1**	1	**1**	1				
		1		4		2		4		1		12	12	3	36
FNS of area		100		116		132		148		164			1584		804
Mults. area		100		464		264		592		164		1584			120
Levers		4		3		2		1		0					
FNS for mts.		400		1392		528		592		0		2912			

Fig. 3.30 A simplified displacement sheet

There will frequently be an appendage below the lowest convenient waterline which must be added to the main body calculations. Areas of this appendage at each ordinate are measured by planimeter and positions of centres of areas either spotted or, if the appendage is considerable, measured by integrator. The effects of the appendage are then included, as described in Chapter 2.

With the offsets for each waterplane so conveniently presented by the displacement table, it is a simple matter to calculate by the methods described in Chapter 2, area, position of CF and the second moments of area for each waterplane. Displacements and CB positions are calculated by adding or subtracting slices from the nearest waterplane calculations.

If Tchebycheff spacings of ordinates are adopted, the Simpson multipliers are, of course, dispensed with in one direction. A form of displacement table in which this is done is not unusual.

The lengthy process of direct calculation has been replaced by computer processing. The process clearly comprises a large number of simple arithmetical steps performed in a logical sequence and repeated for different input data. Offsets may be orthogonal as in Fig. 3.30 or radial. This type of calculation is very readily adaptable to digital computing and programs are available for most common computers often combined with the calculations of later chapters.

THE METACENTRIC DIAGRAM

As has already been discussed, the positions of B and M are dependent only on the geometry of the ship and the waterplane at which it is floating. They can, therefore, be determined without any knowledge of the actual condition of ship loading which causes it to float at that waterline.

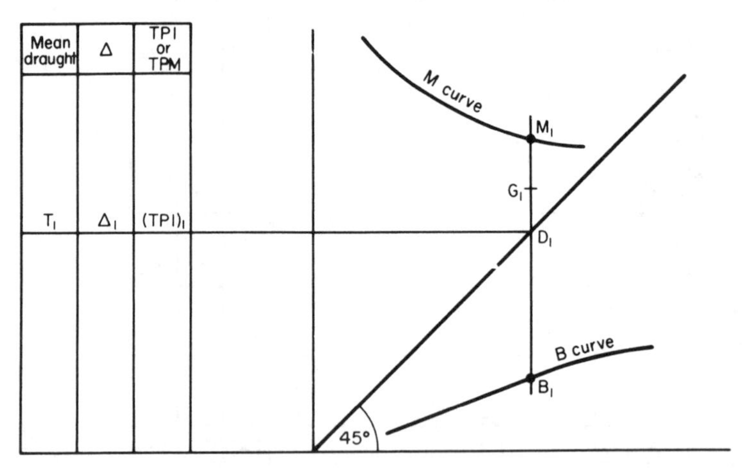

Fig. 3.31 Metacentric diagram

The metacentric diagram is a convenient way of defining variations in relative heights of B and M for a series of waterlines parallel to the design or load waterplane. Such a diagram is shown in Fig. 3.31 and it is constructed as follows. The vertical scale is used to represent draught and a line is drawn at 45 degrees to this scale. For a given draught T_1 a horizontal line is drawn intersecting the 45 degree line in D_1 and a vertical line is drawn through D_1. On this vertical line, a distance $\overline{D_1M_1}$ is set out to represent the height of the metacentre above the waterplane and $\overline{D_1B_1}$ to represent the depth of the centre of buoyancy below the waterplane. This process is repeated a sufficient number of times to define adequately the loci of the metacentre and centre of buoyancy. These two loci are termed the M and B curves. As drawn, M_1 lies above the waterplane but this is not necessarily always the case and the M curve may cross the 45 degree line. A table is constructed to the left of the diagram, in which are listed the displacements and TPI values for each of a number of draughts corresponding to typical ship conditions. These conditions must, strictly, be reduced to a common trim so that the loci of M and B will be continuous. In practice, however, not much error is involved in using the diagram for conditions other than the standard condition provided that the difference in trim is not excessive and provided that the draught is that at the centre of flotation.

In addition to the M and B curves, it is usual to show the positions of G for the conditions chosen. These positions can be obtained by methods discussed in Chapter 4.

Since $\overline{KB}$ is approximately proportional to draught over the normal operating range, the B curve is usually nearly straight for conventional ship shapes. The M curve, on the other hand, usually falls steeply with increasing draught at shallow draught then levels out and may even begin to rise at very deep draught.

Worked examples

EXAMPLE 1. Construct a metacentric diagram for a body with constant triangular cross-section floating with its apex down and at level draught. Dimensions of cross-section are shown in the figure.

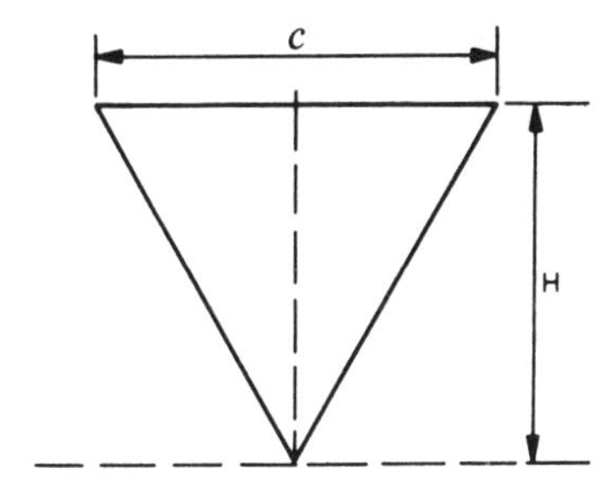

Fig. 3.32

Solution:

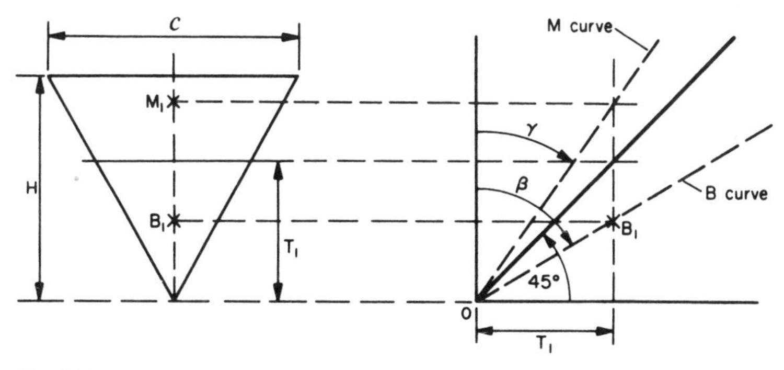

Fig. 3.33

For a given draught T_1, $\overline{KB} = \frac{2}{3}T_1$. Hence, the B curve is a straight line passing through 0, at an angle β to the vertical such that

$$\tan \beta = \frac{T_1}{\frac{2}{3}T_1}$$

i.e.

$$\beta = \tan^{-1} 1{\cdot}5$$

Breadth of waterplane at draught $T_1 = c\dfrac{T_1}{H}$

$$\therefore \quad \overline{B_1M_1} = \frac{I}{\nabla} = \frac{\frac{1}{12}(cT_1/H)^3 L}{\frac{1}{2}(cT_1/H)T_1 L}$$

$$= \frac{1}{6}\frac{c^2 T_1}{H^2}$$

$$\therefore \quad \overline{KM_1} = \overline{KB_1} + \overline{B_1M_1} = \frac{2}{3}T_1 + \frac{1}{6}\left(\frac{c^2}{H^2}\right)T_1$$

$$= \frac{T_1}{6}\left(4 + \frac{c^2}{H^2}\right)$$

Hence, in this case, since c and H are constants, the M curve is also a straight line through 0 but at an angle γ to the vertical such that

$$\tan\gamma = \frac{T_1}{T_1(4 + c^2/H^2)/6} = \frac{6H^2}{4H^2 + c^2}$$

EXAMPLE 2. Construct a metacentric diagram for a vessel of constant circular cross-section floating at uniform draught.

Solution:

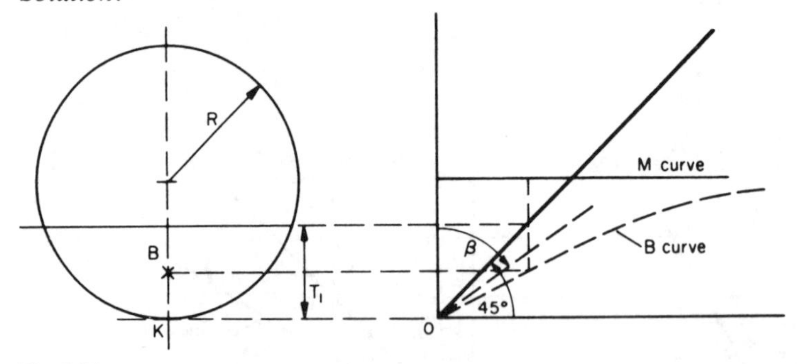

Fig. 3.34

For any inclination at any draught the buoyancy force must, from geometric considerations, pass through the centre of the circular cross-section. Hence, the M curve is a horizontal straight line.

The expression for $\overline{KB}$ is more complicated, but general considerations show that the curve must become tangential to the M curve when $T_1 = 2R$ and that its general form will be as indicated. If it is assumed that for small draughts the circle can be approximated to by a parabola, it can be shown that the tangent to the B curve at 0 is given by

$$\beta = \tan^{-1} \tfrac{5}{3}$$

EXAMPLE 3. When the tide falls, a moored barge grounds in soft homogeneous mud of specific gravity 1·35. With the keel just touching the mud, the barge was drawing a level 1·65 m of salt water of reciprocal weight density 0·975 m³/tonnef. The tide falls a further 18 cm.

Assuming that the barge is box shaped 25 m long and 5 m in breadth, calculate the draught after the tide has fallen. What weight must be removed to free the barge from the mud?

Solution: When floating free

$$\Delta = \frac{1{\cdot}65 \times 25 \times 5}{0{\cdot}975}\,\text{m}^3\,\frac{\text{tonnef}}{\text{m}^3} = 211{\cdot}5 \text{ tonnef.}$$

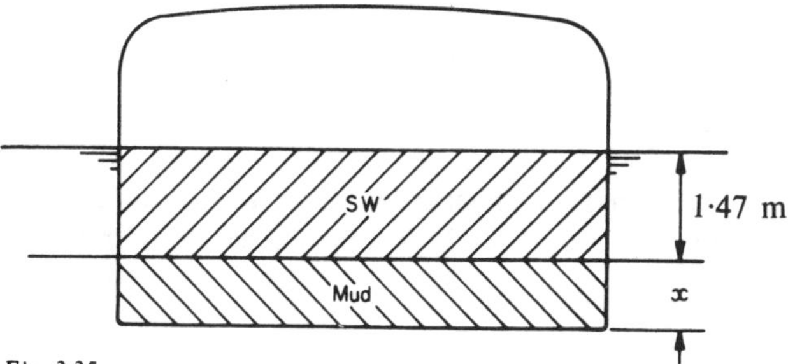

Fig. 3.35

When in mud,

$$\text{buoyancy supplied by salt water} = \frac{1{\cdot}47}{1{\cdot}65} \times 211{\cdot}5 = 188{\cdot}4 \text{ tonnef}$$

$$\text{buoyancy needed from mud} = 211{\cdot}5 - 188{\cdot}4 = 23{\cdot}1 \text{ tonnef}$$

$$\text{buoyancy provided by mud} = x \times 25 \times 5 \times 1{\cdot}35 = 23{\cdot}1$$

$$\therefore \quad x = \frac{23{\cdot}1}{25 \times 5 \times 1{\cdot}35} = 0{\cdot}14 \text{ m}$$

$$\therefore \quad \text{draught when tide has fallen} = 1{\cdot}61 \text{ m}$$

$$\text{weight to be removed to clear mud} = 23{\cdot}1 \text{ tonnef}$$

(the draught x could also have been obtained by taking a proportion of the 18 cm fall in tide in the ratio of the S.Gs of mud and salt water, $(1{\cdot}35 \times 0{\cdot}975)\, x = 0{\cdot}18$.

EXAMPLE 4. Hydrostatic data for a ship show the following

Draught	19 ft 6 in.	21 ft 9 in.	24 ft 0 in.
Displacement (tonf)	8421	9927	11447
TPI	50·56	52·08	52·82

Calculate the displacement at a draught of 22 ft 8 in.

Solution: The TPI curve gives values of TPI at 22 ft $2\frac{1}{2}$ in. and 22 ft 8 in. of 52·27 and 52·44 by plotting

WL	TPI	SM	f(V)
22 ft 8 in.	52·44	1	52·44
22 ft $2\frac{1}{2}$ in.	52·27	4	209·08
21 ft 9 in.	52·08	1	52·08
			313·60

$$\text{displacement of addition} = \frac{5\cdot5}{3} \times 313\cdot6 \text{ in.} \frac{\text{tonf}}{\text{in.}}$$

$$= 574\cdot9 \text{ tonf}$$

$$\therefore \quad \Delta \text{ to 22 ft 8 in. waterline} = 10{,}502 \text{ tonf}$$

(the best obtainable from interpolation of the displacement curve would have been a straight line relationship giving $\Delta = 9927 + \frac{11}{27} \times (11{,}447 - 9927) = 10{,}546$ tonf). We would have obtained a sufficiently accurate answer in this case by assuming the TPI to be constant at its 22 ft $2\frac{1}{2}$ in. value, linearly interpolated between the two top waterplanes, i.e.

$$52\cdot08 + \frac{5\cdot5}{27} \times 0\cdot74 = 52\cdot23$$

then

$$\Delta = 9927 + 11 \times 52\cdot23 = 10{,}502 \text{ tonf}$$

EXAMPLE 5. During a voyage, a cargo ship uses up 320 tonnef of consumable stores and fuel from the fore peak, 85 m forward of midships.

Before the voyage, the forward draught marks 7 m aft of the forward perpendicular, recorded 5·46 m and the after marks, 2 m aft of the after perpendicular, recorded 5·85 m. At this mean draught, the hydrostatic data show the ship to have tonnef per centimetre = 44, one metre trim moment BP = 33,200 tonnef m BP, CF aft of midships = 3 m. Length BP = 195 m. Calculate the draught mark readings at the end of the voyage, assuming that there is no change in water density.

Solution:

$$\text{Parallel rise} = \frac{320}{4400} = 0\cdot073 \text{ m}$$

$$\text{Moment trimming} = 320(85+3) \text{ tonnef m.}$$

$$\text{Trim BP (195 m)} = \frac{320 \times 88}{33{,}200} = 0\cdot848 \text{ m.}$$

$$\text{Draught change at FM due to trim} = \frac{93\cdot5}{195} \times 0\cdot848 = 0\cdot406 \text{ m.}$$

$$\text{Draught change at AM due to trim} = \frac{96 \cdot 5}{195} \times 0 \cdot 848 = 0 \cdot 420 \text{ m.}$$

$$\text{New draught at FM} = 5 \cdot 46 - 0 \cdot 07 - 0 \cdot 41 = 4 \cdot 98 \text{ m.}$$

$$\text{New draught at AM} = 5 \cdot 85 - 0 \cdot 07 + 0 \cdot 42 = 6 \cdot 20 \text{ m.}$$

A clear sketch is essential for a reliable solution to problems of this sort.

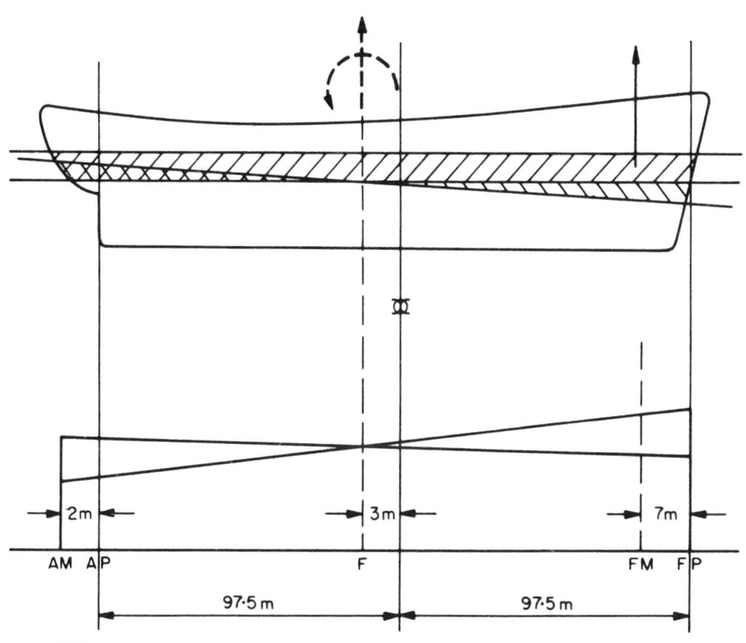

Fig. 3.36

EXAMPLE 6. A fishing vessel grounds on a rock at a point 13·4 m from the FP when the depth of water is 0·3 m above low tide. Before grounding, the vessel was drawing 1·60 m at the FP and 2·48 m at the AP. At this mean draught, TPI in MN per metre immersion = 3·5, one metre trim moment = 13 MN m and CF is 1 m forward of midships. The ship is 49 m BP. Calculate the force at the keel and the new draughts at low tide.

Solution: Let the force at the rock be P MN and replace it by a force P MN and a moment $10 \cdot 1P$ MN m at the CF.

$$\text{Parallel rise due to } P = \frac{P}{3 \cdot 5} \text{ m}$$

$$\text{Change of trim due to } 10 \cdot 1P = \frac{10 \cdot 1P}{13} \text{ m BP}$$

$$\text{Change of draught at rock due to trim} = \frac{10 \cdot 1}{49} \times \frac{10 \cdot 1P}{13} \text{ m}$$

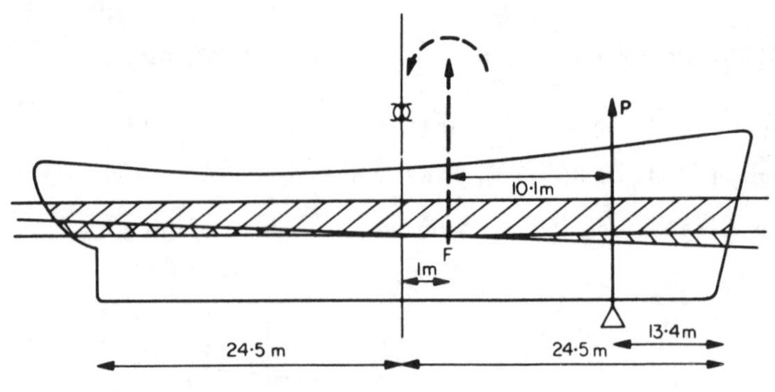

Fig. 3.37

$$\text{Total change of draught at rock} = \frac{P}{3{\cdot}5} + \frac{10{\cdot}1}{49} \times \frac{10{\cdot}1P}{13}\ \text{m}$$

$$= \text{fall of tide} = 0{\cdot}3\ \text{m}$$

$$\therefore \quad P(0{\cdot}286 + 0{\cdot}160) = 0{\cdot}3$$

$$\therefore \quad P = 0{\cdot}673\ \text{MN}$$

Then parallel rise due to $P = 0{\cdot}192$ m.

Change of trim due to $10{\cdot}1P = 0{\cdot}523$ m.

$$\text{Change of draught at FP due to trim} = \frac{23{\cdot}5}{49} \times 0{\cdot}523 = 0{\cdot}251\ \text{m}$$

$$\text{Change of draught at AP due to trim} = \frac{25{\cdot}5}{49} \times 0{\cdot}523 = 0{\cdot}272\ \text{m.}$$

New draught at FP $= 1{\cdot}600 - 0{\cdot}192 - 0{\cdot}251 = 1{\cdot}157$ m.

New draught at AP $= 2{\cdot}480 - 0{\cdot}192 + 0{\cdot}272 = 2{\cdot}560$ m.

(Because the final trim is large, it is unlikely that the answer can be relied upon to within 10 cm.)

EXAMPLE 7. The hydrostatic curves (prepared for water of reciprocal weight density 35 ft^3/tonf) for a guided missile destroyer show it to have the following particulars at a mean draught of 17 ft 2 in. at which it is floating with a 12 in. stern trim in water of reciprocal weight density 35·9 ft^3/tonf:

Length BP	560 ft
Designed trim	2 ft 6 in. by stern
Displacement	7940 tonf
CB abaft amidships	2·4 ft
CF abaft amidships	20·4 ft
TPI	69·4
MCT1″	1720 tonf ft

Calculate (*a*) the true displacement, (*b*) the new draught and trim when the ship puts to sea in water of reciprocal weight density 35·2 ft³/tonf.

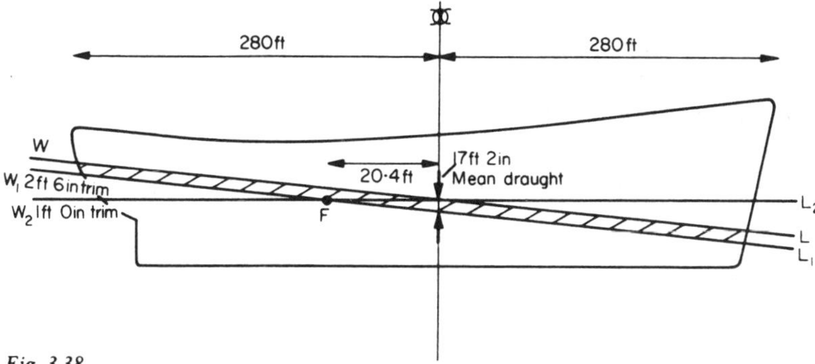

Fig. 3.38

Solution: As shown in Fig. 3.38, because the hydrostatic curves assume a 2 ft 6 in. trim, they overestimate the displacement by the shaded layer of thickness

$$\frac{20{\cdot}4}{560} \times 18 \text{ in.} = 0{\cdot}66 \text{ in.}$$

This layer is composed of 35·9 ft³/tonf water, so that the relevant TPI is

$$69{\cdot}4 \times \frac{35}{35{\cdot}9}$$

and the displacement of the layer

$$= 0{\cdot}66 \times 69{\cdot}4 \times \frac{35}{35{\cdot}9} = 44{\cdot}5 \text{ tonf}$$

$$\text{Now the displacement to WL} = 7940 \times \frac{35}{35{\cdot}9}$$

$$= 7940 - 7940\left(\frac{35{\cdot}9 - 35}{35{\cdot}9}\right)^*$$

$$= 7741 \text{ tonf}$$

$\therefore$ displacement to W_1L_1 (and W_2L_2) $= 7741 - 44{\cdot}5 = 7696$ tonf. (a)

In moving from 35·9 to 35·2 water, which is more buoyant, there is a parallel rise:

$$\text{Volume of layer} = 7696(35{\cdot}9 - 35{\cdot}2) = 5387 \text{ ft}^3$$

This is a layer of 35·9 water, weight

$$\frac{5387}{35{\cdot}9} \text{ tonf} = 149{\cdot}6 \text{ tonf}$$

*(Note the important artifice used to obtain the displacement to WL by calculating the difference by pocket calculator.)

$$\text{TPI in 35·9 water} = 69{\cdot}4 \times \frac{35}{35{\cdot}9}$$

$$\therefore \quad \text{layer thickness} = \frac{5387}{69{\cdot}4 \times 35} = 2{\cdot}2 \text{ in.}$$

$$\text{Moment causing trim} = 149{\cdot}6(20{\cdot}4 - 2{\cdot}4) = 2695 \text{ tonf ft.}$$

$$\text{Trim occurs in 35·2 water for which MCT1}'' = 1720 \times \frac{35}{35{\cdot}2}$$

$$\therefore \quad \text{trim by stern} = \frac{2695}{1720} \times \frac{35{\cdot}2}{35} = 1{\cdot}58 \text{ in.}$$

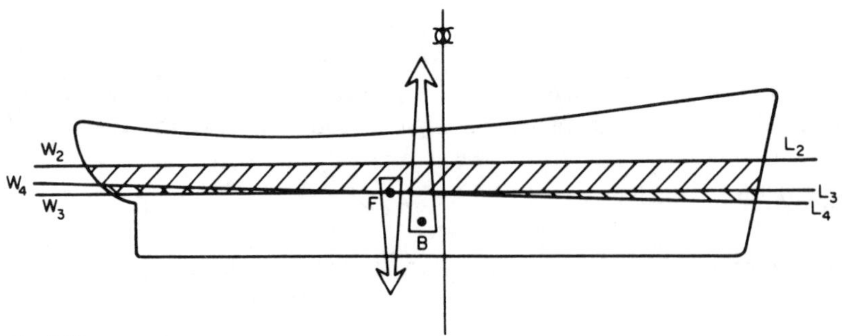

Fig. 3.39

Change in mean draught due to 1·58 in. trim by stern

$$= -\frac{20{\cdot}4}{560} \times 1{\cdot}58 \text{ in.} = -0{\cdot}06 \text{ in.}$$

total reduction in mean draught = 2·26 in.

∴ in water of RWD 35·2 ft^3/tonf

mean draught = 16 ft $11\frac{3}{4}$ in. (b)

stern trim = $13\frac{1}{2}$ in.

Problems

1. (*a*) Calculate the reciprocal weight density of petrol in m^3/MN, given that its specific gravity is 0·70.

(*b*) A pressure of 3·75 MPa is experienced by a submarine. If 1 lb = 0·4536 kg and 1 in. = 2·54 cm, convert this pressure to lbf/in^2. Put g = 9·81 m/s^2.

2. A floating body has a constant triangular section, vertex downwards and has a constant draught of 12 m in fresh water, the breadth at the waterline being 24 m. The keel just touches a quantity of mud of specific gravity 2. The water level now falls 6 m. How far will the body sink into the mud?

3. A wallsided ship goes aground on a mudbank so that its draught marks record a mean draught of 7 ft. Before going aground, the draught of the ship in salt water was 8 ft 10 in. The area of the waterplane is 4180 ft^2. What volume of oil of specific gravity 0·92 should be pumped out to free the ship?

Supposing, before pumping can begin, the tide falls a further foot and the ship sinks a further 6 in. into the mud, displacing a total of 1900 ft^3 of mud. What would be the reciprocal weight density of the mud?

4. A homogeneous wooden pontoon of square section, having a specific gravity of 0·5 is made to float in fresh water with its sides vertical by covering one of its faces completely with a strip of metal having a specific gravity of 8·0.

Show that G coincides with M if the thickness of the strip is approximately one ninetieth ($\frac{1}{90}$) of its width.

5. A model of a ship is built to float in water of 62·5 lbf/ft^3 weight density. It has the following geometric properties:

Length BP, 6·0 ft;	Area of waterplane, 691 in^2
Draught, 5·5 in.;	Trans. MI of waterplane, 12,450 in^4
$\overline{KB}$, 3·0 in.;	Volume of displacement, 4150 in^3

The centre of gravity as built is 7·2 in. above the keel. Where must a 25 lbf weight be placed to bring G below M, presuming the model to be wallsided at the waterline?

6. The TPC (tonnef per centimetre) of a wallsided ship floating at a draught of 2·75 m is 9·40. Its displacement at this draught is 3335 tonnef and the VCB is 1·13 m below the waterline. The centre of gravity of the ship is 3·42 m above the keel. The transverse second moment of area of the waterplane is 7820 m^4.

Calculate the shifts of B and G and the new position of M when a load of 81·2 tonnef is placed 3·05 m off the centreline on a deck 8·54 m above the keel and by constructing the movements on squared paper, estimate, to the nearest degree, the heel of the ship.

7. The half ordinates of the waterplane of a barge, 50 m long, are 0·0, 3·0, 4·5, 5·0, 4·2 and 1·0 m. Its volume of displacement is 735 m^3 and the CG is 2·81 m above the keel. The draught is 3·0 m.

A block of SG 7·0, having a volume of 10 m^3 has to be raised from the deck in a cage. Using an approximate formula for the position of B, estimate the height to which the weight can be raised before G is higher than M.

8. The tonf per inch immersion of a ship at the waterlines, which are 5 ft apart, are given below. In addition, there is an appendage having a displacement of 1480 tonf with its centre of buoyancy 3 ft 3 in. below 7 WL.

Waterline	1	2	3	4	5	6	7
TPI	195	184	169	153	135	108	46

Calculate the displacement of the ship and the vertical position of the centre of buoyancy of the ship.

9. The TPIs of the waterlines of a vessel, which are 7 ft apart, are:

WL	6	$5\frac{1}{2}$	5	4	3	2	1
TPI	20	44	64	88	108	124	132

If there is a 110 tonf appendage whose c.b. is 3 ft below 6WL, find the displacement and the position of the CB of the whole vessel.

10. A yacht of mass displacement 9·5 tonnes to 1 WL in fresh water has a CB 0·25 m below 1 WL. If the distance between WLs is 0·15 m, plot a curve of VCB against draught. The areas of A, 1, 2 and 3 WLs are respectively 28·4, 28·0, 26·8 and 25·2 m². Quote VCB at $2\frac{1}{2}$ WL.

11. A guided missile destroyer is 510 ft long BP and displaces 6130 tonf in salt water when the draught mark readings are 14 ft $10\frac{1}{4}$ in. forward and 15 ft $3\frac{1}{2}$ in. aft. The distances of the draught marks from amidships are 230 ft forward and 275 ft aft. The second moment of area of the waterplane is $82{\cdot}3 \times 10^6$ ft^4 about the centre of flotation which is 5 ft abaft amidships. The waterplane area is 17,500 ft^2.

Calculate the new draughts when 140 tonf of missiles are embarked at a mean distance of 190 ft abaft amidships. Illustrate on a bold diagram, the forces and movements involved and state any assumptions made.

12. The longitudinal moment of inertia about the CF of the waterplane of a depot ship floating in fresh water is 190×10^6 ft^4. The length BP is 684 ft and the draught marks are 32 ft abaft the FP and 8 ft abaft the AP. The area of the waterplane is 48,000 ft^2. Draught marks read 17 ft $3\frac{1}{4}$ in. forward and 22 ft $2\frac{1}{2}$ in. aft.

When 444 tonf of oil with its c.g. 200 ft abaft amidships are pumped out, the reading of the after marks is observed to change to 16 ft $10\frac{1}{2}$ in.

Deduce the position of the CF of the waterplane and the reading of the forward marks.

13. For a mean draught of 4·0 m with a 0·5 m trim by the stern, the ship's book shows a destroyer to have the following particulars:

displacement, mass	=	2300 tonne
TPC	=	9·10
MCT BP	=	3010 tonnef m
CF abaft amidships	=	6·85 m
CB abaft amidships	=	2·35 m

The ship is 102 m BP and is floating where the water has an SG = 1·02 at draughts of 3·50 m forward and 4·50 m aft.

Find, to the nearest half centimetre the draught at the propellers, which are 13 m forward of the AP, when the ship has moved up river to water of SG = 1·01.

14. A ship, displacement 7620 tonnef, floats at draughts of 4·10 m forward and 4·70 m aft. The forward marks are 2·1 m aft of the FP and the after marks are 19·8 m forward of the AP. The ship is 170 m between perpendiculars, has a tonnef per centimetre of 22·4 and the CF is 5·8 m aft of amidships. Determine the new draughts at the marks if a weight of 100 tonnef is placed 18·3 m forward of amidships.

One metre trim moment BP = 20,700 tonnef m.

15. The following are the particulars of an aircraft carrier: Displacement = 44,000 tonf, mean draught BP = 38·4 ft, trim by stern between marks = 7·3 ft, TPI = 84·2, MCT1″ BP = 2550 tonf ft. CF abaft amidships = 40 ft, LBP = 720 ft, forward marks abaft FP = 33 ft, after marks before AP = 52 ft.

Calculate the new draughts at the marks when 530 tonf of aircraft are embarked with their c.g. 260 ft before amidships.

16. A box-shaped barge of length 20 m beam 5 m and mean draught 1·45 m floats with a trim of 30 cm by the stern in water of specific gravity 1·03. If it moves into water of SG = 1·00, what will be its new draughts forward and aft?

17. The buoy mooring cable on a badly moored vessel is vertical and just taut when it has an even draught of 3 m. The vessel is 50 m long and has a TPM = 250 and CF 4 m abaft amidships. MCT is 1300 tonnef m. The mooring cable passes over the bow bullring 5 m before the FP.

Calculate the new draughts at the perpendiculars when the tide has risen 0·8 m.

18. A cargo ship, displacing 5500 tonf, floats at draughts of 15 ft 0 in. forward and 16 ft 6 in. aft, measured at marks 230 ft forward of and 170 ft abaft amidships.

The vessel grounds on a rock 75 ft forward of amidships. What is the force on the bottom when the tide has fallen 1 ft and what are then the draughts at the marks? TPI = 52; CF is 15 ft abaft amidships; MCT1″ BP is 1300 tonf ft; LBP is 480 ft.

19. The draught marks of a ship, 61 m long BP, are 6·1 m aft of the FP and 12·2 m before the AP. The MN per m immersion is 3·570 and the one metre trim moment BP is 7·66 MN m. The CF is 2·45 m abaft amidships and the draught mark readings forward and aft are respectively 2·67 and 3·0 m.

The vessel goes aground on a ridge of rock 9·15 m from the FP. If the tide is falling at an average rate of 81 cm/hr, how long will it be before the freeboard at the AP is reduced by 1·22 m?

What will then be the draught mark readings?

20. A ship 360 ft long, has a design trim of 2 ft by the stern, an after cut-up 75 ft from the after perpendicular and a $\overline{KG}$ of 16 ft. When the after cut-up just touches level blocks, the trim out of normal is measured as 3 ft by the stern and the water level in the dock is 13·28 ft above the blocks.

Calculate (i) the initial displacement and LCG; (ii) the force on the after cut-up and the water level above the blocks when the ship just sues all along.

Hydrostatic data for design trim:

Mean draught (ft)	Displacement (tonf)	LCB (ft aft)	VCB (ft)	LCF (ft aft)	$\overline{BM}_L$ (ft)
12·0	2220	7·62	7·4	21·88	943
11·5	2079	6·66	7·1	21·43	983
11·0	1939	5·64	6·8	20·76	1026
10·5	1800	5·58	6·5	19·61	1070

21. A new class of frigates has the following hydrostatic particulars at a mean draught of 4 m as calculated for water of 1025 kg/m^3: TPC = 7·20, MCT =

3000 tonnef m, CF abaft amidships = 4 m, CB abaft amidships = 1 m, displacement = 2643 tonnef.

When one of the class was launched at Southampton, where the mass density of water is 1025 kg/m^3, it took the water with a mean draught of 3·98 m and a stern trim of 0·6 m.

Supposing it to be in an exactly similar condition, how would you expect a sister ship to float after launch on the Clyde, where the water has a speciffic gravity of 1·006?

All draughts are at the perpendiculars 120 m apart.

22. A ship of 5000 tonf and 500 ft between perpendiculars, has its centre of gravity 5 ft abaft amidships, floats in salt water of 35 ft^3/tonf and has draughts of 10 ft 3 in. at the forward perpendicular and 14 ft 6 in. at the after cut-up, which is 30 ft before the after perpendicular.

Estimate the draughts before docking in water of reciprocal weight density 35·7 ft^3/tonf.

The TPI is originally 50, MCT1″ is 625 tonf ft increasing at the rate of 50 tonf ft per foot increase of draught. The centre of flotation is 25 ft abaft amidships.

23. A Great Lakes bulk carrier has hydrostatic curves drawn for water of reciprocal weight density 36 ft^3/tonf, which show the following information:

Mean draught BP	28 ft	32 ft
Displacement	24,259 tonf	28,032 tonf
Trim by stern	12 in.	12 in.
TPI	78·4	78·8
MCT1″ BP	775 tonf ft	779 tonf ft
CF forward of midships	3·8 ft	2·2 ft
CB aft of midships	12 ft	8·0 ft

The draughts forward and aft in river water of reciprocal weight density 35·8 ft^3/tonf are respectively 30 ft 6 in. and 31 ft 6 in. The length between perpendiculars is 580 ft.

Calculate the draughts when the ship reaches the open sea.

24. The following are the half ordinates in m of a vessel for which the waterlines are 1·75 m apart and ordinates 25 m apart. Determine its volume and the position of the centre of volume below 1WL.

	WLs				
Ords.	5	4	3	2	1
1	0	0	0	0	0
2	1	2	3	4	6
3	2	3	5	7	8
4	3	5	7	8	9
5	2	4	6	8	9
6	1	2	4	6	6
7	0	0	0	0	0

25. Given the half breadths of the waterlines of a ship as shown below, work out the displacement of the main body up to the 1 WL. The WLs are 4 ft 6 in. apart and the ordinates 72 ft.

	Half breadths in feet at ordinates				
	FP	2	3	4	AP
1 WL	1·1	21·3	30·2	27·6	4·4
2 WL	0·8	20·7	29·9	26·7	1·1
3 WL	0·5	18·9	29·0	24·9	1·8
4 WL	0·0	16·2	26·6	21·0	2·9

26. It is desired to fit a gun mounting weighing 20 tonf to a ship without altering the draught at the propellers which are 30 ft forward of the after perpendicular. The draughts are 8 ft forward and 9 ft aft and other particulars of the ship before the gun is added are: LBP = 140 ft, displacement = 260 tonf, MCT1″ = 65 tonf ft, CF abaft amidships = 7 ft, TPI = 5·6.

Where should the gun be mounted? What are the resulting draughts forward and aft?

27. Construct a metacentric diagram for a vessel of rectangular cross-section.

28. Construct metacentric diagrams for prisms, apex downwards,

(*a*) having a parabolic cross-section 6 m deep by 3 m wide at the top;
(*b*) having a section of an isosceles triangle, half apex angle 15 degrees and height 8 m.

4 Stability

The term stability refers to the tendency of a body or system to return to its original state after it has suffered a small disturbance. In this chapter we are concerned at first with a very specific example of static stability whereby a ship floating upright is expected to return to the upright after it has been buffeted for example by wind or wave. If a floating body is very stable it will return quickly to the upright and may engender motion sickness; if it is just stable a disturbance which is not small may cause it to capsize. The stability therefore must be just right in the range of conditions in which a vessel may find itself during its operation and life, even damaged or mishandled.

EQUILIBRIUM AND STABILITY

A rigid body is said to be in a state of equilibrium when the resultant of all the forces acting on it is zero and the resultant moment of the forces is also zero. If a rigid body, subject to a small disturbance from a position of equilibrium, tends to return to that state it is said to possess positive stability or to be in a state of stable equilibrium. If, following the disturbance, the body remains in its new position, then it is said to be in a state of neutral equilibrium or to possess neutral stability. If, following the disturbance, the excursion from the equilibrium position tends to increase, then the body is said to be in a state of unstable equilibrium or to possess negative stability.

The ship is a complex structure and is not in the mathematical sense a rigid body. However, for the purpose of studying stability it is permissible so to regard it. Throughout this chapter the ship will be regarded as a rigid body in calm water and not underway. If the ship is in waves or is underway there are hydrodynamic forces acting on the ship which may affect the buoyancy forces. This problem is discussed briefly in later sections.

Nevertheless, for stability purposes, it is usual to ignore hydrodynamic forces except for high-speed craft, including hydrofoils. For fast motor boats (Ref. 8), the hydrodynamic forces predominate in assessing stability.

For *equilibrium*, the buoyancy force and weight must be equal and the two forces must act along the same straight line. For a floating body this line must be vertical.

DISTURBANCE FROM STATE OF EQUILIBRIUM

Any small disturbance can be resolved into three components of translation and three of rotation with reference to the ship's body axes. The conventional axes are defined as in Fig. 4.1, the positive directions being indicated by the arrows.

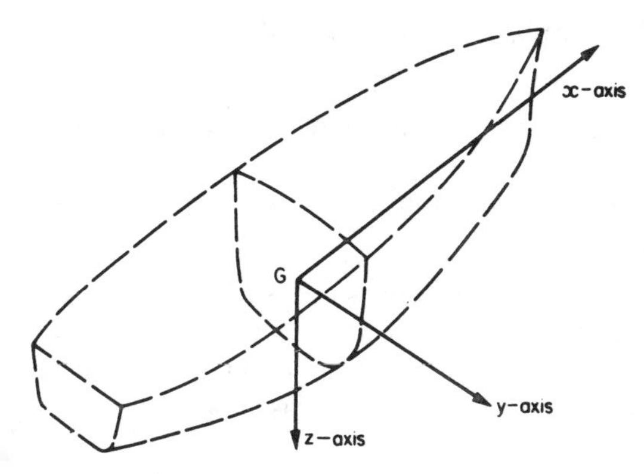

Fig. 4.1 Body axes

Consider each component of a disturbance in turn.

Translation along the x-axis, which is horizontal, leads to no resultant force so that the ship is in neutral equilibrium for this type of disturbance.

Translation along the y-axis also leads to no resultant force so that the ship is in neutral equilibrium for this type of disturbance.

For a floating body, translation along the z-axis in the positive direction results in an augmented buoyancy force which will tend to move the ship in the negative sense of z, i.e. it tends to return the ship to its state of equilibrium and the ship is thus stable for this type of disturbance. The special case of a totally submerged body is considered later.

Rotation about the x-axis, heel, results in a moment acting on the ship about which no generalization is possible and the ship may display stable, neutral or unstable equilibrium.

Rotation about the y-axis, trim, leads to a condition similar to that for heel.

Rotation about the z-axis, yaw, results in no resultant force or moment so that the ship is in neutral equilibrium for this type of disturbance.

The above results are summarized in Table 4.1.

It is clear from Table 4.1 that the only disturbances which herein demand study are the rotations about the x- and y-axes. In principle, there is no distinction between these two but, for the ship application, the waterplane characteristics are such that it is convenient to study the two separately. Also, for small

Table 4.1
Summary of equilibrium conditions

Axis	Equilibrium condition for	
	Translation along:	Rotation about:
x	Neutral	Stable, neutral or unstable
y	Neutral	Stable, neutral or unstable
z	Stable	Neutral

rotational disturbances at constant displacement any skew disturbance can be regarded as compounded of

(*a*) the stability exhibited by the ship for rotations about the *x*-axis—referred to as its transverse stability;
(*b*) the stability exhibited by the ship for rotations about the *y*-axis—referred to as its longitudinal stability.

Note that, because the shape of the heeled waterplane is not usually symmetrical, pure rotation in the *yz* plane must cause a tendency to a small trim. While this is usually of a second order and safely ignored, at large angles it becomes important and computer programs must cater for this cross coupling.

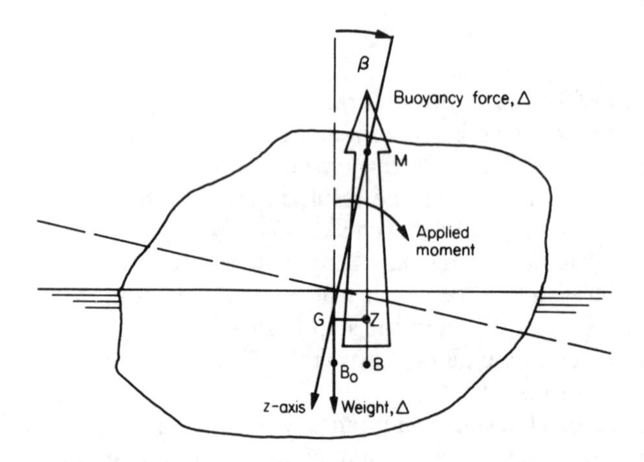

Fig. 4.2 Action of buoyancy force and weight for small rotational disturbance

Initial stability

Consider the irregular shaped body shown in Fig. 4.2 floating in a state of equilibrium. In the equilibrium state, the centre of buoyancy, B_0, and the centre of gravity, G, must lie on the same vertical line. If the body is now subjected to a rotational disturbance, by turning it through a small angle β at constant displacement, the centre of buoyancy will move to some new position, B. For convenience, in Fig. 4.2 the angular disturbance is shown as being caused by a pure moment about G, but it will be realized that the condition of constant displacement will, in general, require translational movements of G by forces as well.

The weight and buoyancy forces continue to act vertically after rotation but, in general, are separated so that the body is subject to a moment $\Delta\ \overline{GZ}$ where Z is the foot of the normal from G on to the line of action of the buoyancy force. As drawn, this moment tends to restore the body to the original position. The couple is termed the *righting moment* and $\overline{GZ}$ is termed the *righting lever*.

Another way of defining the line of action of the buoyancy force is to use its point of intersection, M, with the *z*-axis. As the angle β is indefinitely diminished

M tends to a limiting position termed the *metacentre.* For small values of β it follows that

$$\overline{GZ} = \overline{GM} \sin \beta$$
$$\simeq \overline{GM}\, \beta$$

The distance $\overline{GM}$ is termed the *metacentric height* and is said to be positive when M lies above G. This is the condition of stable equilibrium, for should M lie below G the moment acting on the body tends to increase β and the body is unstable. If M and G coincide the equilibrium is neutral. (See Fig. 4.3.)

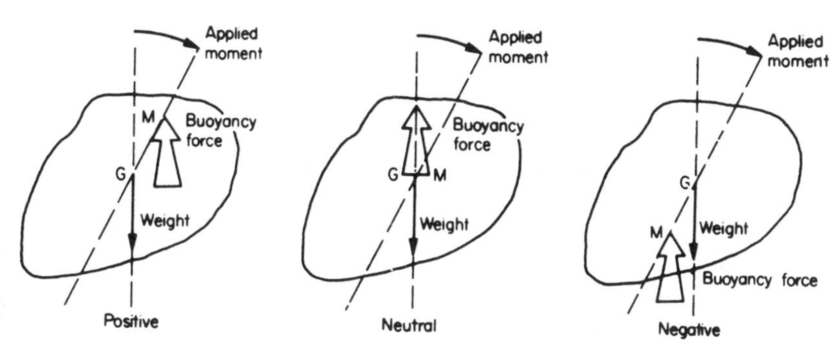

Fig. 4.3 The three stability conditions

ADJUSTMENT OF TRANSVERSE METACENTRIC HEIGHT BY SMALL CHANGES OF DIMENSIONS

The ship's form can be altered by increasing ordinates in the x, y and z directions such that all dimensions in any one direction are altered in the same ratio. This ratio need not be the same for all directions. This method of form changing has the advantage of maintaining the coefficients of fineness unaltered and is often used in the early stages of ship design. For instance, it may be necessary to decrease draught at the expense of the beam to ensure that the ship will be able to negotiate canals or harbour entrances. In this case,

$$\Delta = (\text{constant}) \times L \times B \times T$$

Hence

$$\log \Delta = \log(\text{constant}) + \log L + \log B + \log T$$

and differentiating

$$\frac{d\Delta}{\Delta} = \frac{dL}{L} + \frac{dB}{B} + \frac{dT}{T}$$

Thus, if the percentage changes in the main dimensions are small, their sum will provide the percentage change in the displacement.

Similarly as shown in Chapter 3,

$$\overline{\mathrm{BM}} = \frac{I_T}{\nabla} = (\text{constant}) \times \frac{B^2}{T}$$

and by the same process

$$\frac{\mathrm{d}\overline{\mathrm{BM}}}{\overline{\mathrm{BM}}} = 2\frac{\mathrm{d}B}{B} - \frac{\mathrm{d}T}{T}$$

Also

$$\frac{\mathrm{d}\,\overline{\mathrm{BM}}}{\overline{\mathrm{BM}}} = \frac{\mathrm{d}\,\overline{\mathrm{BG}} + \mathrm{d}\,\overline{\mathrm{GM}}}{\overline{\mathrm{BM}}}$$

$$= \frac{\mathrm{d}\,\overline{\mathrm{BG}} + \mathrm{d}\,\overline{\mathrm{GM}}}{\overline{\mathrm{BG}} + \overline{\mathrm{GM}}}$$

Some special cases are worth considering.

(*a*) Where it is desired to change beam only, the increase in displacement is given by

$$\mathrm{d}\Delta = \Delta\frac{\mathrm{d}B}{B}$$

and the effect on $\overline{\mathrm{GM}}$ is deduced as follows:

$$\frac{\mathrm{d}\,\overline{\mathrm{BG}} + \mathrm{d}\,\overline{\mathrm{GM}}}{\overline{\mathrm{BM}}} = 2\frac{\mathrm{d}B}{B}$$

If it is assumed that $\overline{\mathrm{KG}}$ remains unaltered then $\mathrm{d}\,\overline{\mathrm{BG}} = 0$ and

$$\mathrm{d}\,\overline{\mathrm{GM}} = 2\overline{\mathrm{BM}} \cdot \frac{\mathrm{d}B}{B}$$

Since, in general, $\overline{\mathrm{BM}}$ is greater than $\overline{\mathrm{GM}}$ the percentage increase in metacentric height is more than twice that in beam.

(*b*) If it is desired to keep the displacement and draught unaltered, the length must be decreased in the same ratio as the beam is increased, that is

$$-\frac{\mathrm{d}L}{L} = \frac{\mathrm{d}B}{B}$$

Provided the assumption that $\overline{\mathrm{KG}}$ is unchanged remains true the change in metacentric height is as above.

(*c*) To maintain constant displacement and length the draught must be changed in accord with

$$\frac{\mathrm{d}T}{T} = -\frac{\mathrm{d}B}{B}$$

This change in draught may, or may not, cause a change in $\overline{\mathrm{KG}}$. There is a number of possibilities two of which are dealt with below.

(i) $\overline{KG} = (\text{constant}) \times T$. But $\overline{KB} = (\text{constant}) \times T$ also, i.e. $\overline{BG} = \overline{KG} - \overline{KB} = (\text{constant}) \times T$.

It should be noted that these constants of proportionality are not necessarily the same, hence

$$\frac{d\,\overline{BG}}{\overline{BG}} = \frac{dT}{T} = -\frac{dB}{B}$$

It has been shown that

$$\frac{d\,\overline{BG} + d\,\overline{GM}}{\overline{BG} + \overline{GM}} = 2\frac{dB}{B} - \frac{dT}{T} = \frac{3\,dB}{B}$$

Hence

$$-\overline{BG}\frac{dB}{B} + d\,\overline{GM} = 3(\overline{BG} + \overline{GM})\frac{dB}{B}$$

and

$$d\,\overline{GM} = (4\,\overline{BG} + 3\,\overline{GM})\frac{dB}{B}$$

(ii) $\overline{KG}$ = constant. Since

$$\overline{BG} = \overline{KG} - \overline{KB}$$

in this case, the change in $\overline{BG} = d\,\overline{BG}$

$$= -(\text{change in } \overline{KB})$$

$$= -\overline{KB}\frac{dT}{T} = \overline{KB}\frac{dB}{B}$$

As before

$$d\,\overline{BG} + d\,\overline{GM} = 3(\overline{BG} + \overline{GM})\frac{dB}{B}$$

$$d\,\overline{GM} = (3\,\overline{BG} + 3\,\overline{GM} - \overline{KB})\frac{dB}{B}$$

$$d\,\overline{GM} = (4\,\overline{BG} + 3\,\overline{GM} - \overline{KG})\frac{dB}{B}$$

EXAMPLE 1. A ship has the following principal dimensions: $L = 360$ m, $B = 42$ m, $T = 12$ m, $\Delta = 1200$ MN, $\overline{KB} = 7$ m, $\overline{KM} = 21$ m, $\overline{KG} = 18$ m.

It is desired to reduce draught to 11 m, keeping length and displacement constant and coefficients of fineness constant. Assuming $\overline{KG}$ remains unaltered, calculate the new $\overline{GM}$ and the new beam.

Solution:

$$\frac{d\Delta}{\Delta} = 0 = \frac{dL}{L} + \frac{dB}{B} + \frac{dT}{T}$$

Hence

$$\frac{dB}{B} = \frac{1}{12} \qquad \therefore \quad dB = \frac{42}{12} = 3{\cdot}5 \text{ m.}$$

i.e.

$$\text{new beam} = 45{\cdot}5\,\text{m.}$$

$$\frac{d\,\overline{BM}}{\overline{BM}} = 2\frac{dB}{B} - \frac{dT}{T} = 2\times\frac{1}{12}+\frac{1}{12} = \frac{1}{4}$$

$$d\,\overline{BM} = 14\times\frac{1}{4} = 3{\cdot}5 \text{ m.}$$

Because $\overline{KB}$ is proportional to T

$$\frac{d\,\overline{KB}}{\overline{KB}} = \frac{dT}{T} = -\frac{1}{12}$$

and

$$d\,\overline{KB} = -\frac{7}{12}$$

Now

$$\overline{GM} = \overline{KB}+\overline{BM}-\overline{KG}$$

$$\therefore \quad \text{new } \overline{GM} = 3+(-\tfrac{7}{12})+3{\cdot}5$$

$$= 5{\cdot}92 \text{ m.}$$

EFFECT OF MASS DENSITY

It is clear that a critical feature in the study of initial stability is the vertical position of the centre of gravity. It is instructive to consider a homogeneous rectangular block of material floating freely as shown in Fig. 4.4.

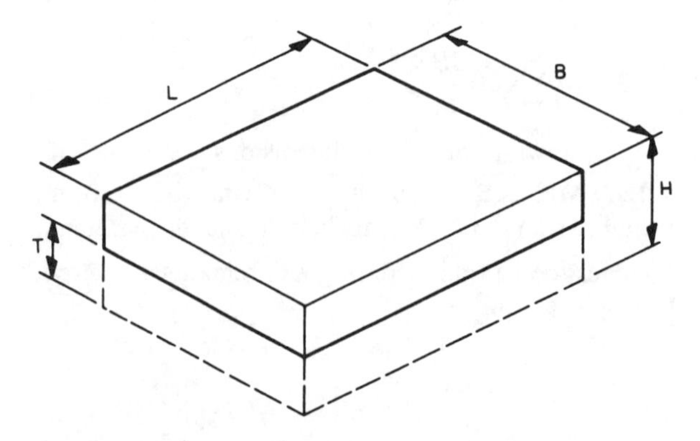

Fig. 4.4 Homogeneous block floating freely

The density of the block must be less than that of the water in which it floats, or else it would sink. If the density is C times that of the water, equating weight and buoyancy shows that

$$C \times LBH = LBT$$

i.e.

$$C = \frac{T}{H}$$

Now for a rectangular block

$$\overline{\mathrm{KB}} = \frac{T}{2}$$

$$\overline{\mathrm{BM}} = \frac{I}{\nabla} = \frac{1}{12}\frac{B^3L}{BLT} = \frac{B^2}{12T}$$

$$\overline{\mathrm{KG}} = \frac{H}{2}$$

and

$$\overline{\mathrm{GM}} = \overline{\mathrm{KB}} + \overline{\mathrm{BM}} - \overline{\mathrm{KG}} = \frac{T}{2} + \frac{B^2}{12T} - \frac{H}{2}$$

$$= \frac{CH}{2} + \frac{B^2}{12CH} - \frac{H}{2}$$

$$= \frac{H}{2}(C-1) + \frac{B^2}{12CH}$$

The condition for positive stability is that

$$\overline{\mathrm{GM}} > 0$$

i.e.

$$\frac{H}{2}(C-1) + \frac{B^2}{12CH} > 0$$

Similar expressions can be derived for homogeneous blocks of other cross-sectional shapes.

EXAMPLE 2. A rectangular block of homogeneous material is 20 m long, 5 m wide and 5 m deep. It is floating freely in fresh water with its longest dimension horizontal. Calculate the range of specific gravity over which the block will be unstable transversely.

Solution: As derived above, the block will be unstable if

$$\frac{H}{2}(C-1) + \frac{B^2}{12CH} < 0$$

i.e.

$$\frac{5}{2}(C-1)+\frac{25}{12C\times 5} < 0$$

$$C-1+\frac{1}{6C} < 0$$

$$6C^2-6C+1 < 0$$

When $C = 0$, the block has zero draught and the inequality is not satisfied, i.e. the block is stable.

When $C = \frac{1}{2}$, the left-hand side $= \frac{6}{4}-3+1$. This is less than zero so that the inequality is satisfied.

When $C = 1$, the other possible extreme value, the inequality is not satisfied.

The block is therefore stable at extreme values of C but unstable over some intermediate range. This range will be defined by the roots of the equation

$$6C^2-6C+1 = 0$$

$$C = \frac{6\pm\sqrt{(36-24)}}{12} = \frac{6\pm\sqrt{12}}{12}$$

i.e.

$$C = 0{\cdot}211 \quad \text{or} \quad 0{\cdot}789$$

Hence the block is unstable if the specific gravity lies within the range 0·211 to 0·789.

EFFECT OF FREE SURFACES OF LIQUIDS

It is quite a common experience to find it difficult to balance a shallow tray or tin containing water. Stability is essentially a problem of balance, and it is therefore necessary to investigate whether a tank of liquid in a ship affects the ship's initial stability. The problem is considered below in relation to transverse stability, but similar considerations would apply to the case of longitudinal stability.

In practice, a ship has tanks with several different liquids—fresh water for drinking or for boilers, salt water for ballast, fuels of various types and lubricating oils—so the problem must be examined with allowance made for the fact that the density of the liquid may not be the same as that of the sea water in which the ship is floating.

In the following, the subscript s is used to denote quantities associated with the ship or the sea water in which she floats and the subscript ℓ to denote the liquid or the tank in which it is contained.

Let the ship be floating initially at a waterline WL and let it be heeled through a small angle ϕ to a new waterline W_1L_1. Since by definition the surface of the liquid in the tank is free, it will also change its surface inclination, relative to the tank, by the same angle ϕ. (See Fig. 4.5.)

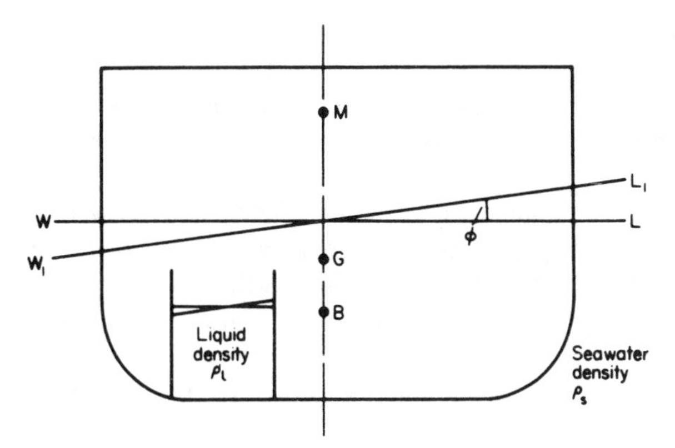

Fig. 4.5 Liquid free surface

In Chapter 3, it was shown that for small angles of inclination the transfer of buoyancy is given approximately by the expression

$$\Delta_s \frac{I_s}{\nabla_s}\phi = \rho_s I_s \phi g$$

It follows that, by similar arguments, the transfer of weight due to the movement of the liquid in the tank is approximately

$$\Delta_\ell \frac{I_\ell}{\nabla_\ell}\phi = \rho_\ell I_\ell \phi g$$

This being a transfer of weight opposes the righting moment due to the transfer of buoyancy and results in the effective righting moment acting on the ship being reduced to

$$\Delta_s \overline{\mathrm{GM}}_s \phi - \rho_\ell I_\ell \phi g$$

Thus, if $\overline{\mathrm{GM}}_\mathrm{F}$ is the effective metacentric height allowing for the action of the liquid free surface

$$\Delta_s \overline{\mathrm{GM}}_\mathrm{F} \phi = \Delta_s \overline{\mathrm{GM}}_s \phi - \rho_\ell I_\ell \phi g$$

and therefore

$$\overline{\mathrm{GM}}_\mathrm{F} = \overline{\mathrm{GM}}_s - \frac{\rho_\ell I_\ell}{\Delta_s} g$$

$$= \overline{\mathrm{GM}}_s - \left(\frac{\rho_\ell}{\rho_s}\right)\frac{I_\ell}{\nabla_s}$$

It should be noted that the effect of the free surface is independent of the position of the tank in the ship; the tank can be at any height in the ship, at any position along its length and need not be on the middle-line. The effect is also independent of the amount of liquid in the tank provided the second moment of area of the free surface is substantially unchanged when inclined. A tank which is almost empty or almost full can suffer such a change in second moment of area and it is for this reason that if, during an inclining experiment (see later), tanks cannot be completely full or empty they are specified to be half full.

It is usual to regard the free surface effect as being a virtual rise of the centre of gravity of the ship, although it will be appreciated that this is merely a matter of convention.

A similar effect can arise from the movements of granular cargoes stowed in bulk such as grain. Here, however, there is no simple relation between the angle of inclination of the ship and the slope of the surface of the cargo. At small angles of heel the cargo is not likely to shift so there is no influence on initial stability.

The effect of a free liquid surface on stability at larger angles of heel is discussed later.

EFFECT OF FREELY SUSPENDED WEIGHTS

It is more difficult to balance a pole with a weight freely suspended from its top than it is to balance the same pole with the same weight lashed securely to the pole. It is of interest to the naval architect to know what effect weights suspended freely in a ship have on the stability of the ship. Such a practical situation arises when a ship unloads her cargo using her own derricks. Consider the moment the cargo load is lifted from the hold before any translation of the load has taken place. The effect on transverse stability will normally be more critical than that on longitudinal stability, although the principles involved are the same.

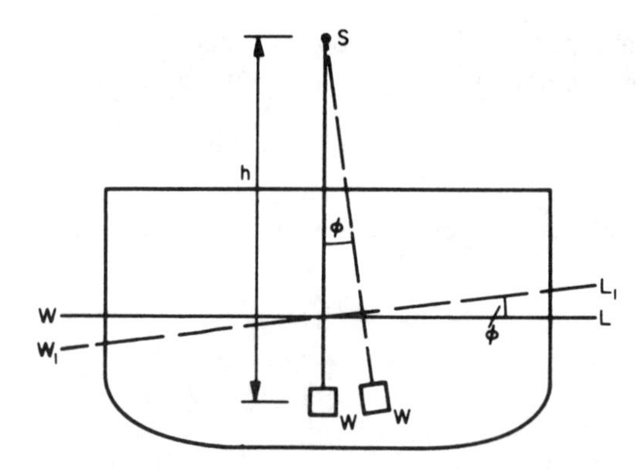

Fig. 4.6 Freely suspended weight

Referring to Fig. 4.6, let the weight W be suspended freely from a point S a distance h above the centroid of mass of the weight. Let the ship be floating at a waterline WL and consider it being heeled through a small angle ϕ to a new waterline W_1L_1. The weight, being freely suspended, will move until it is again vertically below S, i.e. its suspending wire will move through an angle ϕ. Thus, as far as the ship is aware the line of action of the weight of the cargo being lifted always passes through S. Hence, the effect on the ship is as though the weight W were placed at S.

It will be noted that this result is independent of the magnitude of the angle of heel provided the movement of the suspended weight is not restricted in any way.

THE WALL-SIDED FORMULA

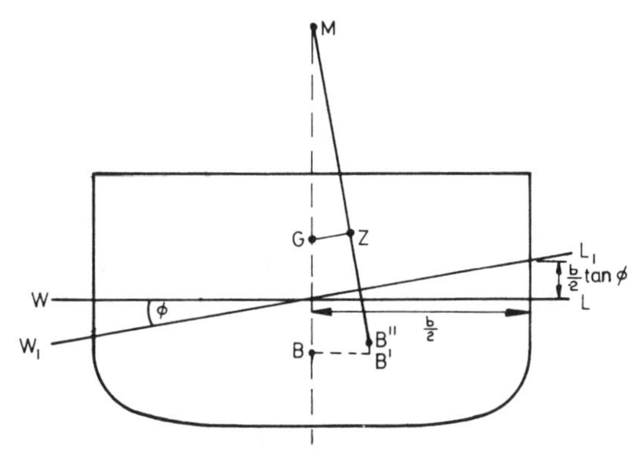

Fig. 4.7 Wall-sided formula

A ship is said to be wall-sided if, for the angles of inclination to be considered, those portions of the outer bottom covered or uncovered by the moving water-plane are vertical with the ship upright. No practical ships are truly wall-sided, but many may be regarded as such for small angles of inclination—perhaps up to about 10 degrees.

Referring to Fig. 4.7, let the ship be inclined from its initial waterline WL to a new waterline W_1L_1 by being heeled through a small angle ϕ. Since the vessel is wall-sided, WL and W_1L_1 must intersect on the centre line.

The volume transferred in an elemental wedge of length δL where the beam is b is

$$\delta L\left(\frac{1}{2}\times\frac{b}{2}\times\frac{b}{2}\tan\phi\right) = \left(\frac{b^2}{8}\tan\phi\right)\delta L$$

Moment of transfer of volume for this wedge in a direction parallel to WL

$$= \frac{b^2}{8}\tan\phi\frac{2b}{3}\,\delta L$$

Hence, for the whole ship, moment of transfer of volume is

$$\int_0^L \frac{b^3}{12}\tan\phi\,\mathrm{d}L$$

Hence, horizontal component of shift of B, $\overline{\mathrm{BB}'}$ is given by

$$\overline{\mathrm{BB}'} = \frac{1}{\nabla}\int_0^L \frac{b^3}{12}\tan\phi\,\mathrm{d}L$$

$$= \frac{I}{\nabla}\tan\phi = \overline{\mathrm{BM}}\tan\phi$$

Similarly, vertical shift $\overline{B'B''}$ is given by

$$\overline{B'B''} = \frac{1}{\nabla}\int_0^L \frac{b^2}{8}\tan\phi\,\tfrac{1}{3}b\tan\phi\,dL$$

$$= \frac{I}{2\nabla}\tan^2\phi = \frac{\overline{BM}}{2}\tan^2\phi$$

By projection on to a plane parallel to W_1L_1

$$\overline{GZ} = \overline{BB'}\cos\phi + \overline{B'B''}\sin\phi - \overline{BG}\sin\phi$$

$$= \overline{BM}\left[\sin\phi + \frac{\tan^2\phi}{2}\sin\phi\right] - \overline{BG}\sin\phi$$

$$= \sin\phi\left[\overline{BM} - \overline{BG} + \frac{\overline{BM}}{2}\tan^2\phi\right]$$

i.e.

$$\overline{GZ} = \sin\phi\left[\overline{GM} + \frac{\overline{BM}}{2}\tan^2\phi\right]$$

For a given ship if $\overline{GM}$ and $\overline{BM}$ are known, or can be calculated, $\overline{GZ}$ can readily be calculated using this formula.

Wall-sided vessel containing a tank with vertical sides containing liquid

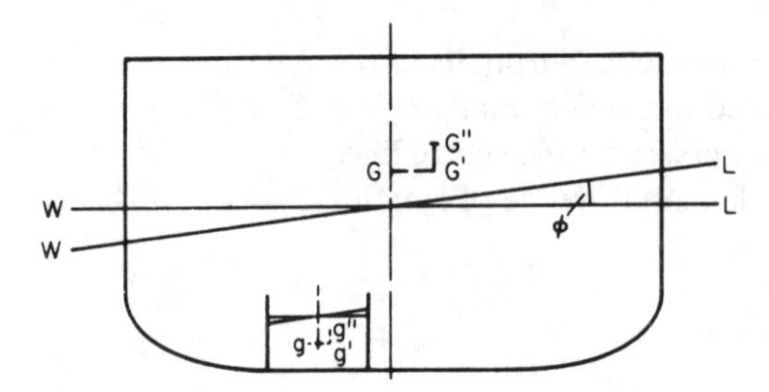

Fig. 4.8 Wall-sided ship with liquid contained in a wall-sided tank

Let the density of the liquid in the tank be ρ times that of the water in which the ship is floating.

Referring to Fig. 4.8, horizontal and vertical shifts of B will be as calculated for the ship without a tank. In addition, however, the centroid of mass of the liquid in the tank will suffer movements

$$\overline{gg'} = \frac{I_t}{\nabla_t}\tan\phi$$

and

$$\overline{g'g''} = \frac{I_t}{2\nabla_t}\tan^2\phi$$

Hence the centre of gravity of the ship, G, will suffer movements of

$$\overline{GG'} = \frac{I_\ell}{\nabla_\ell}\tan\phi \times \rho\frac{\nabla_\ell}{\nabla_s} = \rho\frac{I_\ell}{\nabla_s}\tan\phi$$

and

$$\overline{G'G''} = \rho\frac{I_\ell}{2\nabla_s}\tan^2\phi$$

Thus, the presence of the free surface will effectively reduce $\overline{GZ}$ to $\overline{GZ}_F$ where

$$\overline{GZ}_F = \sin\phi\left[\left(\overline{GM}-\rho\frac{I_\ell}{\nabla_s}\right)+\left(\overline{BM}-\rho\frac{I_\ell}{\nabla_s}\right)\frac{\tan^2\phi}{2}\right]$$

EXAMPLE 3. Calculate the ordinates of the curve of $\overline{GZ}$ for a wall-sided ship up to 15 degrees of heel given that $\overline{GM} = 3$ ft and $\overline{BM} = 18$ ft.

Solution:

$$\overline{GZ} = \left[\overline{GM}+\frac{\overline{BM}}{2}\tan^2\phi\right]\sin\phi$$

$$\overline{GZ} = [3+9\tan^2\phi]\sin\phi$$

Values of $\overline{GZ}$ for varying values of ϕ can be computed as in the table below.

ϕ (deg)	$\tan\phi$	$\tan^2\phi$	$9\tan^2\phi$	$3+9\tan^2\phi$	$\sin\phi$	$\overline{GZ}$ (ft)
0	0	0	0	3	0	0
3	0·052	0·0027	0·0243	3·024	0·052	0·157
6	0·105	0·0110	0·0990	3·099	0·105	0·325
9	0·158	0·0250	0·2250	3·225	0·156	0·503
12	0·213	0·0454	0·4086	3·409	0·208	0·709
15	0·268	0·0718	0·6462	3·646	0·259	0·944

Complete stability

CROSS CURVES OF STABILITY

So far, only stability at small angles of inclination has been discussed. In practice, it is necessary to have a knowledge of stability at large angles of inclination, particularly in the transverse plane. In the following sections, transverse stability at large angles is considered. In principle, the concepts are equally applicable to longitudinal stability but, in practice, they are not normally required because of the relatively small angles of trim a ship can accept for reasons other than stability.

Referring to Fig. 4.9, let the ship be floating initially at a waterline WL. Now let it be heeled through some angle ϕ to a new waterline W_1L_1 such that the displacement remains constant. The buoyancy force Δ will act through B_1 the new position of the centre of buoyancy, its line of action being perpendicular to W_1L_1. If Z_1 is the foot of the perpendicular dropped from G on to the line of action of the buoyancy force then the righting moment acting on the ship is given by $\Delta\overline{GZ}_1$.

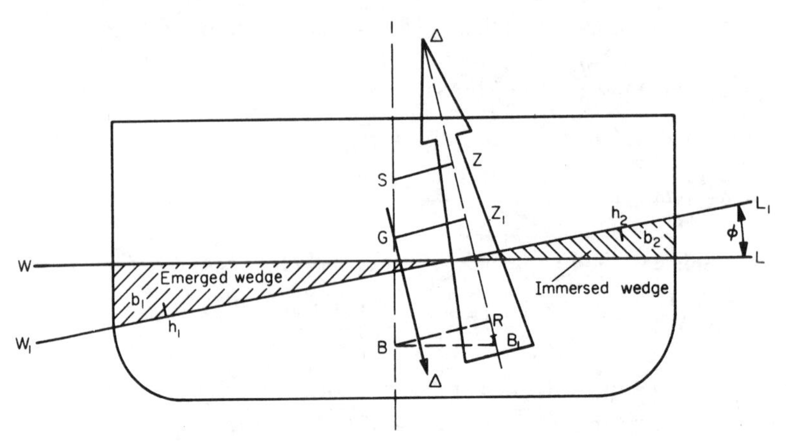

Fig. 4.9 Stability at large angles

This is similar to the concept applied to stability at small angles of heel but, in this case, it is not possible to make use of the concept of a metacentre as the buoyancy force does not intersect the ship's middle line plane in a fixed point.

It should be noted that, in general, a ship when heeled will also trim to maintain its longitudinal equilibrium. This can usually be ignored but where it is significant, the points shown in Fig. 4.9 must be regarded as projections of the true points on to a transverse plane of the ship.

Since the displacement to W_1L_1 is the same as that to WL, it follows that the volumes of the immersed and emerged wedges are equal.

Let

δ = buoyancy force associated with immersed wedge

b_1, b_2 = centroids of volume of the emerged and immersed wedges respectively

h_1, h_2 = feet of perpendiculars from b_1, b_2 on to W_1L_1

R = Foot of the perpendicular from B on to the line of action of the buoyancy force through B_1.

Then

$$\Delta\overline{BR} = \delta\overline{h_1h_2}$$

and righting moment acting on the ship is given by

$$\Delta\overline{GZ} = \Delta\overline{BR} - \Delta\overline{BG}\sin\phi$$

$$= \Delta\left[\frac{\delta}{\Delta}\overline{h_1h_2} - \overline{BG}\sin\phi\right]$$

This formula is sometimes known as *Atwood's Formula*, although it was known and used before Atwood's time.

Unfortunately, G depends upon the loading of the ship and is not a fixed position. It is therefore more convenient to think in terms of an arbitrary, but fixed, pole S and its perpendicular distance $\overline{SZ}$ from the line of action of the buoyancy force. $\overline{SZ}$ then depends only on the geometry of the ship and can be calculated for various angles of heel and for various values of displacement without reference to a particular loading condition. Since the position of S is known, when G has been calculated for a particular condition of loading, $\overline{GZ}_1$ can always be calculated from the equation

$$\overline{GZ}_1 = \overline{SZ} + \overline{SG} \sin \phi \quad \text{(S above G)}$$

Where a whole range of ship conditions is required, it is usual to plot values of $\overline{SZ}$ against displacement for each of a number of angles of inclination as shown in Fig. 4.10. These curves are known as *Cross Curves of Stability*.

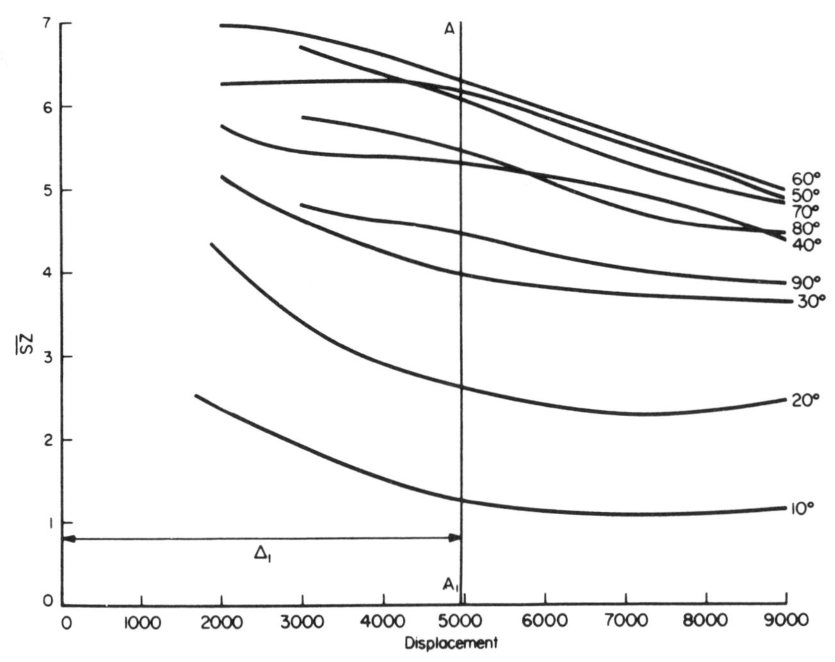

Fig. 4.10 Cross curves of stability

This form of plotting avoids the need to determine inclined waterplanes for precise displacements. It is convenient usually to adopt a set of waterlines which intersect at the same point on the ship's middle line. Then, if a plot of $\overline{SZ}$ against angle of inclination is required at a constant displacement Δ_1, this can be obtained by reading off the $\overline{SZ}$ values along a vertical line AA_1 in Fig. 4.10.

If a whole range of ship conditions is not required, the $\overline{SZ}$ curve for constant displacement can be obtained directly as follows:

Referring to Fig. 4.11, let WL be the upright waterline, W_1L_1 the inclined waterline at the same displacement and W_2L_2 the waterline at the inclination of W_1L_1 passing through the intersection of WL with the centre line of the ship.

Let

δ = bouyancy of immersed or emerged wedge between WL and W_1L_1

δ_1 = emerged buoyancy between WL and W_2L_2

δ_2 = immersed buoyancy between WL and W_2L_2

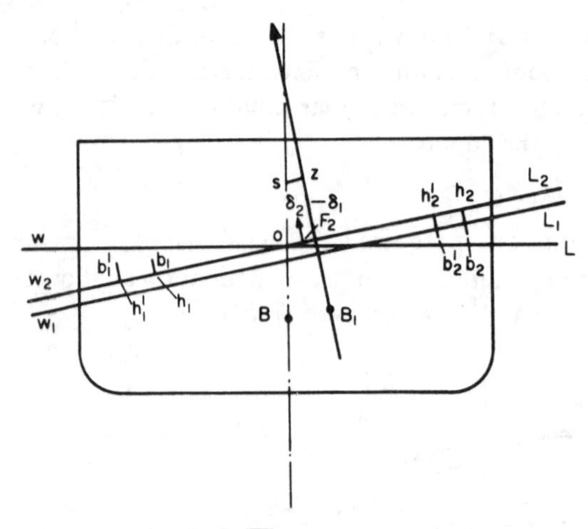

Fig. 4.11 Derivation of $\overline{SZ}$ values for constant displacement

Then, buoyancy of layer between W_1L_1 and W_2L_2 is $\delta_2 - \delta_1$ and let this force act through point F_2 in W_2L_2. If the layer is thin, then F_2 may be regarded as the centroid of the waterplane W_2L_2.

Let

b_1, b_2 be the centroids of the wedges between WL and W_1L_1
b'_1, b'_2 be the centroids of the wedges δ_1 and δ_2
h_1, h_2 projections of b_1, b_2 on to W_2L_2
h'_1, h'_2 projections of b'_1, b'_2 on to W_2L_2

Then

$$\delta \overline{h_1 h_2} = \delta_1 \overline{Oh'_1} + \delta_2 \overline{Oh'_2} - (\delta_2 - \delta_1) . \overline{OF_2}$$

DERIVATION OF CROSS CURVES OF STABILITY

It is now usual to derive data for the plotting of cross curves from a digital computer. A glimpse at manual methods and machines now to be found only in museums is worthwhile because they illustrate the principles on which computer programs are written.

Integrator methods

The integrator was essentially a machine which measured the area lying within a closed curve and the first and second moments of that area about a datum line—the axis of the integrator. The integrator could be used in two ways.

The 'all-round' method

A body plan for the ship is prepared to as large a scale as is convenient for the integrator to work with. On this body plan is marked the position of the selected pole S, and radiating from S a series of lines representing the angles of heel for which $\overline{SZ}$ is to be measured. Then, on a separate sheet of tracing paper is marked off a series of parallel lines to represent waterlines, spaced so as to cover adequately the range of displacement over which it is desired to study the ship's stability. A line normal to the waterlines is drawn to represent the centre line of the ship when upright.

For each angle to be studied, the tracing paper is set up over the body plan with the 'centre line' passing through S and the 'waterlines' at the correct angle and covering the required range of draught. The integrator is then set up so that its axis lies along the 'centre line' drawn on the tracing paper—see Fig. 4.12.

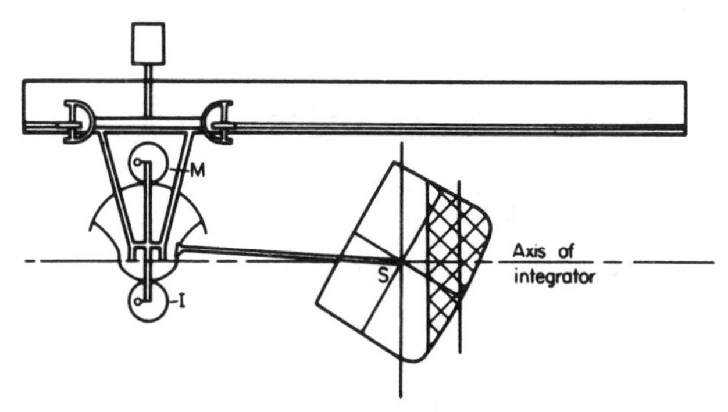

Fig. 4.12 Use of integrator in 'all-round' method

Then, provided the scaling factors to be applied to the dial readings have been accurately assessed, the integrator can be used to determine the area shown cross-hatched in Fig. 4.12 and the moment of that area about the axis of the integrator. This can be repeated for each section of the ship. If a Simpson body plan is used, the areas and moments must be set out in tabular form, multiplied by the appropriate factor and summated to give the total volume of the ship up to the chosen waterline and the moment about the line through S. Displacement and $\overline{SZ}$ follow and the process can be repeated for each waterline and then for other angles of heel having adjusted the position of the body plan below the tracing paper. All the data required for plotting the cross curves of stability are now available except insofar as it may be necessary to allow for appendages.

If a Tchebycheff body plan is used then, since the multiplying factor is the same for all sections of the ship, it is possible to move the integrator round all sections in turn and note only the difference between the first and final readings on the dials. In the same way, the number of readings involved with a Simpson body plan can be reduced by dealing with all sections having the same multiplier without stopping.

The 'figure of eight' method

This method makes use of the fact that the only differences between the upright and an inclined condition are the immersed and emerged wedges. By passing the integrator point round these two wedges in opposite senses—i.e. in a figure of eight motion as in Fig. 4.13—the moment of transfer of buoyancy and the change in displacement are obtained.

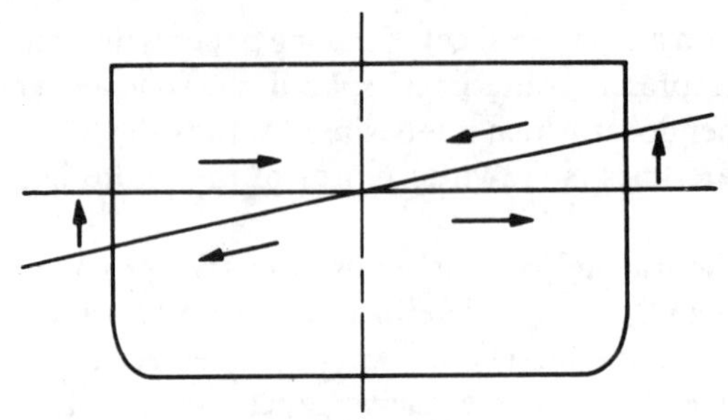

Fig. 4.13 'Figure of eight' method

Barnes's method

This method makes use of the general formula derived above for the transfer of buoyancy, viz.

$$\delta \overline{h_1 h_2} = \delta_1 \overline{\mathrm{Oh}'_1} + \delta_2 \overline{\mathrm{Oh}'_2} - (\delta_2 - \delta_1)\overline{\mathrm{OF}}_2$$

and the various terms are evaluated using radial integration

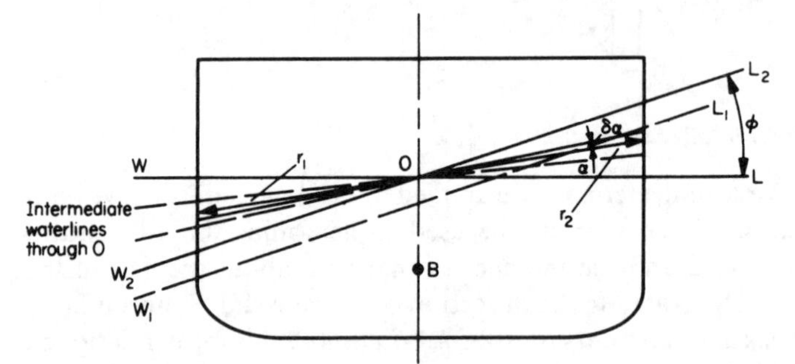

Fig. 4.14 Radial integration for Barnes's method of deriving statical stability data

By considering the element of angle $\delta\alpha$ in Fig. 4.14 and applying the principles of radial integration

$$\delta_1 \overline{\mathrm{Oh}'_1} + \delta_2 \overline{\mathrm{Oh}'_2} = \int_0^L \int_0^\phi \left[\tfrac{1}{2} r_1^2 \, \mathrm{d}\alpha \frac{2r_1}{3} + \tfrac{1}{2} r_2^2 \, \mathrm{d}\alpha \frac{2r_2}{3}\right] \cos(\phi - \alpha) \, \mathrm{d}x$$

$$= \int_0^L \int_0^\phi \left(\frac{r_1^3 + r_2^3}{3}\right) \cos(\phi - \alpha) \, \mathrm{d}x \, \mathrm{d}\alpha$$

This double integral can be evaluated by drawing intermediate radial waterlines at equal increments of heel and measuring the offsets r_ϕ at the appropriate stations. The same process can be followed to evaluate

$$\delta_2 = \int_0^L \int_0^\phi \tfrac{1}{2} r_2^2 \, \mathrm{d}\alpha \, \mathrm{d}x$$

$$\delta_1 = \int_0^L \int_0^\phi \tfrac{1}{2} r_1^2 \, \mathrm{d}\alpha \, \mathrm{d}x$$

The area and first moment of area of waterplane W_2L_2 are given by the expressions

$$\text{Area} = \int_0^L r_\phi \, dx$$

First moment of area about fore and aft axis through 0

$$= \int_0^L \frac{r_\phi^2}{2} dx$$

By dividing the moment by the area the value of $\overline{OF}_2$ is found.
Substituting in Atwood's formula

$$\overline{SZ} = \left(\frac{\delta_1\overline{Oh}'_1 + \delta_2\overline{Oh}'_2 - (\delta_2 - \delta_1)\overline{OF}_2}{\Delta}\right) - \overline{BS}\sin\phi$$

Evaluating this expression for various values of ϕ gives the values of $\overline{SZ}$ at constant displacement.

An adaptation of Barnes's method, if a whole range of ship displacements are to be covéred, would be to find the displacement Δ_ϕ corresponding to waterline W_2L_2 through O and the corresponding $\overline{SZ}$, viz:

$$\Delta_\phi = \Delta + \delta_2 - \delta_1$$

$$\overline{SZ}_\phi = \left(\frac{\delta_1\overline{Oh}'_1 + \delta_2\overline{Oh}'_2}{\Delta_\phi}\right) - \overline{BS}\sin\phi$$

These values correspond to those obtained by the integrator method.

Reech's method

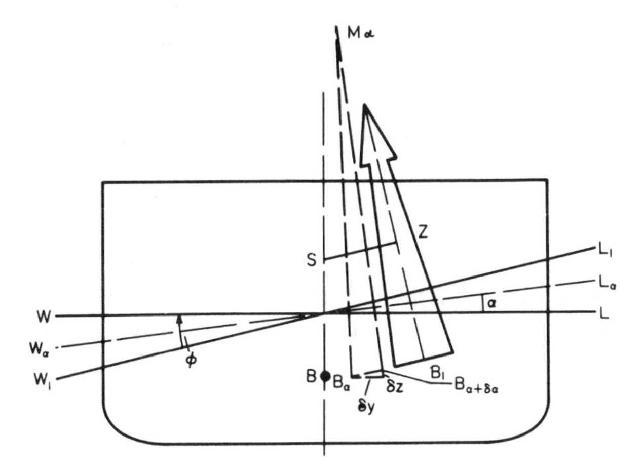

Fig. 4.15 Reech's method

Let B_α and M_α be the positions of the centre of buoyancy and metacentre corresponding to inclined waterplane $W_\alpha L_\alpha$. From the definition of pro-metacentre (see p. 20), if B_α moves to $B_{\alpha+\delta\alpha}$ for a very small additional inclination $\overline{B_\alpha B_{\alpha+\delta\alpha}} = \overline{B_\alpha M_\alpha}\delta_\alpha$, and $\overline{B_\alpha B_{\alpha+\delta\alpha}}$ is parallel to $W_\alpha L_\alpha$

$$\delta y = \overline{B_\alpha M_\alpha} \cos\alpha\, \delta\alpha$$

$$\delta z = -\overline{B_\alpha M_\alpha} \sin\alpha\, \delta\alpha$$

It follows by radial integration that from waterplane W_1L_1 the position of B_1 relative to B can be expressed as (y, z) where

$$y = \int_0^\phi \overline{B_\alpha M_\alpha} \cos\alpha\, d\alpha$$

$$z = -\int_0^\phi \overline{B_\alpha M_\alpha} \sin\alpha\, d\alpha$$

The negative sign for z signifies only that B_1 is 'higher' than B.
Hence

$$\overline{SZ} = y\cos\phi - z\sin\phi - \overline{BS}\sin\phi$$

$$= \cos\phi \int_0^\phi \overline{B_\alpha M_\alpha} \cos\alpha\, d\alpha + \sin\phi \int_0^\phi \overline{B_\alpha M_\alpha} \sin\alpha\, d\alpha - \overline{BS}\sin\phi$$

To evaluate this expression, it is necessary to determine a series of inclined waterlines at constant displacement and for each waterplane obtain the value of $\overline{BM}$ by finding the values of the second moment of area and dividing by the volume of displacement.

Use of model to obtain curve of statical stability

It has been noted that one limitation of the methods described above to determine the statical stability of a ship is the neglect of trim induced by the heeling of the ship. This can be overcome by finding the true equilibrium waterplane at each inclined waterplane, but this leads to considerable complication. One method of automatically allowing for the effects of trim is to use a scale model of the ship, applying to it known heeling moments and noting the resulting angles of heel.

Normally, the expense of making an accurately scaled model is not justified, but the method can be used where a model already exists for other purposes such as the conduct of ship tank tests. The method is particularly useful where there are large changes in waterplane shape for small changes in heel. Such changes can occur with forms which have large rise of floor, very rounded bilges, and/or long well-rounded cut-ups.

Prohaska's method

Reference 1 considers the stability lever $\overline{GZ}$ as composed of two parts, viz:

$$\overline{GZ} = \overline{GM}\sin\phi + \overline{MS}$$

The quantity $\overline{MS}$, which, in a projection on a transverse plane, is the distance from the upright metacentre to the line of action of the buoyancy force, is termed the residuary stability lever (see Fig. 4.16). For convenience in non-dimensional plotting, a coefficient C_{RS} is employed for which

$$C_{RS} = \frac{\overline{MS}}{\overline{BM}_{upright}}$$

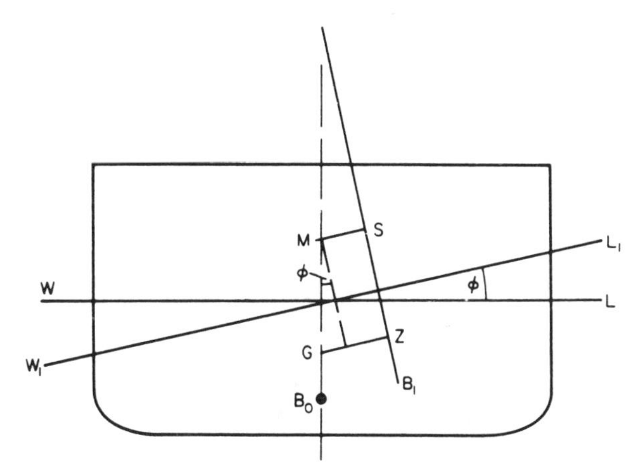

Fig. 4.16

By the use of displacement sheets, the co-ordinates of the centres of buoyancy for 90 and 180 degrees inclination can be calculated relative to those for the upright ship. Also, the metacentric radii for 0, 90 and 180 degrees can be calculated. Using this data and other geometric form data, Prohaska reduces the derivation of the residual stability curve to a calculation in tabular form. The tables used are based on calculations performed on forty-two forms having form characteristics covering the usual merchant ship field, backed up by mathematical treatments assuming that the stability curve can be approximated by a trigonometric series. By its nature the method is approximate, but for general ship forms lying within the range of parameters considered, accuracy comparable with that achieved by the so-called 'exact' methods is obtained.

CURVES OF STATICAL STABILITY

It was explained earlier, that it is usually convenient to derive cross curves of stability from purely geometric considerations by making use of an arbitrary pole about which to measure righting moments. It was also seen, that it simplifies the work in deriving stability curves if displacement is allowed to vary as the ship is inclined and if both displacement and righting lever are calculated for each waterline.

For practical applications, however, it is necessary to present stability in the form of righting moments or levers about the centre of gravity, as the ship is heeled at constant displacement. Such a plot will in general appear as in Fig. 4.17 and is known as a *statical stability curve.*

The curve of statical stability, or $\overline{GZ}$ curve as it is commonly called, is derived from the cross curves of stability by setting up a vertical line such as AA_1

for displacement Δ_1 shown on Fig. 4.10. The $\overline{SZ}$ value corresponding to each angle of inclination plotted can be read off and then, using Fig. 4.9 and the $\overline{SG}$ known for the particular loading of the ship,

$$\overline{GZ} = \overline{SZ} - \overline{SG} \sin \phi$$

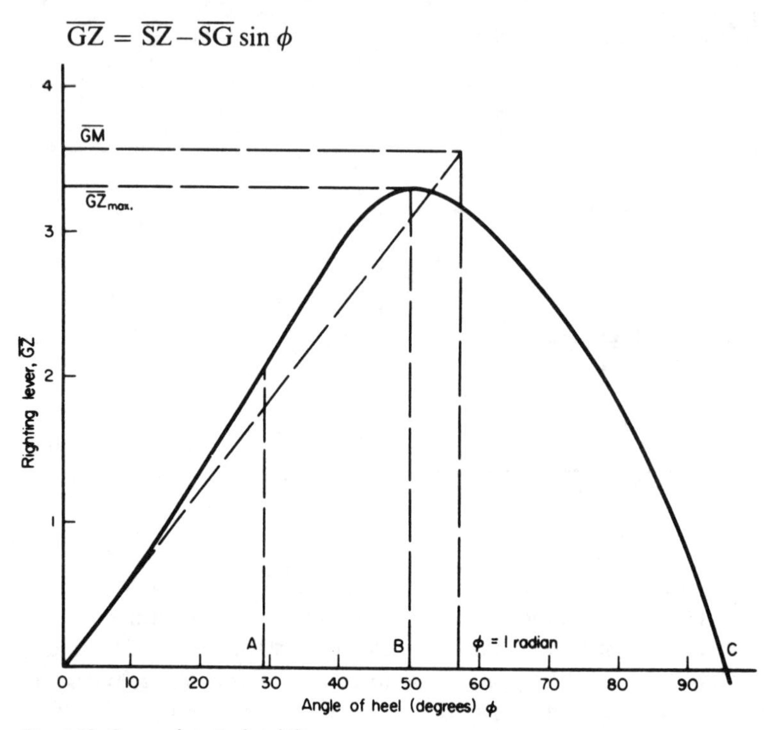

Fig. 4.17 Curve of statical stability

MAIN FEATURES OF THE $\overline{GZ}$ CURVE

Certain features of the $\overline{GZ}$ curve are of particular significance and are useful parameters with which to define the stability possessed by a given design. They include:

(*a*) Slope at the origin. For small angles of heel, the righting lever is proportional to the angle of inclination, the metacentre being effectively a fixed point. It follows, that the tangent to the $\overline{GZ}$ curve at the origin represents the metacentric height.

(*b*) Maximum $\overline{GZ}$. This is proportional to the largest steady heeling moment that the ship can sustain without capsizing, and its value and the angle at which it occurs are both important.

(*c*) Range of stability. At some angle, often greater than 90 degrees, the $\overline{GZ}$ value reduces to zero and becomes negative for larger inclinations. This angle is known as the *angle of vanishing stability* and the range of angle (OC in Fig. 4.17) for which $\overline{GZ}$ is positive is known as the *range of stability*. For angles less than this, a ship will return to the upright state when the heeling moment is removed.

(*d*) Angle of deck edge immersion. For most ship forms, there is a point of inflexion in the curve (shown in Fig. 4.17 as occurring at angle OA) corresponding roughly to the angle at which the deck edge becomes immersed.

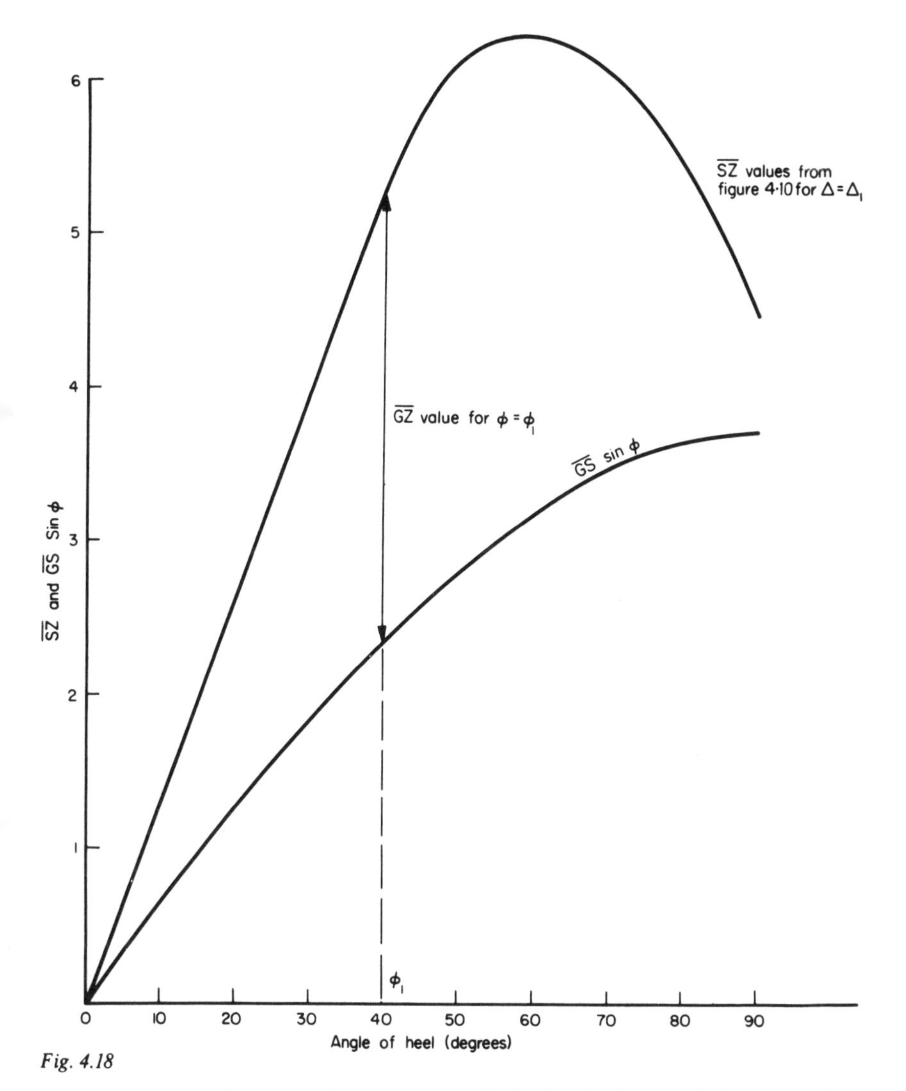

Fig. 4.18

In general, of course, the angle at which the deck edge is immersed varies along the length, but is within a fairly narrow band for the larger sections amidships which exert most influence upon the stability. This point is not so much of interest in its own right as in the fact that it provides guidance to the designer upon the possible effect on stability of certain design changes.

(*e*) Area under the curve. The area under the curve represents the ability of the ship to absorb energy imparted to it by winds, waves or any other external agency. This concept is developed more fully later.

ANGLE OF LOLL

A special case arises when $\overline{GM}$ is negative but $\overline{GZ}$ becomes positive at some reasonable angle of heel. This is illustrated in Fig. 4.19 as ϕ_1. If the ship is

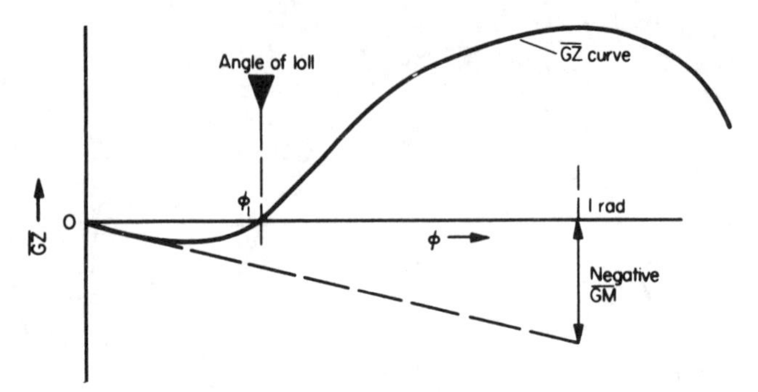

Fig. 4.19 Angle of loll ϕ_1

momentarily at some angle of heel less than ϕ_1, the moment acting due to $\overline{GZ}$ tends to increase the heel. If the angle is greater than ϕ_1, the moment tends to reduce the heel. Thus the angle ϕ_1 is a position of stable equilibrium. Unfortunately, since the $\overline{GZ}$ curve is symmetrical about the origin, as ϕ_1 is decreased, the ship eventually passes through the upright condition and will then suddenly lurch over towards the angle ϕ_1 on the opposite side and overshoot this value before reaching a steady state. This causes an unpleasant rolling motion which is often the only direct indication that the heel to one side is due to a negative $\overline{GM}$ rather than to a positive heeling moment acting on the ship.

As a special case, consider a wall-sided vessel with negative $\overline{GM}$. In this case,

$$\overline{GZ} = \sin\phi(\overline{GM}+\tfrac{1}{2}\overline{BM}\tan^2\phi)$$

$\overline{GZ}$ is zero when $\sin\phi = 0$. This merely demonstrates that the upright condition is one of equilibrium. $\overline{GZ}$ is also zero when $\overline{GM}+\frac{1}{2}\overline{BM}\tan^2\phi = 0$. i.e. when

$$\tan\phi = \pm\sqrt{\left(-\frac{2\overline{GM}}{\overline{BM}}\right)}$$

Also, in this case, the slope of the $\overline{GZ}$ curve at ϕ_1 will be given by

$$\begin{aligned}\frac{\mathrm{d}\,\overline{GZ}}{\mathrm{d}\,\phi} &= \cos\phi(\overline{GM}+\tfrac{1}{2}\overline{BM}\tan^2\phi)+\sin\phi\,\overline{BM}\tan\phi\sec^2\phi \\ &= 0+\overline{BM}\tan^2\phi_1/\cos\phi_1 \quad (\text{putting } \phi = \phi_1) \\ &= -2\overline{GM}/\cos\phi_1\end{aligned}$$

EFFECT OF FREE LIQUID SURFACES ON STABILITY AT LARGE ANGLES OF INCLINATION

It was shown earlier, that the effect of a free liquid surface on initial stability is equivalent to a reduction in the metacentric height of

$$\left(\frac{\rho_\ell}{\rho_s}\right)\frac{I_\ell}{\nabla_s}$$

This was derived by considering the transfer of weight of liquid which occurred in opposition to the righting moment induced by the transfer of buoyancy at the waterplane. The same principle can be applied to any form of tank at any angle of heel.

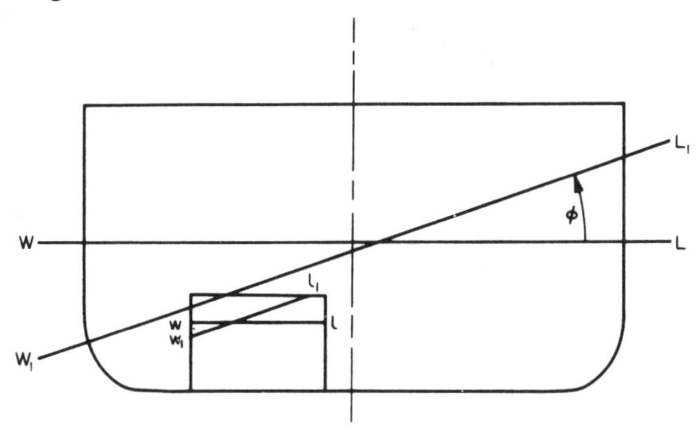

Fig. 4.20 Effect on stability of free surface at large angles of heel

Referring to Fig. 4.20, let a ship initially floating at waterline WL be inclined, at constant displacement, through angle ϕ. Then if wl is the upright free surface of liquid in the tank shown, the surface when inclined becomes w_1l_1 also at angle ϕ to wl.

Knowing the shape of the tank, the transfer of weight can be computed and translated into an effective reduction in the righting moment due to transfer of buoyancy in the wedges between WL and W_1L_1. It follows, by analogy with the upright case, that the reduction in *slope* of the $\overline{GZ}$ curve at angle ϕ will be given by

$$\left(\frac{\rho_\ell}{\rho_s}\right)\cdot\frac{(I_\ell)_\phi}{\nabla_s}$$

where $(I_\ell)_\phi$ is the moment of inertia of the liquid surface at angle ϕ.

If the tank is not a simple geometric shape, the transfer of weight can be computed using an integrator or any of the other methods discussed on p. 109 *et seq.* for the derivation of cross curves of stability. Even so, it is fairly tedious to evaluate for sufficient angles of inclination in order to define the effective $\overline{GZ}$ curve precisely, and it is common for some simplifying assumption to be made.

It is common practice to compute the metacentric height and the $\overline{GZ}$ value for 45 degrees heel allowing for the free surface. The effective $\overline{GZ}$ curve up to 45 degrees inclination is then constructed by drawing the curve through the 45 degree spot, following the general character of the uncorrected curve and fairing into the modified tangent at the origin. For angles greater than 45 degrees, the reduction of $\overline{GZ}$ at 45 degrees is applied as a constant correction. This method is illustrated in Fig. 4.21.

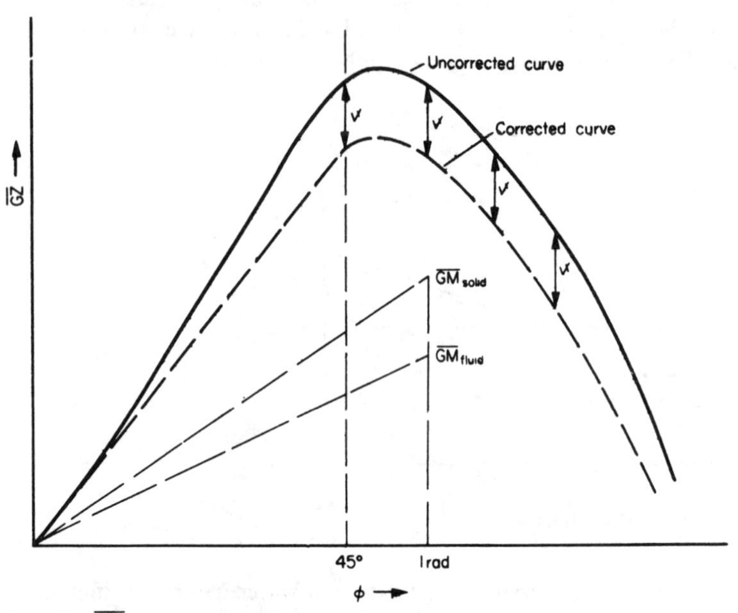

Fig. 4.21 $\overline{GZ}$ *curve corrected for free surface*

SURFACES OF B, M, F AND Z

When a body floats in water, there are unique positions of the centre of buoyancy, metacentre, centre of flotation of the waterplane and Z, the foot of the perpendicular from the c.g. on to the line of the buoyancy force. If the body is rotated at constant displacement, these points will, in general, move to new positions. Considering all possible inclinations B, M, F and Z trace out surfaces which are fixed relative to the body. The simplest case to consider is that of a sphere as in Fig. 4.22.

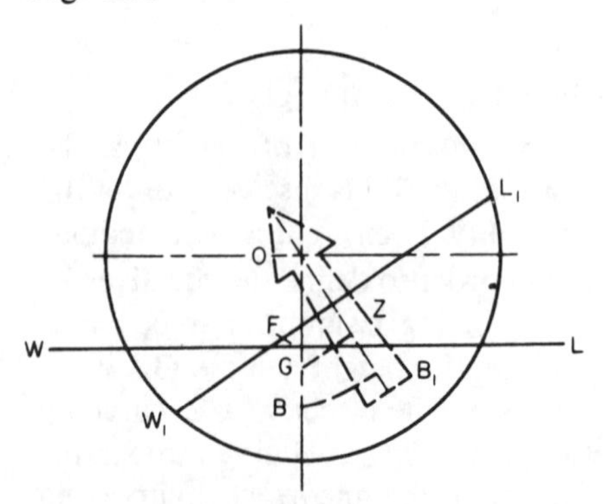

Fig. 4.22 Sphere floating in fluid at WL

Let the sphere float at a waterline WL as shown (the argument is unaffected if WL is above the centre of the sphere, O). Then, by arguments of symmetry it follows that for any inclination, OB will be constant, OF will be constant, F being the foot of the perpendicular from O on to WL, and M will always be at O. Thus, in this case, M is a single point and the surfaces traced out by B and F are spheres of radius OB and OF, respectively. As the sphere rotates, the angle OZG is always a right angle. Hence, the locus of Z is a sphere with $\overline{OG}$ as diameter.

Interest, as far as a ship is concerned, is usually centred on transverse stability, i.e. on rotation at constant displacement about a fore and aft axis. In this case, B, M, F and Z trace out lines across the surfaces discussed above. The projections of these traces on to a plane normal to the axis of rotation are termed the *curves of* B, M, F *and* Z and it is of interest to study certain of their characteristics. For a symmetrical ship form, of course, all curves will be symmetrical about the middle line of the ship.

B *curve or curve of buoyancy*

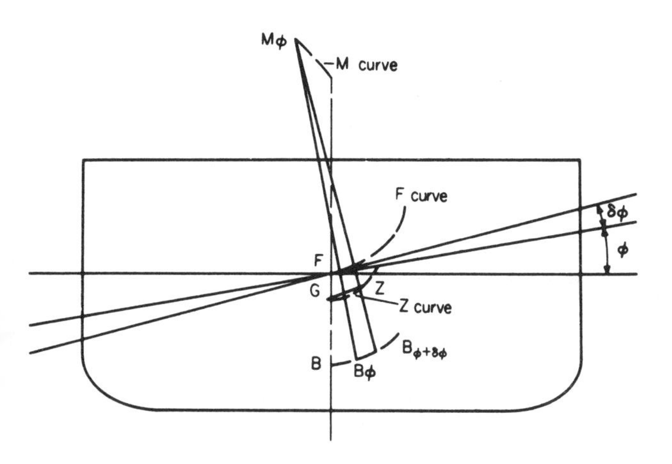

Fig. 4.23 Curves of B, M, F *and* Z

Let B_ϕ and $B_{\phi+\delta\phi}$ be the positions of the centre of buoyancy corresponding to inclinations ϕ and $\phi+\delta\phi$. Then $\overline{B_\phi B_{\phi+\delta\phi}}$ is parallel to the line joining the centroids of volume of the two wedges between the waterplanes at ϕ and $\phi+\delta\phi$. In the limit, $\overline{B_\phi B_{\phi+\delta\phi}}$ becomes the tangent to the curve of buoyancy and this tangent must be parallel to the corresponding waterplane.

The projections of the lines of action of buoyancy through B_ϕ and $B_{\phi+\delta\phi}$ intersect in a point which, in the limit, becomes M_ϕ, called the *prometacentre* and

$$\overline{B_\phi M_\phi} = \frac{I_\phi}{\nabla}$$

The co-ordinates of the point on the curve of buoyancy corresponding to any given inclination ϕ have already been shown to be

$$y = \int_0^{\phi} \overline{B_\alpha M_\alpha} \cos \alpha \, d\alpha$$

$$z = -\int_0^{\phi} \overline{B_\alpha M_\alpha} \sin \alpha \, d\alpha$$

The M *curve or metacentric curve*

The metacentric curve is the projection of the locus of pro-metacentres, for inclinations about a given axis, on to a plane normal to that axis.

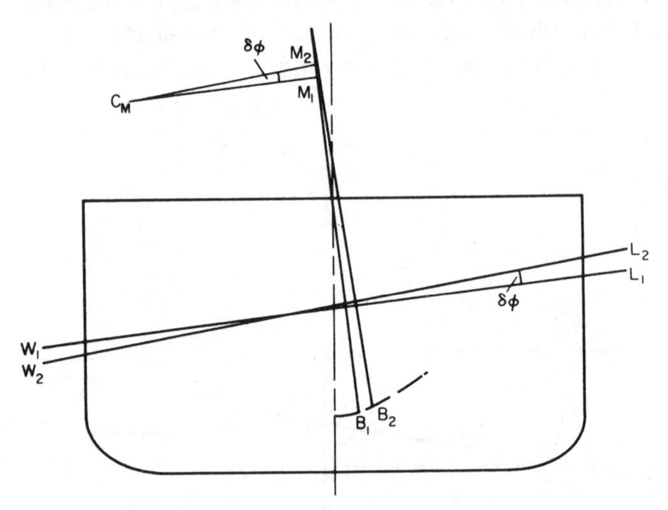

Fig. 4.24 Metacentric curve

Let W_1L_1 and W_2L_2 be two waterlines at relative inclination of $\delta\phi$ and let B_1 and B_2, M_1 and M_2 be the corresponding positions of the centre of buoyancy and pro-metacentre. Let C_M be the centre of curvature of the metacentric curve at M_1. Then $\overline{C_M M_1}$ is normal to $\overline{B_1 M_1}$ and therefore parallel to W_1L_1.

Referring to Fig. 4.24

$$\overline{M_1 M_2} = \delta(\overline{B_\phi M_\phi}) = \overline{C_M M_1}\, \delta\phi$$

and in the general case

$$\overline{C_M M_\phi} = \frac{d\overline{B_\phi M_\phi}}{d\phi} = \frac{1}{\nabla}\frac{dI_\phi}{d\phi}$$

The F *curve or curve of flotation*

For very small changes in inclination successive waterplanes intersect in a line through their centres of flotation. In the limit, therefore, the waterplane at any angle is tangential to the curve of flotation at the point corresponding to that inclination.

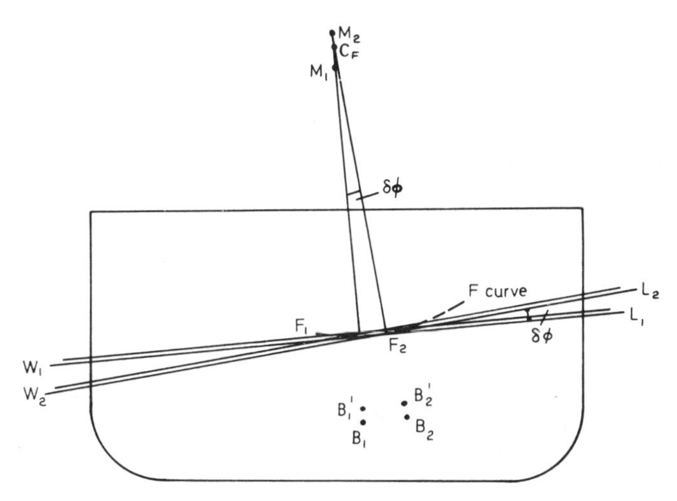

Fig. 4.25 Curve of flotation

Let W_1L_1 and W_2L_2 be waterplanes at a relative inclination $\delta\phi$ and B_1 and B_2, F_1 and F_2 and M_1 and M_2 their respective centres of buoyancy and flotation and pro-metacentres. Let the lines through F_1 and F_2 normal to W_1L_1 and W_2L_2 intersect in C_F which, in the limit, as $\delta\phi$ diminishes, will become the centre of curvature of the F curve.

Now, consider a small layer added to each waterline, the added volume of buoyancy in each case being $\delta\nabla$. Then let B_1B_2 take up new positions $B_1'B_2'$ and so on. Effectively, $\delta\nabla$ is added at F_1 and F_2 for the two inclinations and B_1, B_1' and F_1 will lie on a straight line as will B_2, B_2' and F_2.

By taking moments

$$\frac{\overline{B_1B_1'}}{\overline{B_1F_1}} = \frac{\delta\nabla}{\nabla+\delta\nabla} = \frac{\overline{B_2B_2'}}{\overline{B_2F_2}}$$

Also $\overline{B_1B_2}$, $\overline{B_1'B_2'}$ and $\overline{F_1F_2}$ are parallel, and

$$\overline{B_1B_2} = \overline{B_1M_1}\,\delta\phi = \frac{I_1}{\nabla}\,\delta\phi, \text{ if } \delta\phi \text{ is small.}$$

Hence

$$\frac{\overline{B_1F_1}}{\overline{B_1B_1'}} = \frac{\overline{B_1B_2}-\overline{F_1F_2}}{\overline{B_1B_2}-\overline{B_1'B_2'}}$$

By substitution, this becomes

$$\frac{\nabla+\delta\nabla}{\delta\nabla} = \frac{\dfrac{I_1}{\nabla}\delta\phi-\overline{C_FF_1}\,\delta\phi}{\dfrac{I_1}{\nabla}\delta\phi-\dfrac{I_1+\delta I}{\nabla+\delta\nabla}\delta\phi}$$

From which it follows that, in the limit,

$$C_F F_1 = \frac{dI}{d\nabla}$$

This is known as Leclert's theorem.

The Z curve

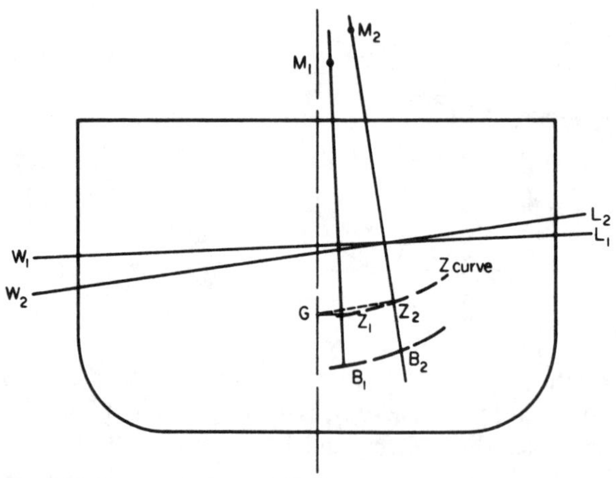

Fig. 4.26 Z curve

The Z curve is the projection on to a transverse plane of the foot of the perpendicular from G on to successive lines of action of the buoyancy force.

Curves of B, M, F *and* Z *for a typical ship*

The form of the B, M, F and Z curves is illustrated in Fig. 4.27 for a 20,000 tonf depot ship. The principal features of this design are:

	20,000 tonf depot ship	
Deep displacement, tonf (MN)	19,300	(192)
L.B.P., ft (m)	675	(206)
Beam, ft (m)	75·8	(23·1)
Mean draught, ft (m)	22·2	(6·8)
Block coefficient	0·595	
Prismatic coefficient	0·617	
Waterplane coefficient	0·755	

INFLUENCE OF SHIP FORM ON STABILITY

It has been shown how a designer, faced with a given form and loading condition, can assess the stability characteristics of his ship. It is unlikely that his first choice of form will satisfy his requirements completely. Initial stability, maximum $\overline{GZ}$ and range may all be too great or too small. Perhaps one feature alone may need modification. How then is the designer to modify his form to produce the required result? The effect of small changes in principal dimensions

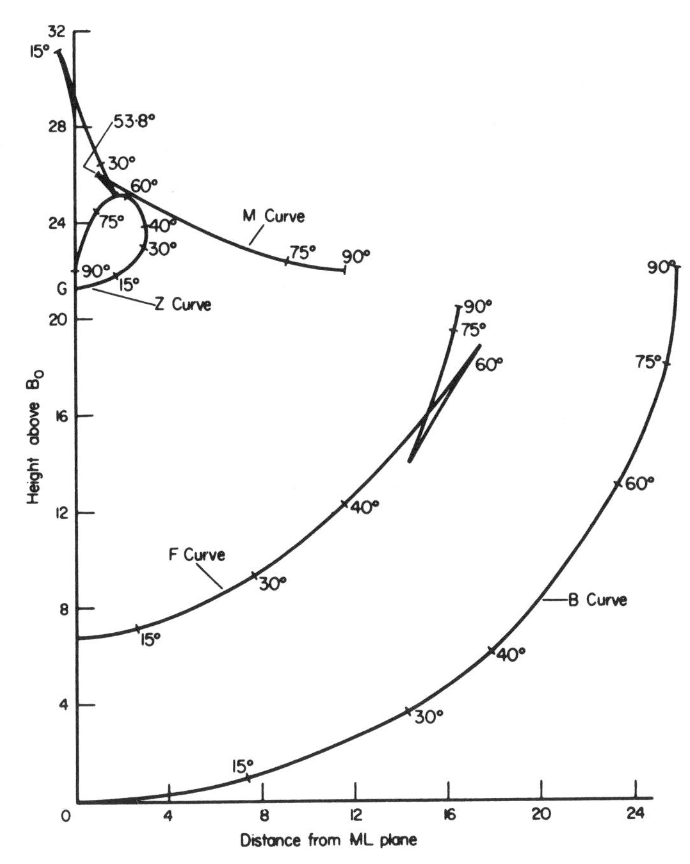

Fig. 4.27

on initial stability has already been discussed, and if only such a change is required this method might be adopted provided that the consequential changes in stability at large angles are acceptable.

In general, however, there must be a certain amount of 'trial and error' in the process. That is to say, the designer must make certain changes, recalculate the stability and assess what additional changes may be necessary. The electronic computer is a great help in this type of investigation as, by its use, extensive stability data can be obtained in a very short time. Indeed, if the range of forms the designer has to deal with is limited, the computer can produce a 'methodical series' of forms from which the designer can select a suitable design.

Although the designer cannot predict precisely the steps necessary to effect a desired change in stability characteristics, it is essential that he know the general effect of changing the ship form in a specified manner. The following comments are intended as a general qualitative guide to the changes that can be expected. It will be appreciated that, in general, it is impossible to vary one parameter alone without some consequential change being required in some other factor.

Length

If length is increased in proportion to displacement, keeping beam and draught constant, the transverse $\overline{KB}$ and $\overline{BM}$ are unchanged. In general, increasing length increases $\overline{KG}$ so reducing initial stability although this will not be the case for a truly geometrically similar ship.

If length is increased at the expense of beam, there will be a reduction in stability over the full range. If length is increased at the expense of draught, there will, in general, be an increase in initial stability but a reduction at larger angles.

Beam

Of the form parameters which can be varied by the designer, the beam has the greatest influence on transverse stability. It has been shown that

$$\overline{BM} \propto \frac{B^2}{T}$$

Thus $\overline{BM}$ will increase most rapidly if beam is increased at the expense of the draught. If freeboard remains constant, the angle of deck edge immersion will decrease and stability at larger angles will be reduced. If the total depth of ship remains constant, then stability will be reduced by the increase in $\overline{BG}$ and, although this is offset by the increased beam at small angles, the stability at larger angles will suffer.

Draught

Reduction in draught in proportion to a reduction in displacement increases initial stability, leads to greater angle of deck immersion but reduces stability at large angles.

Displacement

Changes in length, beam and draught associated with displacement are discussed above. If these parameters are kept constant then displacement increases lead to a fatter ship. In general, it can be expected that the consequential filling out of the waterline will more than compensate for the increased volume of displacement and $\overline{BM}$ will increase. Fattening of the ship in this way is also likely to lead to a fall in G. These changes will also, in general, enhance stability at all angles.

Centre of gravity

The best way, but often the most difficult, of achieving improved stability at all angles is to reduce the height of the centre of gravity above the keel. This is clear from a consideration of the formulae used to derive $\overline{GZ}$ in earlier sections of this chapter. What is perhaps not so immediately obvious, but which cannot be too highly emphasized, is that too high a value of $\overline{KG}$ results in poor stability at larger angles no matter what practical form changes are made.

The effect of change in $\overline{KG}$, if not associated with any change of form, is given directly by the relationship

$$\overline{GZ} = \overline{SZ} - \overline{SG} \sin \phi$$

For a more thorough review of the influence of form changes, the student is advised to consult papers published in the various transactions of learned bodies such as Refs. 2 and 3. The broad conclusion from Prohaska's work, is that the ratios of beam to draught and depth of ship have a comparatively greater influence on stability than that caused by the variation of the fullness parameters of the forms.

Using Prohaska's basic approach, Ref. 4 reports the results of applying regressional analysis to data from 31 ships to obtain expressions for C_{RS} (see p. 111) in terms of other hull parameters. It was shown that the following expressions give reasonable estimates at 30 degrees of heel

$$C_{RS} = 0{\cdot}8566 - 1{\cdot}2262\, \overline{KB}/T - 0{\cdot}035\, B/T$$

$$C_{RS} = -0{\cdot}1859 - 0{\cdot}0315\, B/T + 0{\cdot}3526\, C_M$$

STABILITY OF A COMPLETELY SUBMERGED BODY

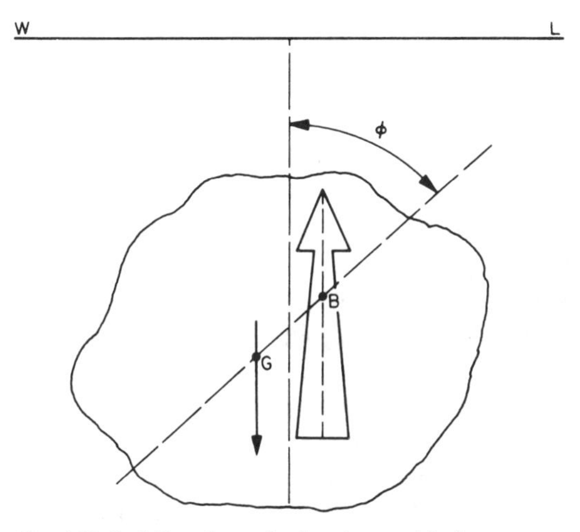

Fig. 4.28 Stability of completely submerged body

As in the case of surface ships there will be a position of rotational equilibrium when B and G are in the same vertical line. Let us determine the condition for stable equilibrium for rotation about a horizontal axis.

Since the body is completely submerged, the buoyancy force always acts through B. Viewing this another way, in the absence of any waterplane $\overline{BM}$ must be zero so that B and M become coincident. If the body is inclined through an angle ϕ, the moment acting on it is given by

$$\Delta \overline{BG} \sin \phi$$

It is clear from Fig. 4.28, that this will tend to bring the body upright if B lies above G. This then is the condition for stable equilibrium. If B and G coincide, the body has neutral stability and if B lies below G it is unstable.

Both the transverse and longitudinal stability of a submerged body are the same and the $\overline{GZ}$ curve becomes a sine curve as in Fig. 4.29. A submarine is typical although the relative positions of B and G, and therefore the stability in the two directions, may be changed by flooding or emptying tanks.

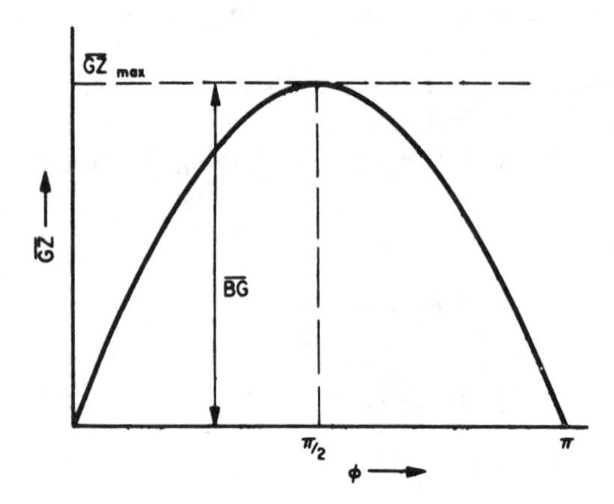

Fig. 4.29 Statical stability curve for submarine

Another mode of stability of a completely submerged body which is of interest is that associated with depth of submersion. If the body in Fig. 4.28 is initially in equilibrium, and is disturbed so as to increase its depth of submersion it is subject to greater hydrostatic pressure. This causes certain compression of the body which, if the body is more readily compressed than water, leads to a reduction in the buoyancy force. The weight of the body remains constant so a resultant downward force is created. This will tend to increase the depth of submersion still further. In these circumstances—applicable to a submarine—the body is unstable in depth maintenance.

Generally, in the case of a submarine the degree of instability is no great embarrassment and, in practice, the effect is usually masked by changes in water density associated with temperature or salinity as depth changes. It becomes important, however, with deep submergence vehicles.

Dynamical stability

The *dynamical stability* of a ship at a given angle of heel is defined as the work done in heeling the ship to that angle very slowly and at constant displacement, i.e., ignoring any work done against air or water resistance.

Consider a ship with a righting moment curve as shown in Fig. 4.30.

Let the righting moment at an angle of heel ϕ be M_ϕ. Then, the work done in heeling the ship through an additional small angle $\delta\phi$ is given approximately by

$M_\phi \, \delta\phi$

Hence, the total work in heeling to an angle Φ is

$$\int_0^\Phi M_\phi \, d\phi$$

$$= \int_0^\Phi \Delta \overline{GZ}_\phi \, d\phi$$

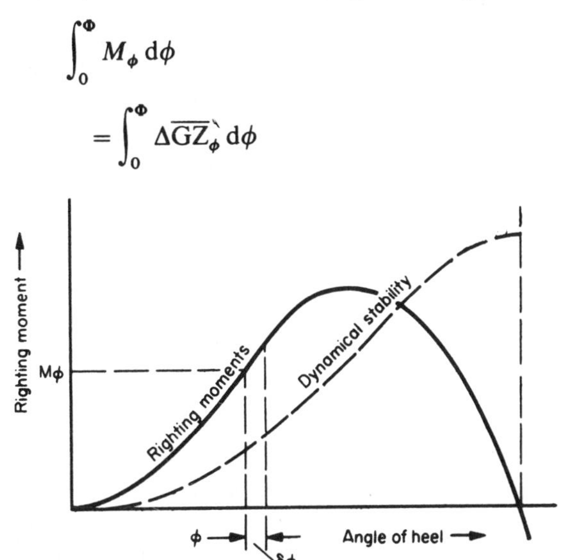

Fig. 4.30

The dynamical stability at any angle, therefore, is proportional to the area under the statical stability curve up to that angle.

There is an alternative way of determining the work done in heeling a ship. Weight and displacement forces remain constant during the process but they are separated vertically, thereby raising the potential energy of the ship.

Referring to Fig. 4.9, let B move to B_1 when ship is heeled from waterline WL to waterline W_1L_1. Increased vertical separation $= \overline{B_1Z_1} - \overline{BG}$. Hence, dynamical stability at angle ϕ is given by

$$\Delta(\overline{B_1Z_1} - \overline{BG})$$

Let v be the volume of the immersed or emerged wedge, let their centroids of volume be at b_1 and b_2 and let h_1 and h_2 be the feet of perpendiculars dropped from b_1 and b_2 on to W_1L_1.
Then

$$\overline{B_1R} = \frac{v(b_1h_1 + b_2h_2)}{\nabla}$$

Now

$$\overline{B_1Z_1} = \overline{B_1R} + \overline{BG} \cos \Phi$$

Therefore

$$\overline{B_1Z_1} = \frac{v(b_1h_1 + b_2h_2)}{\nabla} + \overline{BG} \cos \Phi$$

and the dynamical stability is given by

$$\Delta\left[\frac{v(b_1h_1+b_2h_2)}{\nabla}-\overline{BG}(1-\cos\Phi)\right]$$

This formula is known as *Moseley's Formula.* Although this expression can be evaluated for a series of angles of heel, and the curve of dynamical stability drawn, its application is laborious without a computer. Dynamical stability should not be confused with dynamic stability of course which is discussed in Chapter 13.

Stability assessment

STABILITY STANDARDS

Having shown how the stability of a ship can be defined and calculated, it remains to discuss the actual standard of stability to be aimed at for any particular ship bearing in mind its intended service. For conventional ships, the longitudinal stability is always high and need not be considered here. This may not be true for offshore drilling barges and other less conventional vessels.

What degree of simple static stability is needed? Unless a ship has positive stability in the upright condition, it will not remain upright because small forces —from wind, sea or movements within the ship—will disturb it. Even if it does not actually capsize, i.e. turn right over, it would be unpleasant to be in a ship which lolled to one side or the other of the upright. Thus one need for stability arises from the desire to have the ship float upright. Although, in theory, a very small metacentric height would be sufficient for this purpose, it must be adequate to cover all conditions of loading of the ship and growth in the ship during its life. If the ship is to operate in very cold climates, allowance will have to be made for the topweight due to ice forming on the hull and superstructure.

Next, it is necessary to consider circumstances during the life of the ship which will cause it to heel over. These include

(*a*) The action of the wind which will be most pronounced in ships with high freeboard or large superstructure;
(*b*) the action of waves in rolling the ship. This will be most important for those ships which have to operate in large ocean areas, particularly the North Atlantic;
(*c*) the action of the rudder and hull forces when the ship is manoeuvring;
(*d*) loading and unloading cargoes.

The ship may be subject to several of the above at the same time. In addition, the ship may, on occasion, suffer damage leading to flooding or she may suffer from shifting cargo in very rough weather. These aspects of the design are dealt with in more detail in the next chapter.

The standards of stability aimed at by various navies are not generally available but Ref. 5 discusses the standards specified for US naval surface ships. The authors point out that if a ship experiences a heeling moment in perfectly calm water it would be sufficient if the curve of statical stability had a lever in excess of that represented by the heeling moment. In practice, it is also necessary to allow a certain reserve of dynamical stability to enable the ship to absorb the energy imparted to it by waves or by a gusting wind.

Certain relaxations are made for ships operating in coastal waters and for ships in service, but for new design intact ships to operate in ocean areas the following criteria are specified.

(*a*) *Beam winds and rolling*

A wind heeling arm curve is calculated and is superimposed on the statical stability curve for the intact ship as in Fig. 4.31. A simple formula for calculating the heeling arm due to wind is

$$\text{Heeling arm} = \frac{0{\cdot}0035V^2Al\cos^2\phi}{2240\Delta}\ \text{ft}$$

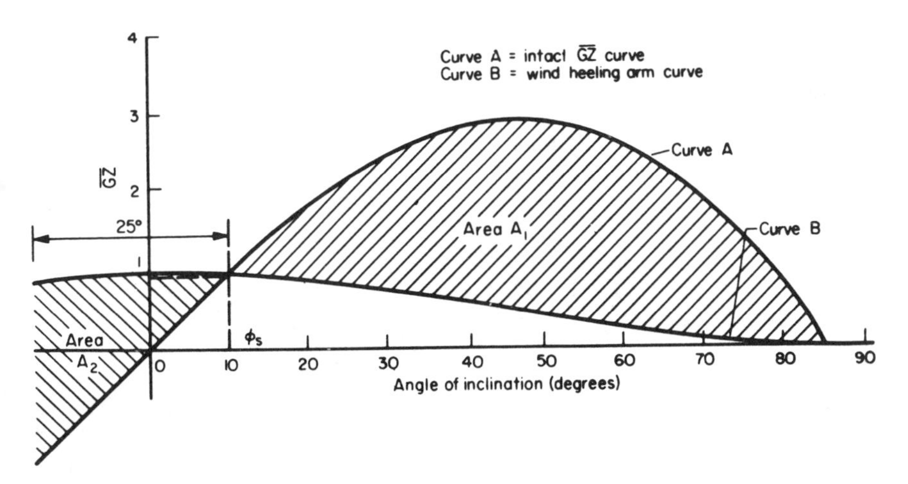

Fig. 4.31 Stability criteria for beam wind and rolling

where V = nominal wind velocity, knots (taken as 100 knots). A = projected area (ft^2) of ship above waterline and l = lever arm from half draught to the centroid of the projected area. Δ = displacement in tonf.

More accurately, the formula

$$\text{Heeling arm} = \frac{0{\cdot}004V^2Al\cos^2\phi}{2240\Delta}\ \text{ft}$$

can be used when the wind speed V at each height above the sea surface is taken from Fig. 9.22 for a nominal wind velocity of 100 knots. In this case, it is necessary to divide A into a number of convenient horizontal strips and carry out a vertical integration.

If A and l are measured in metres and Δ in MN, the two equations become

$$\text{Heeling arm} = 0{\cdot}17 \times 10^{-6} \frac{AV^2l}{\Delta} \cos^2 \phi \text{ m}$$

$$\text{Heeling arm} = 0{\cdot}19 \times 10^{-6} \frac{AV^2l}{\Delta} \cos^2 \phi \text{ m}$$

The angle ϕ_s at which the two curves cross, is the steady heel angle the ship would take up if the wind were perfectly steady and there were no waves and the corresponding righting arm is $\overline{GZ}_s$. To allow for the rolling of the ship the curves are extended back to a point 25 degrees to windward of ϕ_s and the areas A_1 and A_2 shown shaded in Fig. 4.31 are computed. For stability to be regarded as adequate the following conditions must be met by USN ships:

$\overline{GZ}_s$ not more than $0{\cdot}6 \times \overline{GZ}_{max}$
Area A_1 not less than $1{\cdot}4 \times$ Area A_2.

(*b*) *Lifting of heavy weights over the side*

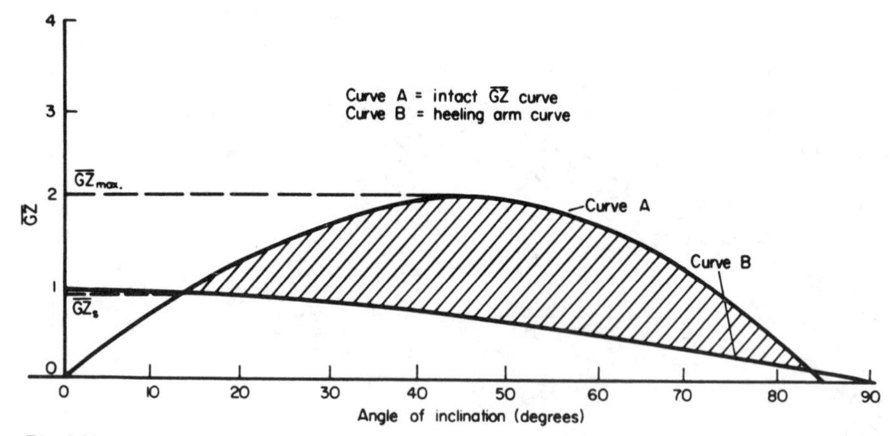

Fig. 4.32

Again, the heeling arm curve is superimposed on the statical stability curve (Fig. 4.32) the heeling arm being taken as

$$\frac{Wa}{\Delta} \cos \phi$$

where W = weight of lift, a = transverse distance from centre line to end of boom and Δ = displacement, including weight of lift.

The stability criteria applied by the USN are

$\overline{GZ}_s$ not more than $0{\cdot}6 \times \overline{GZ}_{max}$
ϕ_s not greater than 15 degrees
Shaded area not less than 40 per cent of the total area under the statical stability curve.

(*c*) *Crowding of passengers to one side*

The same formula and criteria are used as for case (*b*), except that, in this case,

W = weight of passengers
a = distance from centre line of ship to c.g. of passengers.

(*d*) *Heeling during a high speed turn*

When a ship is turning steadily, it must be acted upon by a force directed towards the centre of the circle and of value

$$\frac{\Delta\,(\text{steady speed})^2}{g\,(\text{radius of turn})} = \frac{\Delta V^2}{gR}$$

This problem is discussed in more detail in Chapter 13, but, approximately, the heeling moment acting on the ship is given by

$$\frac{\Delta V^2}{gR}\left(\overline{\text{KG}} - \frac{T}{2}\right)\cos\phi$$

The heeling arm curve is then

$$\frac{V^2}{gR}\left(\overline{\text{KG}} - \frac{T}{2}\right)\cos\phi$$

and can be plotted as in Fig. 4.32. The stability criteria adopted are as for case (*b*) above.

In Ref. 6, the standards of stability adopted by Japan are presented and discussed. As with the USN standards, they are essentially standards for metacentric height, maximum $\overline{\text{GZ}}$ and dynamical stability.

So far, only minimum standards have been mentioned. Why not then make sure and design the ship to have a very large metacentric height? In general, this could only be achieved by considerably increasing the beam partially at the expense of draught. This will increase the resistance of the ship at a given speed, and may have adverse effects on stability at large angles of heel and following damage. It will also lead to more violent motion in a seaway which may be unpleasant to passengers and indeed dangerous to all personnel. In addition, the larger beam may make it impossible to negotiate certain canals or dock entrances. Hence the designer is faced with providing adequate but not excessive stability.

PASSENGER SHIP REGULATIONS

Passenger ships (Ref. 13) must exhibit intact stability which conforms to the following:

(*a*) The area under the $\overline{\text{GZ}}$ curve shall not be less than
  (i) 0.055 m rad up to 30 degrees
  (ii) 0.09 m rad up to 40 degrees or up to the downflooding angle (the least angle where there are openings permitting flooding)
  (iii) 0.03 m rad between 30 degrees and angle (ii)

(*b*) $\overline{\text{GZ}}$ must be greater than 0·20 m at 30 degrees
(*c*) Maximum $\overline{\text{GZ}}$ must be at an angle greater than 30 degrees
(*d*) $\overline{\text{GM}}$ must be at least 0·15 m.

THE INCLINING EXPERIMENT

It has been shown that, although much of the data used in stability calculations depends only on the geometry of the ship, the position of the centre of gravity must be known before the stability can be assessed finally for a given ship condition. Since $\overline{KG}$ may be perhaps ten times as great as the metacentric height, it must be known very accurately if the metacentric height is to be assessed with reasonable accuracy.

$\overline{KG}$ can be calculated for a variety of conditions provided it is accurately known for one precisely specified ship condition. It is to provide this basic knowledge that the *inclining experiment* is carried out. The term 'experiment' is a misnomer in the scientific sense but it is universally used. It would be better called a stability survey. The purposes of the inclining experiment are to determine the displacement and the position of the centre of gravity of the ship in an accurately known condition. It is usually carried out when the ship is as nearly complete as possible during initial building and, in the case of warships, may be repeated at intervals during her life as a check upon the changing armaments, etc. For instance, an additional inclining may be carried out following an extensive modernization or major refit.

The experiment is usually carried out by moving weights across the deck under controlled conditions and noting the resulting angle of heel. The angles are normally kept small so that they are proportional to the heeling moment, and the proportionality factor provides a measure of the metacentric height. The angle of heel is usually recorded using long pendulums rigged one forward and one aft in the ship, perhaps through a line of hatches. The pendulum bobs may be immersed in water or oil to damp their motion.

Conduct of an inclining experiment

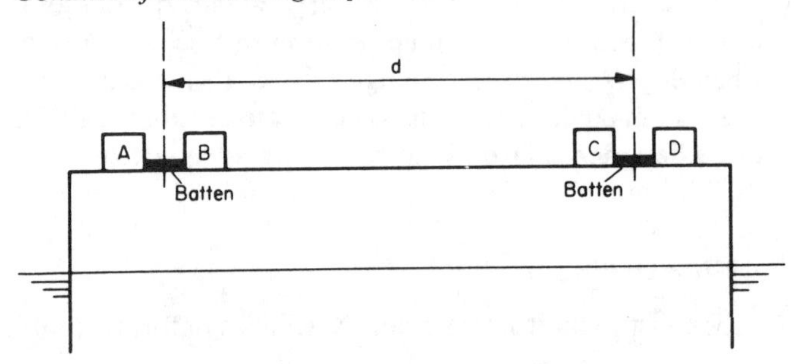

Fig. 4.33 Movement of weights during an inclining experiment

The principal steps in the carrying out of an inclining experiment are:

(*a*) The ship is surveyed to determine weights to be removed, to come on board or be moved for final completion.
(*b*) The state of all tanks is noted accurately.
(*c*) The draughts are accurately read at each set of draught marks, including amidships, on both sides of the ship.
(*d*) The density of the water in which the vessel is floating is measured at a number of positions and depths around the ship.

(*e*) Weights, arranged on the deck in four groups, shown in Fig. 4.33, are moved in the following sequence:
weights A to a position in line with weights C
weights B to a position in line with weights D
weights A and B returned to original positions
weights C to a position in line with weights A
weights D to a position in line with weights B
weights C and D returned to original positions
The weight groups are often made equal.

(*f*) The angle of heel is recorded by noting the pendulum positions before the first movement of weight and after each step given above.

Analysis of results

Due to the shift of a weight w through a distance d, G will move parallel to the movement of the weight to G′ where

$$\overline{GG'} = \frac{wd}{\Delta}$$

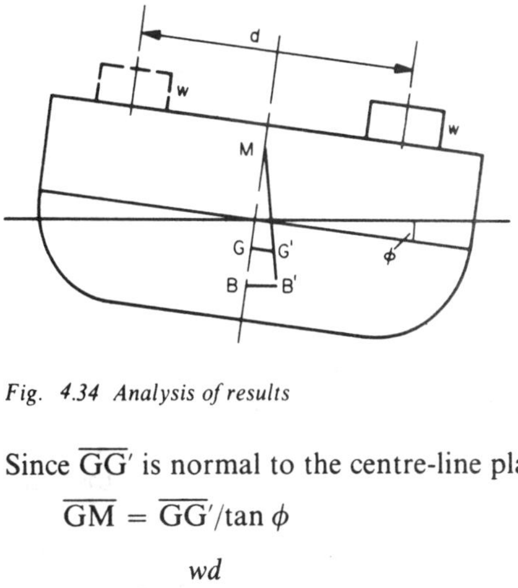

Fig. 4.34 Analysis of results

Since $\overline{GG'}$ is normal to the centre-line plane of the ship

$$\overline{GM} = \overline{GG'}/\tan\phi$$

$$= \frac{wd}{\Delta \tan\phi}$$

If the movement of each pendulum, length l, is measured as y on a horizontal batten

$$\tan\phi = \frac{y}{l}$$

The displacement can be determined from the measurements of draught and water density. The latter are averaged and the former are used to determine the trim of the vessel. The amount of hog or sag present can be determined if the ship is provided with amidship draught marks. When hog or sag is detected it is usual to assume, in calculating the displacement, that the ship takes up a parabolic curve. The mean draught is calculated from the draughts at the ends of the ship and is increased or reduced by 70 per cent of the sag or hog respectively. Similarly, it may be possible to find the extent of any stern drop.

Hence, $\overline{GM}$ can be calculated for each transfer of weight provided the metacentre remains fixed up to the maximum angle of heel used. Typically, this should be of the order of 3 or 4 degrees and presents no problem for most ships. If equal weight transfers are made, this is readily confirmed if the angle changes are equal. If unequal weight transfers are used, an approximate check can be made by dividing the angles by the respective weights and comparing the results. If it is suspected that the relationship between angle and heeling moment is not linear, the results obtained must be plotted as a statical stability curve and the tangent at the origin determined.

The data can be analysed in various ways. One possible method is to produce a table of the asymmetric moment applied at each stage (m_1, m_2, etc.) and the corresponding pendulum deflections (d_1, d_2, etc.). The moment values are squared and the moments and deflections multiplied together as illustrated below:

Weight movement (Fig. 4.33)	Applied moment m	m^2	Measured deflection d	$m \times d$
A to C	m_1	m_1^2	d_1	$m_1 d_1$
B to D	m_2	m_2^2	d_2	$m_2 d_2$
A and B returned	0	0	d_3	0
C to A	$-m_4$	m_4^2	d_4	$-m_4 d_4$
D to B	$-m_5$	m_5^2	d_5	$-m_5 d_5$
C and D returned	0	0	d_6	0
		$\sum m^2$		$\sum md$

The average deflection per unit applied moment is given by $\sum md / \sum m^2$.

It will be noted that, in this method of analysis, the deflections measured when all weights are returned to their original positions play no part in the analysis. They are used at the time of the experiment to indicate whether any loose weights, or liquids in tanks have moved from one side to another.

The datum readings are used in a second method of analysis in which the deflection of the pendulum per unit applied moment is calculated for each shift of weights. This is illustrated below:

Weight moment (Fig. 4.33)	Applied moment, m	Measured deflection, d	Change in deflection per unit change of moment
A to C	m_1	d_1	d_1/m_1
B to D	m_2	d_2	$(d_2 - d_1)/(m_2 - m_1)$
A and B returned	0	d_3	$(d_3 - d_2)/(0 - m_2)$
C to A	$-m_4$	d_4	$(d_4 - d_3)/(-m_4)$
D to B	$-m_5$	d_5	$(d_5 - d_4)/(-m_5 + m_4)$
C and D returned	0	d_6	$(d_6 - d_5)/m_5$

The average deflection per unit change of moment is taken as the mean of the figures in the last column.

EXAMPLE 4. A ship of 6000 tonnef displacement is inclined. The ballast weights are arranged in four equal units of 10 tonnef and each is moved transversely through a distance of 10 m. The pendulum deflections recorded are 31, 63, 1, −30, −62 and 0 cm with a pendulum length of 7 m. Calculate the metacentric height at the time of the experiment.

Solution: Since the moment is applied in equal increments of 100 tonnef m, each such increment can be represented by unity in the table as below, using the first method of analysis:

Applied moment Units of 100 tonnef m m	m^2	Deflection d cm	$m \times d$
1	1	31	31
2	4	63	126
0	0	1	0
−1	1	−30	30
−2	4	−62	124
0	0	0	0
	$\sum m^2 = 10$		$\sum md = 311$

∴ Mean deflection per unit moment = 31·1 cm, i.e. the pendulum is deflected through 31·1 cm for each 100 tonnef m of applied moment

$$\text{Angle of heel} = \frac{\text{Pendulum deflection}}{\text{Pendulum length}}$$

$$= \frac{31{\cdot}1}{700} = 0{\cdot}0444 \text{ radians.}$$

For small angles

$$\text{Applied moment} = \Delta\overline{\text{GM}}\phi$$

$$\therefore \quad 100 = 6000\,\overline{\text{GM}}\,0{\cdot}0444$$

$$\therefore \quad \overline{\text{GM}} = 0{\cdot}375 \text{ m}$$

Applying the second method of analysis would have given:

Change of moment Units of 100 tonnef m	Change in pendulum deflection (cm)	Pendulum deflection per unit change of moment
1	31	31
1	32	32
−2	−62	31
−1	−31	31
−1	−32	32
2	62	31

$$\text{Hence mean pendulum deflection} = \frac{188}{6}$$

$$= 31{\cdot}3 \text{ cm}$$

This represents less than one per cent difference compared with the answer obtained by the first analysis. This is within the general limits of accuracy of the experiment.

Precautions to be taken

The importance of obtaining an accurate result has already been emphasized and certain precautions must be taken to ensure that accuracy is obtained. These are really a matter of common sense and include

(*a*) The ship must be floating upright and freely without restraint from ropes.
(*b*) There should be no wind on the beam.
(*c*) All loose weights should be secured.
(*d*) All cross-connections between tanks should be closed.
(*e*) Tanks should be empty or pressed full. If neither of these conditions is possible, the level of liquid in the tank should be such that the free surface effect is readily calculable and will remain sensibly constant throughout the experiment.
(*f*) The number of men on board should be kept to a minimum and they should go to specified positions for each reading of the pendulums.
(*g*) Any mobile equipment used to move the weights across the deck must return to a known position for each set of readings.

It should be noted that (*c*), (*d*), (*f*) and (*g*) will, to some extent, be checked by the accuracy with which the pendulums return to their zero readings following the return of weights to their original positions half way through the experiment and again at the end. If the pendulum reading fails to take up a steady value but slowly increases or diminishes, it is likely that liquid is passing between tanks through an open cross-connection.

PRECISION OF STABILITY STANDARDS AND CALCULATIONS

The stability characteristics of a given form can be derived with considerable precision and some authorities fall into the trap of quoting parameters such as initial metacentric height with great arithmetic accuracy. This shows a lack of appreciation of the true design concept of stability and the uncertainties surrounding it in the real world.

Even in the very limited case of the stability of a vessel at rest in calm water great precision is illusory since:

(*a*) Ships are not built precisely to the lines plan.
(*b*) Weights and $\overline{\text{KG}}$ of sister ships even in the unloaded state will differ. Apart from dimensional variations, plate thicknesses are nominal.

(*c*) The accuracy of $\overline{KG}$ assessment is limited by the accuracy with which an inclining experiment can be conducted.
(*d*) Many calculations ignore the effects of trim.

However, these inaccuracies are relatively trivial compared with the circumstances appertaining when a vessel is on voyage in a rough sea.

By now the student should appreciate that stability is a balance between forces acting on the hull, trying on the one hand to overturn it and on the other to return it to the upright position. All the major elements in the equation are varying, viz.:

(*a*) Weight of ship and its distribution depend upon initial lading and the way in which consumables are used up during a voyage. The presence or otherwise of free surfaces will be significant.
(*b*) Wind forces will depend upon speed and direction relative to ship. They will vary as the wind gusts and as the ship moves to present a different aspect to the wind, e.g. by rolling.
(*c*) Water forces will depend upon the nett wave pattern compounded by the ship's own pattern interacting with that of the surrounding ocean; also upon the ship's altitude and submergence as it responds to the forces acting on it. To take one example: the shipping of water and the ability of clearing ports to clear that water is important.

A ship's speed and direction relative to the predominant wave direction can be important. Reference 10, for example, shows that at relatively high speed in quartering seas a ship encounters waves of a wide range of lengths at almost exactly the same frequency. If this frequency happened to coincide with the ship's natural roll frequency, then very large roll amplitudes could build up quickly, possibly resulting in capsize.

Thus the concepts of static equilibrium and stability have many limitations in dealing with what is in reality a dynamic situation. To be realistic, a designer must recognise the varying environment and the variations in ship condition. Traditionally he has dealt with this by setting standards based loosely on realistic conditions and experience of previously successful design. Margins provide safety under all but extreme conditions which it would be uneconomical to design for. Whilst this approach can allow different standards for say ocean going or coastal shipping it has a number of drawbacks, viz.:

(*a*) A succession of successful ships may mean they have been overdesigned.
(*b*) The true risk factor is not quantified. Is it a 1 in 10^6 or 1 in 10^9 probability of a ship being lost?

In recent years more attention has been given to the problems of stability and capsizing as dynamic phenomena. Advances have been made in the understanding of the dynamic problem spurred on, unfortunately, by a number of marine disasters such as the loss of the trawler *Gaul* (Ref. 11). This was a modern vessel which met IMO standards for the service in which it was used but which was nevertheless lost. In this case it was concluded in Ref. 11 that the *Gaul* was not lost because of inadequate intact stability or poor seakeeping qualities. It is surmised that she met severe wave and wind conditions at the same time as she suffered from some other unknown circumstance such as internal flooding.

The above remarks will have made the student more aware of the true nature of the problem of stability of ships at sea, and the limitations of the current static approach. However, there is, as yet, no agreed method of dealing with the true dynamic situation. The designer still uses the static stability criteria but with an eye on the other issues. Thus Ref. 11 included quasi-static calculations of stability curves with a ship balanced on a wave. Righting arms are reduced below the still water values when a wave crest is near amidships and increased when a trough is amidships. In the case of *Gaul* the variation in maximum $\overline{GZ}$ was about ± 30 per cent. The changing roll stiffness due to the $\overline{GZ}$ variations as the ship passes through successive crests and troughs can result in a condition of parametric resonance leading to eventual capsize. The effect of the Smith correction on righting arm was found to be insignificant for angles up to 20 degrees but gave changes of 10 per cent at 50 degrees. Reference 11 also shows how valuable model experiments can be for studies in waves.

Another possible approach is outlined in Ref. 3 by which the hull shape is so configured as to retain a constant $\overline{GZ}$ curve in the changing circumstances between the 'as designed' and 'as built' conditions and in motion in a seaway. It discusses how to create a hull form in way of the design draught such that the metacentric height remains constant with changing displacement and trim and extends this to quasi-static sea conditions. Although the approach concentrated on initial stability it was shown that forms developed in accord with the principles exhibited sensibly constant $\overline{GZ}$ curves in waves, and for damaged stability and change of draught up to quite large heel angles. The shaping required did not conflict greatly with other design considerations.

Heeling trials

These trials are mentioned only because they are commonly confused with the inclining experiment. In fact, the two are quite unrelated. Heeling trials are carried out to prove that the ship, and the equipment within her, continue to function when the vessel is held at a steady angle of heel. For instance, pumps may fail to maintain suction, bearings of electric motors may become overloaded, boats may not be able to be lowered by the davits clear of the ship, etc.

For warships, the trials are usually carried out for angles of heel of 5, 10 and 15 degrees although not all equipment is required to operate up to the 15 degree angle.

Tilt test

This is another test carried out in warships which is quite unrelated to the inclining experiment but is sometimes confused with it. The tilt test is carried out to check the alignment of items of armament—gun turrets, missile launchers, directors—with respect to each other and to a reference datum in the ship.

Problems

1. A small weight is added to a wall-sided ship on the centre line and at such a position as to leave trim unchanged. Show that the metacentric height is reduced if the weight is added at a height above keel greater than $T-\overline{GM}$, where T is the draught.

2. Prove that for a wall-sided ship, the vertical separation of B and G at angle of heel ϕ is given by

$$\cos\phi[\overline{BG}+\tfrac{1}{2}\overline{BM}\tan^2\phi]$$

where $\overline{BG}$ and $\overline{BM}$ relate to the upright condition. Show that when the ship has a free surface in a wall-sided tank the corresponding expression is

$$\overline{BG}_V = \cos\phi[\overline{BG}_S+\tfrac{1}{2}(\overline{BM}-\overline{G_SG_F})\tan^2\phi]$$

where subscripts S and F relate to the solid and fluid centres of gravity.

3. Show that the dynamical stability of a wall-sided vessel up to angle ϕ is given by the expression

$$\Delta[\overline{BG}(\cos\phi-1)+\tfrac{1}{2}\overline{BM}\tan\phi\sin\phi]$$

4. Prove the wall-sided formula.

A rectangular homogeneous block 30 m long × 7 m wide × 3 m deep is half as dense as the water in which it is floating. Calculate the metacentric height and the $\overline{GZ}$ at 15 degrees and 30 degrees of heel.

5. The statical stability curve for a cargo ship of 10,000 tonnef displacement is defined by

ϕ (degrees)	0	15	30	45	60	75	90
$\overline{GZ}$ (m)	0	0·275	0·515	0·495	0·330	0·120	−0·100

Determine the ordinate of the dynamical stability curve at 60 degrees and the change in this figure and loss in range of stability if the c.g. of the ship is raised by 0·25 m.

6. A hollow triangular prism whose ratio of base to height is 1 : 2 floats vertex downwards. Mercury of density 13·6 is poured in to a depth of 1 m and water is then poured on top till the draught of the prism is 4 m. Calculate $\overline{GM}$ solid and fluid assuming the prism weightless.

7. A ship has the following principal dimensions: length 120 m, beam 14 m, draught 4 m, displacement 30 MN. The centres of buoyancy and gravity are 2·5 m and 6 m above keel and the metacentre is 7 m above keel.

Calculate the new metacentric height and beam if draught is reduced to 3·6 m, keeping length, displacement, $\overline{KG}$ and coefficients of fineness unaltered.

8. The model of a ship, 5 ft long, is found to be unstable. Its waterplane has an area of 550 in^2 and a transverse MI of 5500 in^4. The draught is 4·6 in. and the volume of displacement is 2200 in^3 in fresh water.

If G is originally 6 in. above the keel, how far above the keel must a weight of 20 lb be put to just restore stability? $\overline{KB}$ = 2·5 in.

9. A hollow prism has a cross section in the form of an isosceles triangle. It floats apex down in fresh water with mercury, of specific gravity 13·5, dropped into it to make it float upright.

If the prism may be assumed weightless by comparison and has an included angle of 90 degrees, express the fluid metacentric height in terms of the draught T.

10. A ship is floating in fresh water at a draught of 14 ft when a weight of 420 tonf is placed 42 ft above the keel and 14 ft to port. Estimate the angle of heel.

In sea water, the hydrostatic curves show that at a draught of 14 ft, TPI = 72; displacement = 36,000 tonf; $\overline{KM}$ = 26 ft; $\overline{KB}$ = 8 ft. Before the weight is added $\overline{KG}$ = 19 ft.

11. The raft shown has to support a weight which would produce a combined height of the centre of gravity above the keel of 3·5 m. The draught is then uniformly 1 m.

What is the minimum value of *d*—the distance apart of the centre lines of the baulks—if the $\overline{GM}$ must not be less than 2 m?

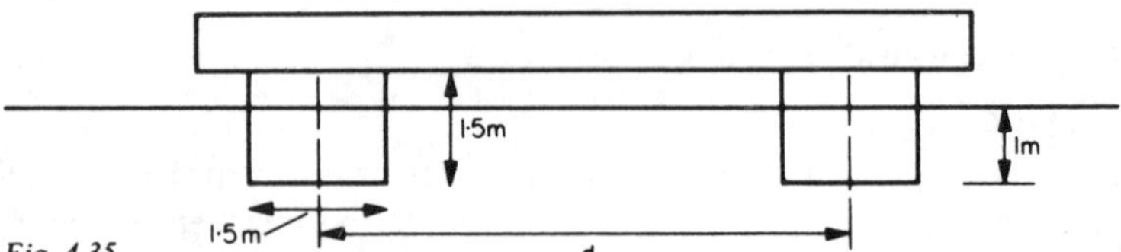

Fig. 4.35

12. A homogeneous solid is formed of a right circular cylinder and a right circular cone of the same altitude *h* on opposite sides of a circular base radius *r*. It floats with the axis vertical, the whole of the cone and half of the cylinder being immersed. Prove that the metacentric height is $3r^2/10h - 21h/80$, so that for equilibrium *r* must be greater than 0·935*h*.

13. A ship of 50 MN (5100 tonnef) displacement has a metacentric height of 1 m. Suppose 1·0 MN of fuel to be shifted so that its centre of gravity moves 6 m transversely and 2 m vertically up, what would be the angle of heel of the vessel, if it were upright before the movement? No free surface effects are produced.

14. A prismatic log of wood, of specific gravity 0·75, whose uniform transverse section is that of an isosceles triangle, floats in fresh water with the base of the section horizontal and vertex upwards. Find the least vertical angle of the section for these conditions to hold.

15. A pontoon raft 10 m long is formed by two cylindrical pontoons 0·75 m diameter spaced 2 m apart between centres and is planked over with wood forming a platform 10 m × 3 m.

When laden, the raft floats with the cylindrical pontoons half immersed in river water and its centre of gravity when laden is 1 m above the water-line. Calculate the transverse and longitudinal metacentric heights.

16. Describe the procedure required to carry out an inclining experiment. The displacement of a vessel as inclined in sea water is 75 MN, 0·3 MN of ballast with a spread of 15 m causes a deviation of 0·2 m on a 4 m pendulum. The net

weight to go on includes 7 MN of fuel (s.g. = 0·85) with a KG of 6 m. The transverse M.I. of the fuel surface is 1000 m^4 units. If the value of $\overline{KM}$ is 9 m and is kept at this value for the increased draught, find the final $\overline{GM}$.

17. Determine the transverse metacentric height from the following data obtained from an inclining experiment:

Length of pendulum	= 20 ft
Mean deflection	= 8·25 in.
Shift of weight	= 20 tonf through 56 ft
Draughts at perpendiculars fwd.	= 17 ft $9\frac{1}{2}$ in.
aft.	= 19 ft $1\frac{1}{4}$ in.
Density of water	= 35·2 ft^3/tonf
Length BP	= 565 ft

The hydrostatic particulars for normal trim of 2 ft by the stern and water 35·0 ft^3/tonf are:

Draught (mean BP) (ft)	Displacement (tonf)	TPI	CF. aft of ⌀ (ft)
18·0	10,524	49·8	30·0
18·5	10,823	50·0	31·1

18. During an inclining experiment on a 60 MN guided missile destroyer weights in units of 0·05 MN are moved 10 m transversely. Deflections at the bottom of a pendulum 4 m long are defined by the following table:

Weight movement	Pendulum reading (m)
0	0
(*a*) 0·05 MN to port	0·023
(*b*) 0·05 MN to port	0·049
(*a*) & (*b*) returned	0·003
(*c*) 0·05 MN to starboard	−0·021
(*d*) 0·05 MN to starboard	−0·045
(*c*) & (*d*) returned	−0·001

Calculate the metacentric height as inclined.

19. A crane in a cargo liner, of 50 MN displacement and $\overline{GM}$ 1 m, lifts a weight of 0·2 MN from a cargo hold through a vertical height of 10 m. If the height of the crane's jib is 20 m above the hold and the radius of the jib is 10 m, calculate the reduction in $\overline{GM}$ caused by lifting the weight.

What is the approximate angle of heel caused by turning the crane through 30 degrees assuming it is initially aligned fore and aft?

20. A fire in the hangar of an aircraft carrier, floating in water of 35 ft^3/tonf, is fought by dockside fire tenders using fresh water. When the fire is under control the hangar, which is 360 ft long and 80 ft broad, is found to be flooded to a depth of 3 ft. If the hangar deck is 23 ft above the original c.g. of the ship,

what is the resultant $\overline{GM}$? The following particulars apply to the ship before the fire:

Displacement	= 43,000 tonf	$\overline{KB}$	= 19 ft
Mean draught	= 39 ft 2 in.	$\overline{BM}$	= 23 ft
TPI	= 120	$\overline{GM}$	= 11·5 ft

21. The following particulars apply to a new cruiser design:

Deep displacement	12,500 tonnef
Length on WL	200 m
$\overline{KG}$	8 m
$\overline{GM_T}$	1·5 m
Area of waterplane	3000 m^2
Draught	6 m
Depth of hull amidships	14 m
Beam	21 m

A proposal is made to increase the armament by the addition of a weight of 200 tonnef, the c.g. 2 m above the upper deck. Find the amount by which the beam must be increased to avoid loss of transverse stability assuming other dimensions unchanged. Assume water of 0·975 m^3/tonne.

22. A pontoon has a constant cross-section in the form of a trapezium, width 6 m at the keel, 10 m at the deck and depth 5 m. At what draught will the centre of curvature of the curve of flotation be at the deck level?

23. Show that a wall-sided vessel with negative $\overline{GM}$ lolls to an angle given by

$$\tan\phi = \pm\left(\frac{2\overline{GM}}{\overline{BM}}\right)^{\frac{1}{2}}$$

A body of square cross-section of side 6 m has a metacentric height of $-0{\cdot}5$ m when floating at a uniform draught of 3 m. Calculate the angle of loll.

24. Derive an expression for the radius of curvature of the metacentric locus.

Prove that in a wall-sided vessel, the radius of curvature of the metacentric locus is given by $3\,\overline{BM}\sec^3\phi\tan\phi$ where $\overline{BM}$ relates to the upright condition.

25. The half-ordinates in metres of a waterplane for a ship of 5 MN, 56 m long are 0·05, 0·39, 0·75, 1·16, 1·63, 2·12, 2·66, 3·07, 3·38, 3·55, 3·60, 3·57, 3·46, 3·29, 3·08, 2·85, 2·57, 2·26, 1·89, 1·48, and 1·03. If $\overline{KB}$ is 1·04 m and $\overline{KG}$ is 2·2 m calculate the value of $\overline{GM}$ assuming the ship is in water of 0·975 m^3/Mg.

26. A computer output gives the following data for plotting cross curves of stability ($\overline{KS}$ = 15 ft).

ϕ (deg)	Dispt. (tonf)	$\overline{SZ}$ (ft)	Dispt. (tonf)	$\overline{SZ}$ (ft)	Dispt. (tonf)	$\overline{SZ}$ (ft)
10	1300	3·12	2350	2·08	3580	1·60
	4890	1·29	6200	1·10	7660	1·08
20	1410	4·89	2430	3·78	3660	3·05
	4950	2·62	6330	2·35	7830	2·30

(continued on p. 141)

(*continued from p. 140*)

ϕ (deg)	Dispt. (tonf)	$\overline{SZ}$ (ft)	Dispt. (tonf)	$\overline{SZ}$ (ft)	Dispt. (tonf)	$\overline{SZ}$ (ft)
30	1580	5·56	2600	4·76	3820	4·31
	5130	3·94	6550	3·79	8100	3·69
40	1810	5·84	2870	5·45	4120	5·38
	5500	5·23	6870	4·98	8350	4·60
50	2100	6·24	3270	6·29	4570	6·23
	5850	5·87	7150	5·48	8450	5·08
60	2530	6·90	3680	6·69	4950	6·30
	6200	5·89	7400	5·49	8600	5·11
70	2950	6·70	4030	6·37	5300	5·93
	6500	5·47	7620	5·13	8770	4·87
80	3250	5·81	4370	5·63	5530	5·26
	6680	4·82	7830	4·58	8970	4·42
90	3670	4·67	4630	4·51	5740	4·32
	6900	4·08	8050	3·91	9170	3·85

Plot the cross curves of stability and deduce the $\overline{GZ}$ curves for the following conditions:

(*a*) Δ = 3710 tonf; $\overline{GM}$ = 1·92 ft, $\overline{KG}$ = 21·84 ft
(*b*) Δ = 4275 tonf; $\overline{GM}$ = 2·80 ft, $\overline{KG}$ = 20·20 ft
(*c*) Δ = 4465 tonf; $\overline{GM}$ = 2·88 ft, $\overline{KG}$ = 19·90 ft
(*d*) Δ = 4975 tonf; $\overline{GM}$ = 3·56 ft, $\overline{KG}$ = 18·70 ft

Plot the $\overline{GZ}$ curves and find the angles of vanishing stability.

27. A ship of displacement 50 MN has a $\overline{KG}$ of 6·85 m. The $\overline{SZ}$ values read from a set of cross curves of stability are as below:

ϕ	10	20	30	40	50	60	70	80	90
$\overline{SZ}$ (m)	0·635	1·300	1·985	2·650	3·065	3·145	3·025	2·725	2·230

Assuming that $\overline{KS}$ for the curves is 5 m, calculate the ordinates of the $\overline{GZ}$ curve and calculate the dynamical stability up to 80 degrees.

28. A passenger liner has a length of 200 m, a beam of 25 m, a draught of 10 m and a metacentric height of 1 m. The metacentric height of a similar vessel having a length of 205 m, beam 24 m, draught of 10·5 m is also 1 m. Assuming that $\overline{KG}$ is the same for the two vessels, calculate the ratio of $\overline{KB}$ to $\overline{BM}$ in the original vessel. What is the ratio of the displacements of the two vessels?

29. A vessel 72 ft long floats at 6 ft draught and has 4·5 ft freeboard, with sides above water vertical. Determine the $\overline{GZ}$ at 90°, assuming $\overline{KB}$ = 3·5 ft and $\overline{KG}$ = 5 ft with the vessel upright. The half-ordinates of the waterplane are 0·8, 3·3, 5·4, 6·5, 6·8, 6·3, 5·1, 2·8 and 0·6 ft. The displacement is 100 tonf and the middle-line plane is rectangular in shape.

30. For a given height of the centre of gravity and a required standard of stability (i.e. $\overline{GM}$), what is the relationship between a ship's length, beam, draught and volume of displacement?

State your assumptions clearly.

A ship is designed to the following particulars:

Displacement 10,000 tonf; draught 24 ft; beam 56 ft, $\overline{GM}$ 3·5 ft; G above WL 4·0 ft; TPI 48.

Alterations involve 100 tonf extra displacement and 1 ft rise of c.g.

What increase in beam will be necessary to maintain $\overline{GM}$, assuming that the depth of the ship remains unaltered and the small alterations in dimensions will not further alter the position of the c.g. or weight?

Assume the ship to be wall-sided and that length, waterplane coefficients and block coefficients are unchanged.

31. A ship of 1500 tonf displacement carries 150 tonf of oil fuel in a rectangular tank 30 ft long, 20 ft wide and 20 ft deep. If the oil is assumed solid the ship has a $\overline{GM}$ of 4·90 ft and the following values of $\overline{GZ}$:

angle (deg)	15	30	45	60	75	90
$\overline{GZ}$ (ft)	1·35	2·72	3·17	2·18	0·60	−1·13

Taking the density of the oil as 40 ft^3/tonf, find the values of $\overline{GZ}$ at 30, 45 and 75 degrees allowing for the mobility of the fuel, and state the initial $\overline{GM}$ (fluid) and the angle of vanishing stability.

32. Prove Atwood's formula for calculating $\overline{GZ}$ for a ship at an angle of heel ϕ.

A simple barge has a constant cross-section throughout its length, a beam B at the waterplane and straight sides which slope outwards at an angle α to the vertical (beam increasing with draught).

If the barge is heeled to an angle ϕ, prove that the heeled waterplane will cut the upright waterplane at a horizontal distance x from the centre line where:

$$x = \frac{B}{2 \tan \phi \tan \alpha}\left[1-\{1-(\tan \phi \tan \alpha)^2\}^{\frac{1}{2}}\right]$$

If $B = 20$ ft, $\alpha = 10°$ and $\phi = 30°$, calculate the value of x, and then, by calculation or drawing, obtain the $\overline{GZ}$ at 30° of heel if $\overline{B_0G} = 4·6$ ft and the underwater section area $= 140$ ft^2.

33. State the purposes of an inclining experiment. Explain the preparations at the ship before the experiment is performed.

Calculate the change in $\overline{GM}$ due to the addition of 150 tonf, 36 ft above the keel of a frigate whose particulars before the addition are given below:

Displacement	= 2700 tonf	$\overline{KG}$	= 21·3 ft
TPI	= 18	$\overline{KB}$	= 8·2 ft
Draught	= 14·0 ft	$\overline{BM}$	= 18·5 ft

34. Derive an expression for the radius of curvature of the curve of flotation in terms of the waterplane characteristics and the volume of displacement (Leclert's theorem).

A homogeneous prism with cross-section as an isosceles triangle, base width B and height H, floats with its apex downwards in a condition of neutral equilibrium. Show that the radius of curvature of the curve of flotation in this condition is

$$B^2H/(4H^2+B^2)$$

References

1. Prohaska, C. W. Residuary Stability, *TINA*, 1947.
2. Prohaska, C. W. Influence of ship form on stability, *TINA*, 1951.
3. Burcher, R. K. The influence of hull shape on transverse stability, *TRINA*, 1980.
4. Brown, D. K. Stabililty at large angles and hull shape considerations, *NA*, Jan. 1979.
5. Sarchin, T. H. and Goldberg, L. L. Stability and buoyancy criteria for US naval surface ships, *TSNAME*, 1962.
6. Yamagata, M. Standard of stability adopted in Japan, *TRINA*, 1959.
7. Welford, S. The development of the self-righting lifeboat, *NA*, July 1974.
8. Dorey, A. L. High speed small craft, *NA*, July 1990.
9. Odabasi, A. Y. Ultimate stability of ships, *TRINA*, 1977.
10. Lewison, G. R. G. and Paulling, J. R. Encountering a congregation of waves, *NA*, Sept. 1980.
11. Morrall, A. The *Gaul* disaster. An investigation into the loss of a large stern trawler, *TRINA*, 1980.
12. RoRo Safety and Vulnerability; International Conferences, *RINA*, 1987, 1991.
13. Merchant Shipping (Passenger Ship Construction) Regulations 1984, SI 1984 No 1216 as amended 1990 (No 892), 1992 (No 2358).

5 Hazards and protection

The naval architect needs to be more precise than the old and loved hymn, because he must ask himself why those on the sea are in peril and what he can do to help them. Hazard to those sailing in well designed ships arises through mishandling, misfortune or an enemy and these can cause flooding, fire, explosion, structural damage or a combination of these. It is the job of the naval architect to ensure that the effects do not result in immediate catastrophe and that they can be counteracted by equipment provided in the ship, at least up to a substantial level of damage. He must further ensure that when the damage is such that the ship will be lost, workable equipment and time are adequate to save life. To this end, the naval architect must anticipate the sort of mishandling and misfortune possible and the probable intent of any enemy, so that he may design his ship to mitigate the effects and render counter action readily available.

This chapter is concerned with this anticipation. It seeks to define the hazards in sufficiently precise terms and to describe the forms of protection and life saving possible. Some of these forms are demanded by international and national regulations; such regulations do not, however, remove the need to understand the problems.

Despite all that anticipation and regulation can achieve, losses at sea continue. In 1990, 230 ships were lost, representing about 0·3 per cent of the 76,000 ships at risk worldwide and 0·2 of world tonnage. Typically about a third are lost by fire or explosion. Whilst not a large percentage, they remain an anxiety to naval architects as well as to masters, owners and safety authorities.

Flooding and collision

WATERTIGHT SUBDIVISION

Excluding loading in excess of the reserve of buoyancy, a ship can be sunk only by letting water in. Water may be let in by collision, grounding, by enemy action or by operation of a system open to the sea. A compartment which has been opened to the sea is said to have been *bilged*. However the water is let in, there is a need to isolate the flooded volume for the following reasons:

(*a*) to minimize the loss of transverse stability,
(*b*) to minimize damage to cargo,
(*c*) to prevent *plunge*, i.e. loss of longitudinal stability,
(*d*) to minimize the loss of reserve of buoyancy.

Ideally, a ship should sustain more and more flooding without loss of stability until it sinks bodily by loss of its reserve of buoyancy. This is *foundering*. The dangerous effect of asymmetric moment on the $\overline{GZ}$ curve was discussed in the last chapter; the ideal way to avoid such a possibility, is to subdivide the entire ship transversely and this is a major aim of the designer. It is not often wholly possible, but it is nowadays a major feature of most ships even at the expense of increasing the volume of water admitted. Free surface can also cause appreciable loss of transverse stability as discussed in Chapter 4. The effects of free surface can be minimized by longitudinal bulkheads which are undesirable and, better, by sills; because the loss of stability is proportional to the cube of the width of free surface, the effect of this type of subdivision is considerable (see Fig. 5.1). The sill reduces the effect of shallow flooding but cannot cure the effect of excessive transverse flooding.

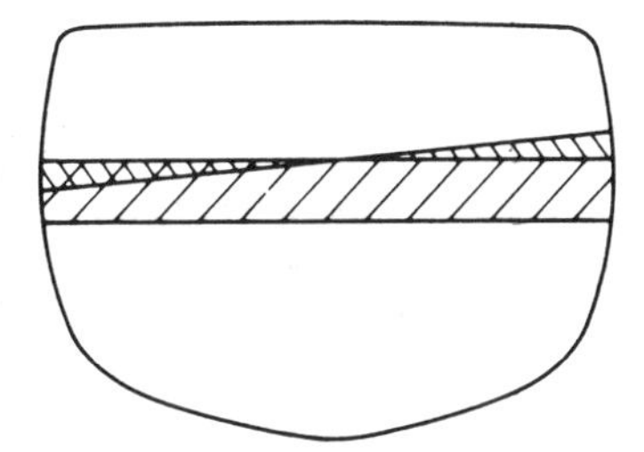
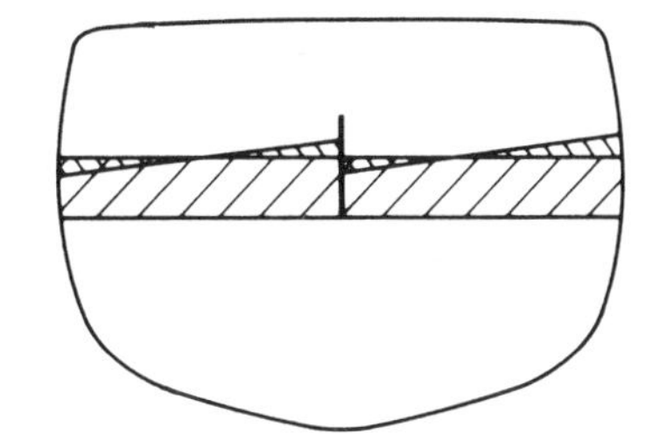

Fig. 5.1 Effect of a sill

Extensive watertight subdivision is an inconvenience to everyone. The ship is larger, more expensive to build, access around the ship is inconvenient and cargo becomes more difficult to stow and more expensive to handle. A compromise between degree of safety and economics must be found. Generally, this results in watertight volumes larger in merchant ships than in warships, and a greater tendency in the former to plunge when damaged. Regulations to minimize this possibility in the behaviour of merchant ships are dealt with later in the chapter under 'Damage Safety Margins'.

Certain watertight subdivision is incorporated into a ship with the direct intention of isolating common and likely forms of damage. Collision, for example, is most likely to let water in at the bow. For this reason, Reference 1 requires most passenger ships to have a collision bulkhead at 5 per cent of the length from the bow. A second bulkhead is required in ships longer than 100 m. Also required are an afterpeak bulkhead and bulkheads dividing main and auxiliary machinery spaces from other spaces. They must be watertight up to the bulkhead deck and able to withstand the water pressures they might be subject to after damage. The collision bulkhead should not be pierced by more than one pipe below the margin line. A watertight double bottom is required and in ships over 76 m in length this must extend from the collision to afterpeak bulkhead. It provides useful tank capacity as well as protection against grounding and also some protection against non-contact underbottom explosions, although to be most effective it should be only partially filled with liquid (usually oil fuel), which causes large free surface losses. The effects of contact explosion.by torpedo or mine are minimized in large warships by the provision at the sides of sandwich protection comprising watertight compartments containing air or fluid (see Fig. 5.2). The basic principle is to absorb the energy of the explosion without allowing water to penetrate to the ship's vitals. Similar usable watertight space is efficacious in RoRo ferries against collision.

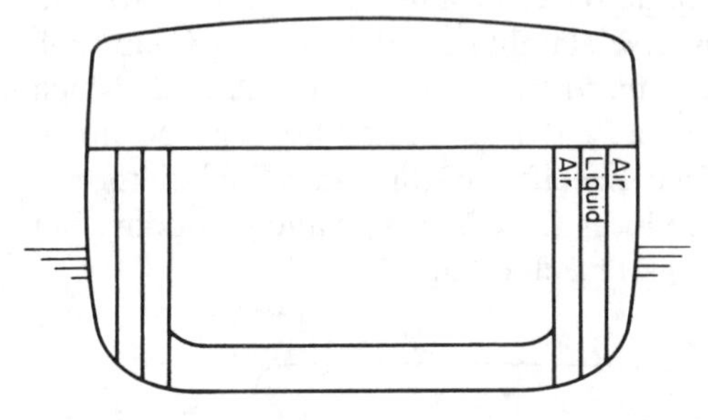

Fig. 5.2 Sandwich protection

Special subdivision is required in ships having a nuclear reactor. The reactor, primary circuit and associated equipment must be in a containment pressure vessel which should be so supported in the ship that it accepts no strain from the ship. It must be protected from damage by longitudinal bulkheads between it and the ship's side, by an extra deep double bottom and by cofferdam spaces longitudinally (Fig. 5.3).

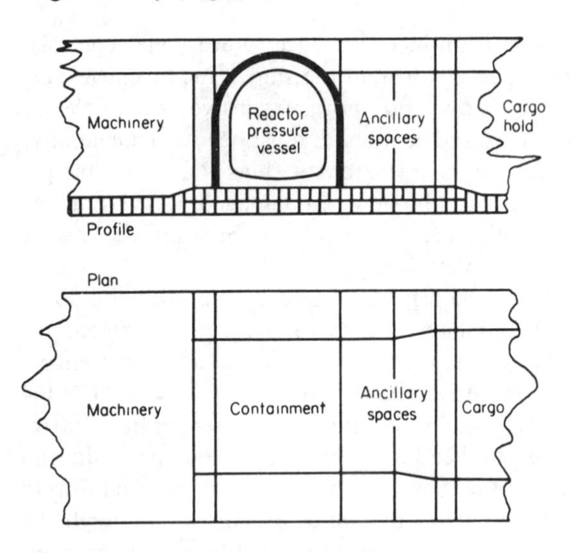

Fig. 5.3 Reactor protection

FLOTATION CALCULATIONS

In order to assess a ship's ability to withstand damage, it is first necessary in the design stages to define the degrees of flooding to be examined. A range of examples will be chosen normally based, for a warship, on an expert assessment of likely weapon damage or, for a merchant ship, on statutory figures given in the United Kingdom by the Dept. of Trade in the Merchant Shipping (Construction) Rules. For each of these examples, it is necessary to discover.

(*a*) the damaged waterline, heel and trim,
(*b*) the damaged stability for which minimum standards are laid down in the same Rules.

Consider, first, a central compartment of a rectangular vessel open to the sea (Fig. 5.4).

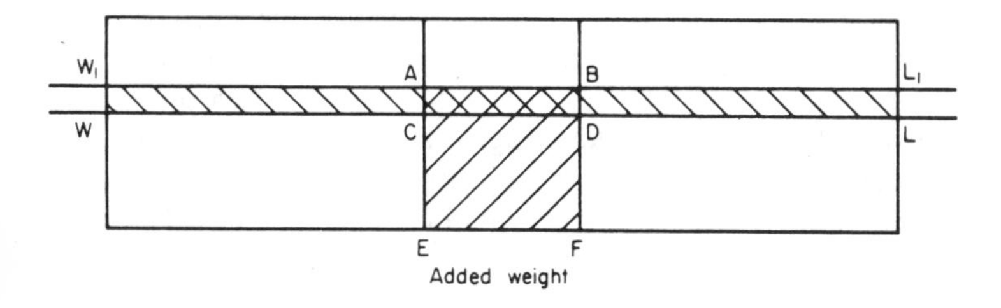

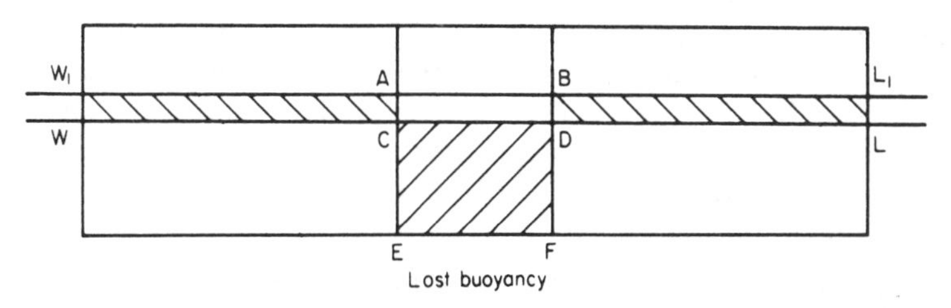

Fig. 5.4 A central compartment open to the sea

As a result of this flooding, the ship sinks from WL to W_1L_1. The amount of weight added to the ship by the admission of water is represented by ABFE and the additional buoyancy required exactly to support it is represented by W_1L_1LW. To calculate the added weight to the depth AE, it is necessary to guess the new waterline and then verify that the guess is correct; trial and error will yield an answer. There is a better way to approach the problem. Instead of thinking of an increase in ship weight, the flooded portion CDFE must be thought of as a loss of buoyancy which must be made up by the buoyancies of W_1ACW and BL_1LD. The lost buoyancy CDFE can be calculated exactly because it is up to the original waterline only, and the additional buoyancy up to W_1L_1 can be calculated from the tonf parallel immersion of the waterplane excluding the portion AB. Weight and buoyancy of the portion ABDC cancel each other out. The ship's weight is unchanged. These two approaches are known generally as the '*added weight method*' and the '*lost buoyancy method*' respectively.

Compartments of ships open to the sea do not fill totally with water because some space is already occupied by structure, machinery or cargo. The ratio of the volume which can be occupied by water to the total gross volume is called the *permeability*. Typical values are

Space	*Permeability* (per cent)
Watertight compartment	97 (warship) 95 (merchant ship)
Accommodation spaces	95
Machinery compartments	85
Dry cargo spaces	70
Coal bunkers, stores, cargo holds	60

Permeabilities should be calculated or assessed with some accuracy. Gross floodable volume should therefore be multiplied by the permeability to give the lost buoyancy or added weight. Formulae for the calculation of permeability for merchant ships are given in the Merchant Shipping (Construction) Rules.

Often, a watertight deck will limit flooding of a compartment below the level of the new waterline. But for a possible free surface effect due to entrapped air, the compartment may be considered pressed full and its volume calculated. In this case, the new waterline, heel and trim are best calculated by regarding the flooding as a known added weight whose effects are computed by the methods described in Chapters 3 and 4.

Where there is no such limitation and the space is free flooding, heel, trim and parallel sinkage are calculated by regarding the flooding as lost buoyancy in a manner similar to that already described for a central compartment. Now, because the ineffective area of waterplane is not conveniently central, the centre of flotation will move and the ship will not heel about the middle line. The procedure is therefore as follows:

(i) calculate permeable volume of compartment up to original waterline;
(ii) calculate TPI, longitudinal and lateral positions of CF for the waterplane with the damaged area removed;
(iii) calculate revised second moments of areas of the waterplane about the CF in the two directions and hence new $\overline{BM}$s;
(iv) calculate parallel sinkage and rise of CB due to the vertical transfer of buoyancy from the flooded compartment to the layer;
(v) calculate new $\overline{GM}$s
(vi) calculate angles of rotation due to the eccentricity of the loss of buoyancy from the new CFs.

This is best illustrated by an example.

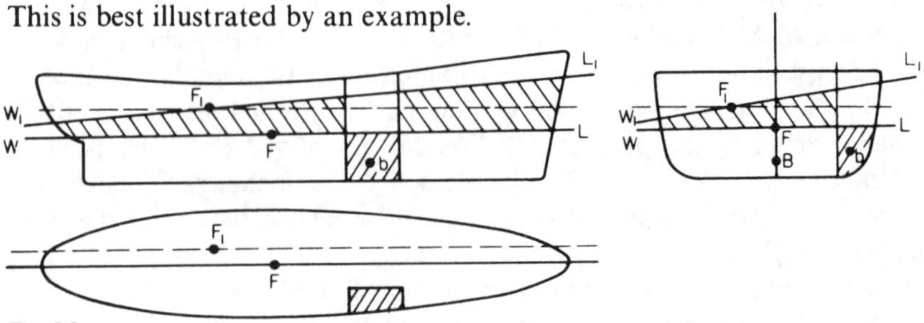

Fig. 5.5

EXAMPLE. A compartment having a plan area at the waterline of 100 m^2 and centroid 70 m before midships, 13 m to starboard is bilged. Up to the waterline obtaining before bilging, the compartment volume was 1000 m^3 with centres of volume 68·5 m before midships, 12 m to starboard and 5 m above keel. The permeability was 0·70.

Before the incident the ship was floating on an even draught of 10 m at which the following particulars obtained

Displacement mass, 30,000 tonnes	$\overline{KM}$ long, 170 m
$\overline{KG}$, 9·40 m	WP area, 4540 m^2
$\overline{KM}$ transverse, 11·40 m	CF, 1 m before midships
$\overline{KB}$, 5·25 m	LBP, 220 m

Calculate the heel and trim when the compartment is bilged.

Solution: Refer to Fig. 5.5. Use the lost buoyancy method.

(i) Permeable volume = 0·70 × 1000 = 700 m^3

(ii) Damaged WP area = 4440 m^2

$$\text{Movement of CF aft} = \frac{100 \times (70 - 1)}{4440} = 1{\cdot}55 \text{ m}$$

$$\text{Movement of to CF port} = \frac{100 \times 13}{4440} = 0{\cdot}29 \text{ m}$$

(iii) Original transverse I

$$I = (11{\cdot}40 - 5{\cdot}25) \times 0{\cdot}975 \times 30{,}000 = 179{,}889 \text{ m}^4$$

Damaged transverse I

$$I = 179{\cdot}89 \times 10^3 - 100 \times 13^2 - 4440 \times (0{\cdot}29)^2 = 162{,}620 \text{ m}^4$$

(ignoring I of the damage about its own axis)

$$\text{Damaged transverse } \overline{BM} = \frac{162{,}620}{0{\cdot}975 \times 30{,}000} = 5{\cdot}56 \text{ m}$$

Original long. I

$$I = (170{\cdot}0 - 5{\cdot}25) \times 0{\cdot}975 \times 30{,}000 = 4{\cdot}819 \times 10^6 \text{ m}^4$$

Damaged long. I

$$I = 4{\cdot}819 \times 10^6 - 100\,(69)^2 - 4440\,(1{\cdot}55)^2 = 4{\cdot}332 \times 10^6 \text{ m}^4$$

Damaged long. $\overline{BM}$

$$\overline{BM} = \frac{4{\cdot}332 \times 10^6}{0{\cdot}975 \times 30{,}000} = 148{\cdot}1 \text{ m}$$

(iv) Parallel sinkage $= \dfrac{700}{4440} = 0{\cdot}16$ m

Rise of B $= \dfrac{700\,(10 + 0{\cdot}08 - 5)}{0{\cdot}975 \times 30{,}000} = 0{\cdot}12$ m

(v) Damaged transverse $\overline{GM} = 5{\cdot}25 + 0{\cdot}12 + 5{\cdot}56 - 9{\cdot}40 = 1{\cdot}53$ m

Damaged long. $\overline{GM} = 5{\cdot}25 + 0{\cdot}12 + 148{\cdot}1 - 9{\cdot}40 = 144{\cdot}1$ m

(vi) Angle of heel $= \dfrac{700 \times 12{\cdot}29}{0{\cdot}975 \times 30{,}000 \times 1{\cdot}53} \times \dfrac{180}{\pi} = 11{\cdot}0°$

Angle of trim $= \dfrac{700 \times (68{\cdot}5 - 1 + 1{\cdot}55)}{0{\cdot}975 \times 30{,}000 \times 144{\cdot}1} = 0{\cdot}01147$ rads

Change of trim $= 0{\cdot}01147 \times 220 = 2{\cdot}52$ m between perps.

Note that displacement and $\overline{KG}$ are unchanged.

The principal axes of the waterplane, as well as moving parallel to their original positions, also rotate. The effects of this rotation are not great until a substantial portion of the waterplane has been destroyed.

This approach to the flotation calculations is satisfactory for trim which does not intersect keel or deck and for heel up to about 10 degrees. Thereafter, it is necessary to adopt a trial and error approach and to calculate the vertical, athwartships and longitudinal shifts of the centre of buoyancy for several trial angles of waterplanes. That angle of waterplane which results in the new centre of buoyancy being perpendicularly under the centre of gravity is the correct one. This is illustrated for the transverse plane in Fig. 5.6, in which the transfer of buoyancy from *b* to the two wedges (one negative) results in a move of the centre of buoyancy of the whole from B to B_1. $\overline{B_1G}$ must be perpendicular to W_1L_1. To ensure that the displacement is unaltered, several waterlines parallel to W_1L_1 will be necessary at each angle as described presently under 'Damaged Stability Calculations', Fig. 5.8. These calculations are performed normally by a digital computer.

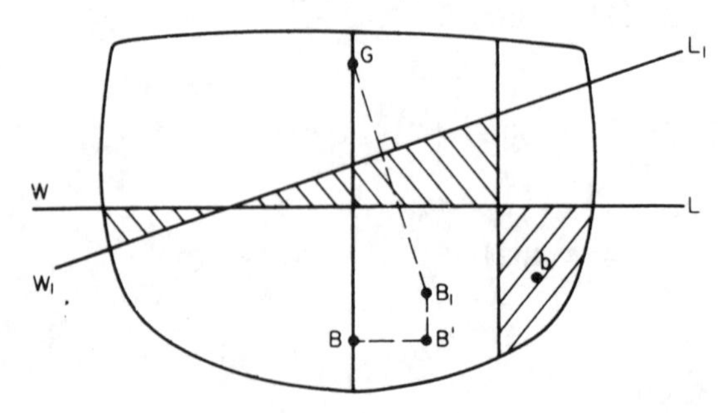

Fig. 5.6

DAMAGED STABILITY CALCULATIONS

Consider, first, the effect of a flooded compartment on initial stability. Let us show that, whether the flooding be considered an added weight or a lost buoyancy, the result will be the same.

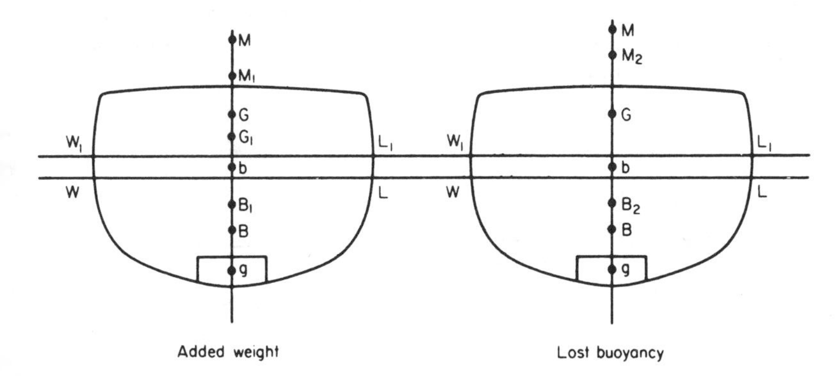

Fig. 5.7

Consider a ship of displacement Δ admitting a weight w having a free surface i as shown in Fig. 5.7. Regarding as an added weight,

$$\text{Rise of B} = \overline{\text{BB}}_1 = \frac{w\overline{\text{Bb}}}{\Delta + w}$$

$$\text{Fall of G} = \overline{\text{GG}}_1 = \frac{w\overline{\text{Gg}}}{\Delta + w}$$

$$\text{New } \overline{\text{GM}} = \overline{\text{B}_1\text{M}_1} - \overline{\text{B}_1\text{G}_1} = \overline{\text{B}_1\text{M}_1} - \overline{\text{BG}} + \overline{\text{BB}}_1 + \overline{\text{GG}}_1$$

i.e.

$$\overline{\text{G}_1\text{M}_1} = \rho\frac{(\text{I}_1 - \text{i})}{\Delta + w} - \overline{\text{BG}} + \frac{w}{\Delta + w}(\overline{\text{Bb}} + \overline{\text{Gg}})$$

$$\begin{aligned}\text{Righting moment} &= (\Delta + w)\overline{\text{G}_1\text{M}_1}\phi = \{\rho(\text{I}_1 - \text{i}) - \Delta\overline{\text{BG}} - w\overline{\text{BG}} \\ &\qquad + w(\overline{\text{Bb}} + \overline{\text{BG}} + \overline{\text{Bg}})\}\phi \\ &= \{\rho(\text{I}_1 - \text{i}) - \Delta\overline{\text{BG}} + w\overline{\text{bg}}\}\phi\end{aligned}$$

Regarding as a loss of buoyancy,

$$\text{Rise of B} = \overline{\text{BB}}_2 = \frac{w\overline{\text{bg}}}{\Delta}$$

$$\text{New } \overline{\text{GM}} = \overline{\text{B}_2\text{M}_2} - \overline{\text{B}_2\text{G}} = \overline{\text{B}_2\text{M}_2} - \overline{\text{BG}} + \overline{\text{BB}}_2$$

$$\overline{\text{GM}}_2 = \rho\frac{(\text{I}_1 - \text{i})}{\Delta} - \overline{\text{BG}} + \frac{w\overline{\text{bg}}}{\Delta}$$

$$\text{Righting moment} = \Delta\overline{\text{GM}}_2\phi = \{\rho(\text{I}_1 - \text{i}) - \Delta\overline{\text{BG}} + w\overline{\text{bg}}\}\phi \text{ as before.}$$

Thus, so far as righting moment is concerned (and this is what the ship actually experiences), it does not matter whether the flooding is regarded as added weight or lost buoyancy although $\overline{GM}$ and $\overline{GZ}$ values will not be the same. The lost buoyancy method is also called the *constant displacement method.*

It is probable that the approximate method of correcting the $\overline{GZ}$ curve for contained liquid given in Chapter 4 will be insufficiently accurate for large volumes open to the sea. In this case, cross curves for the volume of flooding are constructed in the following manner:

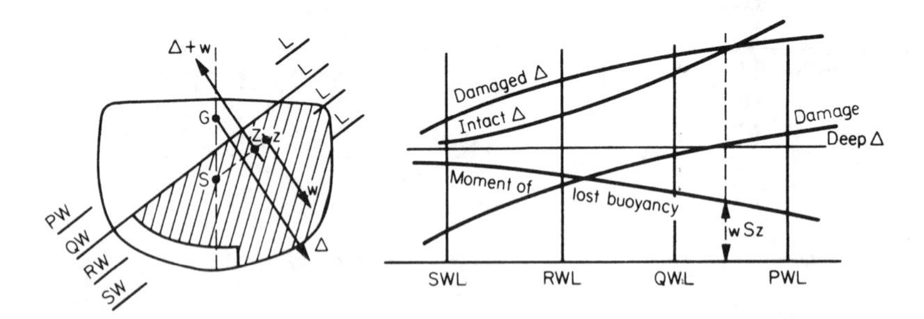

Fig. 5.8

For each angle of heel, a series of parallel waterlines P, Q, R, S, is drawn. For each of these waterlines, the weight of flooding and its lever $\overline{Sz}$ from the pole of the ship's cross curves are calculated by integrator, computer or direct mensuration, i.e. by one of the methods used for the ship itself and described in Chapter 4. The weight of flooding up to each waterline is added to the deep displacement to give a damaged displacement which is plotted on a base of WL together with $w\overline{Sz}$. Intact displacement to the same base will be available from the calculations which led to the cross curves for the ship (if not, they must be calculated). Where these two lines cross is the point of vertical equilibrium for that angle and $w\overline{Sz}$ can be read off. Now the restoring lever $\overline{SZ}$ for the intact ship at displacement $\Delta + w$ can be read directly from the cross curves. Then the damaged righting moment is given by

$$(\Delta + w)\overline{SZ} - \Delta\overline{SG}\sin\phi - w\overline{Sz}$$

A righting lever corresponding to the original displacement Δ is given by dividing this expression by Δ. This procedure is repeated for all angles. The damaged GZ curve will, in general, appear as in Fig. 5.9. The point at which the curve crosses the ϕ-axis, at angle ϕ_1, represents the position of rotational equilibrium; this will be zero when the flooding is symmetrical about the middle line.

With symmetrical flooding, the upright equilibrium position may not be one of stable equilibrium. In this case, the $\overline{GZ}$ curve will be as in Fig. 5.10 where ϕ_2 is the angle of loll. The significance of the angle of loll is discussed in Chapter 4. If the $\overline{GZ}$ curve does not rise above the ϕ-axis the ship will capsize; if little area is left above, capsizing will occur dynamically.

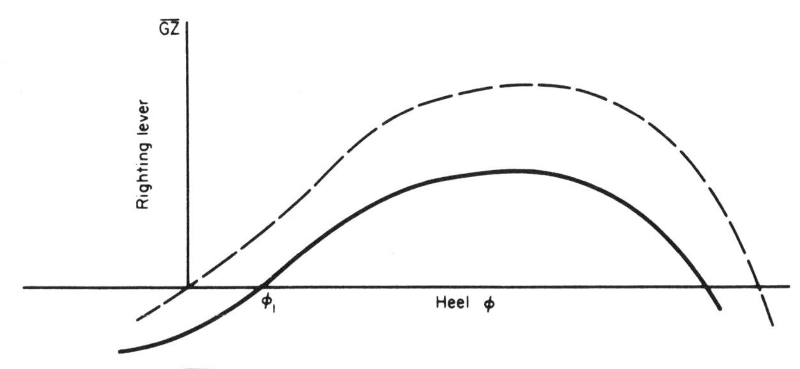

Fig. 5.9 Damaged $\overline{GZ}$ *curve*

Definition of damaged stability is not easy. Reference 1 requires that the margin line shall not be submerged and a residual metacentric height of at least 0·05 m as calculated by the constant displacement method in the standard damaged condition be maintained. Adequacy of range, maximum $\overline{GZ}$ and angle of maximum stability are important, particularly for warships which do not conform to the margin line standards discussed presently. Trim in a warship can cause rapid loss of waterplane inertia and deterioration of damaged stability and this is one reason why the low quarter deck destroyers common in the second world war have been replaced by modern flush deck, high freeboard frigates.

The fact that ships may take up large angles of heel following damage is recognized in the regulations governing ship construction, e.g. watertight doors must be able to be opened up to angles of heel of 15 degrees.

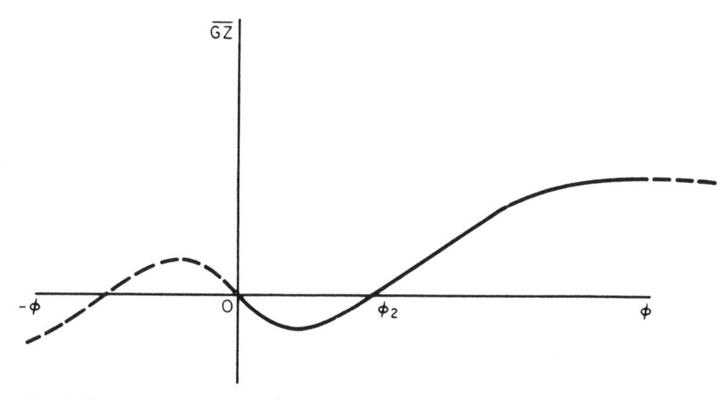

Fig. 5.10

DAMAGE SAFETY MARGINS

Minimum standards of behaviour of merchant ships under damaged conditions are agreed by international convention and enforced by national authority.

For the United Kingdom this authority is the Department of Transport who may delegate certain authority to specific agencies. The regulations which are the immediate concern of this chapter are those relating to damaged stability which has already been discussed, to *floodable length* and to *freeboard*. Floodable length calculations are required to ensure that there is sufficient effective longitudinal waterplane remaining in a damaged condition to prevent plunge, i.e. loss of longitudinal stability. Minimum permitted freeboard is allotted to ensure a sufficient reserve of buoyancy to accommodate damage. It is necessary, as always, to begin with clear definitions; while these embrace a majority of ships, interpretations for unusual vessels may have to be obtained from the full Rules, from the national authority or, even by litigation.

The *bulkhead deck* is the uppermost weathertight deck to which transverse watertight bulkheads are carried.

The *margin line* is a line at least 76 mm below the upper surface of bulkhead deck at side.

The *floodable length* at any point in the length of a ship is the length, with that point as centre, which can be flooded without immersing any part of the margin line when the ship has no list.

Formulae are provided in the Rules for the calculation of a *factor of subdivision* which must be applied to the floodable length calculations. This factor depends on the length of the ship and a *criterion numeral* which is intended to represent the criterion of service of the ship and is calculated from the volumes of the whole ship, the machinery spaces and the accommodation spaces and the number of passengers. Broadly, the factor of subdivision ensures that one, two or three compartments must be flooded before the margin line is immersed and ships which achieve these standards are called one-, two- or three-compartment ships. Compartment standard is an inverse of the factor of subdivision. As indicated in Fig. 5.11, very small ships would be expected to have a one-compartment and large passenger ships a three-compartment standard.

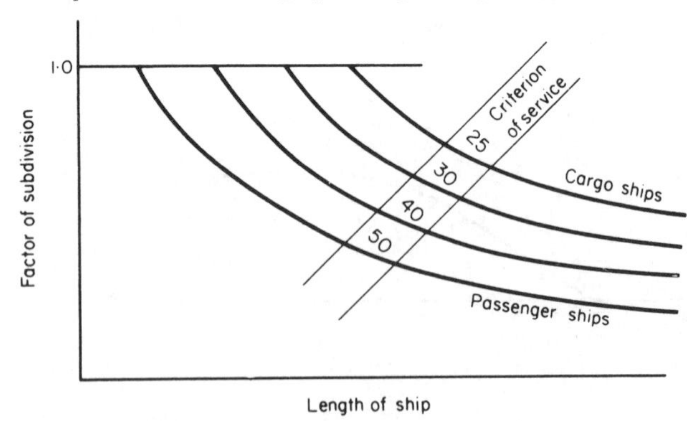

Fig. 5.11 Factor of subdivision

The product of the floodable length and the factor of subdivision gives the *permissible length*. With certain provisos concerning adjacent compartments, a compartment may not be longer than its permissible length. Flooding calculations, either direct or aided by standard comparative parametric diagrams available in the United Kingdom from the Department of Transport can produce curves similar to those shown in Fig. 5.12 for floodable and permissible lengths and in general, a triangle erected from the corners of a compartment with height equal to base must have an apex below the permissible length curve. The base angle of this triangle is $\tan^{-1}2$ if vertical and horizontal scales are the same. Lines at this angle, called the forward and after terminals, terminate the curves at the two ends of the ship. The steps in the curve are due to different permeabilities of compartments; in the initial determination of bulkhead spacing for the ship on this basis, it is necessary to complete sets of permissible length curves for a range of permeabilities.

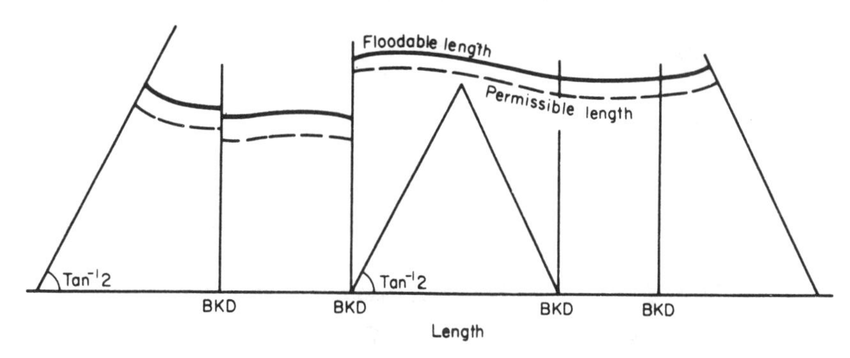

Fig. 5.12 Floodable and permissible length curves

Summarizing, the steps to be taken to carry out the floodable length calculations are as follows:

(*a*) Define bulkhead deck and margin line;
(*b*) calculate factor of subdivision;
(*c*) calculate permeabilities;
(*d*) assess floodable lengths;
(*e*) plot permissible lengths.

Calculations are conveniently carried out in methodical manner in a standard tabulated form recommended by the Department of Transport.

The second important insurance against damage in merchant ships is the allocation of a statutory freeboard. The rules governing the allocation of freeboard are laid down by an international Load Line Convention, of which one was held in 1966, and ratified nationally by each of the countries taking part. Amendments to this were introduced by IMO in 1971, 1975 and 1979. Because the amount of cargo which can be carried is directly related to the minimum permissible freeboard, the authorities are anxious to be fair and exact and this tends to produce somewhat complicated rules. The intention, however, is clear enough. The intention is to provide a simple visual check that a laden ship has sufficient reserve of buoyancy and to relate, therefore, the watertight volume above the laden waterline to a height which can be readily measured. This clearly involves the geometry of the ship, and the statutory calculation of freeboard involves water density, ship's length, breadth, depth, sheer, size of watertight superstructures and other relevant geometrical features of the ship. The Rules require minimum standards of closure in watertight boundaries before allowing them to be counted in the freeboard calculation. Standards for hatch covers, crew protection, freeing port areas, ventilators and scuttles are enforced by these Rules.

The freeboard thus calculated results in a *load line* which is painted boldly on the ship's side in the form shown in Fig. 5.13. Because the density of water affects the reserve of buoyancy and weather expectation varies with season and location, different markings are required for Winter North Atlantic (WNA), Winter (W), Summer (S), Tropical (T), Fresh (F) and Tropical Fresh (TF).

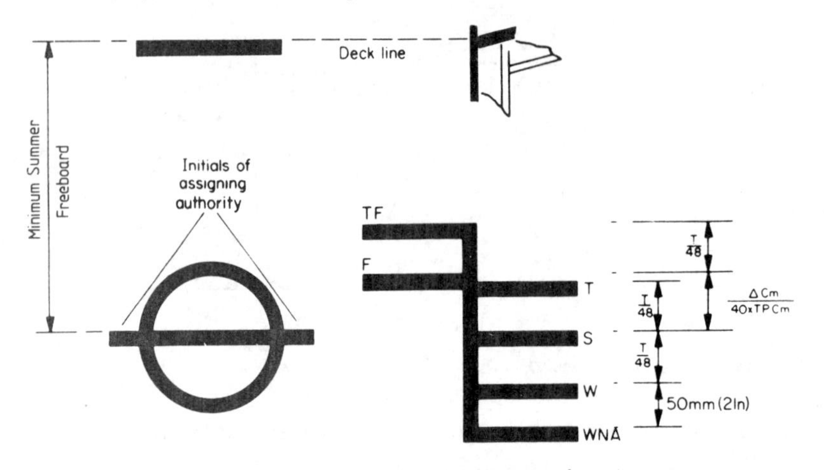

Fig. 5.13 Load line markings (min. freeboard differences shown)

DAMAGED STABILITY STANDARDS FOR PASSENGER SHIPS

Following the loss of the ferry *Herald of Free Enterprise* in 1987, IMO agreed revised standards to be incorporated into SOLAS 74 which are now amendments to the Passenger Ship Construction Rules 1984. They require, *inter alia*, that after damage;

(*a*) the range of the residual $\overline{GZ}$ curve should be 15 degrees beyond the equilibrium point and that the area of the $\overline{GZ}$ curve beyond this point be 0·015 m rad up to the smaller of the downflooding angle or 22 degrees from the upright for one compartment flooding or 27 degrees for two compartment flooding,

(*b*) the residual $\overline{GZ}$ shall be obtained from the worst circumstance of all passengers (at 75 kg each, four per square metre) at the ship's side or all survival craft loaded and swung out or wind force at a pressure of 120 N/m^2 all calculated by:

$$\text{residual } \overline{GZ} = \frac{\text{heeling moment}}{\text{displacement}} + 0{\cdot}04 \text{ m but not less than } 0{\cdot}10 \text{ m}$$

Intermediate stages of flooding must also be considered. The Master must determine and record the state of his ship before sailing and this is readily done with the help of a computer.

LOSS OF STABILITY ON GROUNDING

The upward force at the keel due to docking or grounding, whose magnitude was found in Chapter 3, causes a loss of stability. Let the force at the keel be w. Consider a slightly inclined vessel before and after the application of this force (Fig. 5.14).

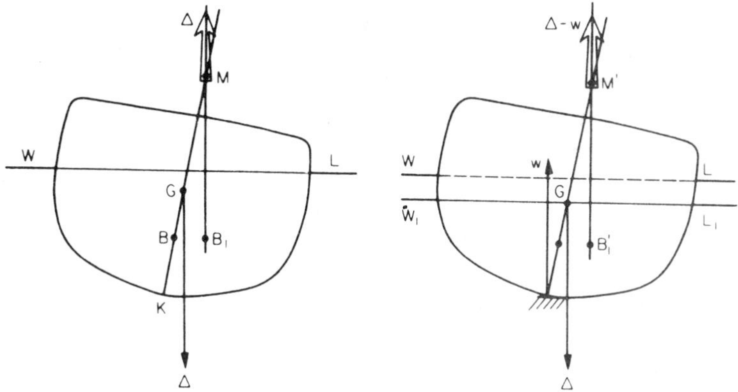

Fig. 5.14

The righting moment at inclination ϕ before the application of w is, of course, $\Delta\overline{GM}\sin\phi$. After application, the righting moment is

$$(\Delta - w)\overline{GM}'\sin\phi - w\overline{KG}\sin\phi = \Delta\left(\overline{GM}' - \frac{w}{\Delta}\overline{KM}'\right)\sin\phi$$

The movement of M to M′ is due, as described in Chapter 3 to (*a*), the fall of B due to the removal of the layer of buoyancy at the waterline, (*b*), the change in $\overline{BM}$ due to the differences in I and volume of displacement ∇.

BERTHING AND ICE NAVIGATION

A ship has no brakes and is slow to respond to the propulsive machinery; unexpected currents and gusts of wind can make a ship awkward to handle. Despite dexterous handling by the Master, there will be occasions when a ship comes alongside heavily and may be indented by fenders or catamarans or tugs. Local internal structural stiffening is often fitted in the vicinity of the waterline to afford some protection, and the waterline strake may be increased in thickness.

Similar stiffening is fitted in ships intended for navigation in ice and Lloyd's Register lays down five degrees of stiffening dependent on the type of ice to be negotiated, making a note against the ship to this effect in the Register Book. This is not to be confused with the stiffening required in an ice breaker which is much more formidable, since to perform its task, the ice breaker is driven to ride up on the ice, allowing its weight to break the ice.

Safety of life at sea

Representatives of almost 150 seafaring nations contribute to the International Maritime Organization (IMO) which is an organization formed in 1959 under the auspices of the United Nations, intended to promote co-operation amongst all countries on the questions relating to ships. This organization promotes discussion on such topics as tonnage measurement, safety of life and standards of construction. It convened the Load Line Convention of 1966, the one previous having been held in 1930 and it has taken over the functions of the International Conferences on the Safety of Life at Sea (SOLAS) five of which have been held, in 1914 (following the *Titanic* disaster), in 1929, 1948, 1960 and 1974. The topics with which the SOLAS Conferences concerned themselves were standards of construction, watertight subdivision, damage behaviour, damage control, fire protection, life-saving appliances, dangerous cargoes, nuclear machinery protection, safety of navigation and many other aspects of safety. Following the 1960 SOLAS Conference, IMO set up a dozen or so sub committees dealing actively with stability and subdivision, fire protection, bulk cargoes, oil pollution, tonnage, signals and other aspects of safety under the auspices of a Maritime Safety Committee.

FIRE

In the design and construction of a ship, it is necessary to provide means to contain fire, the most feared, perhaps, of all the hazards which face the mariner. Large death tolls are to be expected in ship fires due mainly to the effects of smoke. In the passenger ship *Scandinavian Star* in 1990, 158 people died, all but six due to the effects of smoke and toxic fumes. When it is appreciated that these disasters occur in peacetime, with so many modern aids available, the seriousness of fire as a hazard can be seen. The causes of fires are manifold—electrical short circuits, chemical reactions, failure of insulation, ingress of hot particles. Whatever the cause, however, there are two factors about fire which give the clue to its containment,

(*a*) fire cannot be sustained below a certain temperature which depends upon the particular material;
(*b*) fire cannot be sustained without oxygen.

Thus, to put a fire out, it must be cooled sufficiently or it must be deprived of oxygen. A ship must be designed readily to permit both.

The principal method of cooling a fire is to apply water to it and this is most efficiently done by fine spray. Automatic sprinkler systems are often fitted, actuated by bulbs which burst at a temperature of about 80°C, permitting water to pass. Passenger vessels must embody arrangements whereby any point in the ship can be reached by two hoses delivering a specified quantity of water. Dual escape from every point is also needed.

Deprivation of oxygen is effected in a variety of ways. The watertight subdivision of the ship is conveniently adapted to provide vertical zone fire boundaries built to particular standards of fire resistance that can be closed off in the event of a fire. The fire itself can be smothered by gas, fog or foam provided by portable or fixed apparatus, especially in machinery spaces.

Regulations for merchant ships registered in the United Kingdom are contained in the Merchant Shipping (Fire Appliances) Rules and allied publications issued by Government bodies. These specify the constructional materials, method of closing the fire boundaries, the capacities of pumps and the provision of fire-fighting appliances. While the quantities depend upon the particular classes of ship, the principles of fire protection in merchant ships and warships are similar:

(*a*) the ship is subdivided so that areas may be closed off by fire boundaries; this also prevents the spread of smoke, which can asphyxiate and also hampers firefighters in finding the seat of the fire;
(*b*) a firemain of sufficient extent and capacity is provided, with sufficient cross connections to ensure supply under severe damage;
(*c*) fixed foam, or inert gas smothering units are provided in machinery spaces and in most cargo holds;
(*d*) sufficient small portable extinguishers are provided to put out a small fire before it spreads;
(*e*) fire control plans are supplied to the Master and a fire fighting headquarters provided.

Some cargoes contain or generate their own oxygen and are not therefore susceptible to smothering. Cooling by spraying or flooding is the only means of extinguishing such fires and any extensive flooding thus caused brings with it the added hazard of loss of stability.

LIFE-SAVING EQUIPMENT

Ships must conform to the safety regulations of the country with whom they are registered. Standards depend upon the class of ship and the service. Generally sufficient lifeboats and rafts are to be provided for all persons carried. They must be launchable with the ship listing up to 15 degrees and trimming up to 10 degrees. There have been significant advances in life saving appliances in recent years following SOLAS 74. Modern systems now include escape chutes similar to those in aircraft (Ref. 9) while rescue boats at rigs can withstand severe oil fires. Information provided to passengers, particularly in ferries, has been improved and drills for crews are now regularly enforced in all well run ships.

ANCHORING

Lives of the crew and the value of ship and cargo are often entrusted totally to the anchor—an item of safety equipment much taken for granted. Bower anchors carried by ships need to hold the ship against the pull of wind and tide. The forces due to wind and tide enable the naval architect to select the size of cable and an anchor of sufficient holding pull. Classification societies define an equipment number to represent the windage and tide forces and relate this to the cable and to approved anchor designs.

A good anchor should bite quickly and hold well in all types of sea bed. It must be stable (i.e. not rotate out), sufficiently strong and easy to weigh.

Efficiency of the anchor itself is measured by the ratio of holding pull to anchor weight which may nowadays exceed 10 in most sea beds. However, the efficiency of the mooring depends also upon the length of cable veered and the depth of water, the ratio of these being the scope. The scope must be sufficient to ensure a horizontal pull on the anchor shank. Figure 5.15.

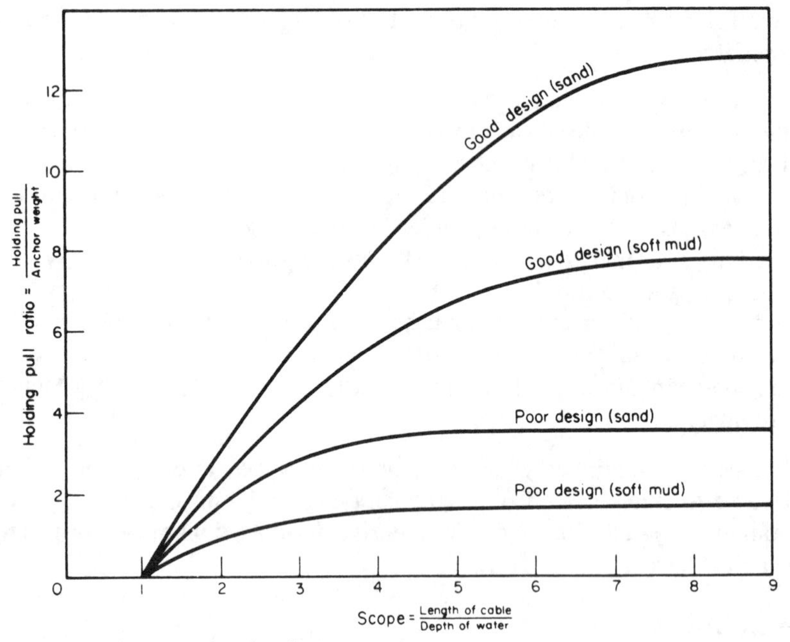

Fig. 5.15 Characteristics of anchoring

Ideally, the holding pull of the anchor should be slightly less than its own proof load and slightly more than the total wind and tide drag on the ship, while the proof load of the cable should be slightly greater than both.

The above applies to anchors which provide horizontal restraint and are used in ships and moorings. A different type is the uplift resisting anchor used for permanent moorings. Modern types depend upon being buried deep in the sea bed. This is often achieved using mechanical aids—water jetting, vibration or explosives.

DAMAGE CONTROL

The importance of a damage control organization is recognized in both merchant shipping and warship circles. Because of its greater complexity and its need to defend itself from nuclear, chemical and bacteriological attack as well as explosion and flooding, the organization in a warship is more complex. However, the Master of a merchant ship is required to be familiar with the damage control characteristics of his ship, to institute and drill an effective organization and to ensure that damage control plans are kept displayed.

The aim is to organize the ship in order rapidly to ascertain the damage, to isolate the incident and to take corrective action. The ship is constructed to make this as easy as possible; watertight subdivision is arranged as already discussed; watertight doors and hatches must be operable at 15 degrees of heel and are marked in levels of importance; minimum standards of damaged stability are met; pumping, flooding and fire-fighting systems are arranged to facilitate action under conditions of damage; electrical supplies are arranged with a view to providing sufficient breakdown capacity and duplication. All important systems are cross connected to ensure continuity of supplies following damage.

In a warship, information and control is concentrated in a central compartment with several subsidiary control posts each controlling one of the sections of the ship into which it is divided for purposes of damage control. Displayed in the headquarters is the flooding board which shows the watertight subdivision of the ship and the approximate heel and trim which would result from the flooding of any watertight compartment. Also displayed is information on the pumping, electrical, ventilation and other systems. A comprehensive communication network is available in the headquarters. With this organization, the damage control officer is able to appreciate the position and order corrective action.

Because a ship is most easily sunk by loss of transverse stability, the most urgent corrective action, once flooding has been contained by the closure of boundaries, is to reduce the angle of heel. This is most effectively done by counterflooding, i.e. flooding a compartment on the side opposite the damage and the extent necessary is indicated by the flooding board. In merchant ships, various regulations govern the limiting angle of heel. Reference 1 stipulates that after corrective action, heel should not exceed 7 degrees or the margin line immersed. Asymmetric flooding should be avoided if possible and cross-flooding systems capable of bringing heel under control in 15 minutes. Following corrective measures, the damage control officer may order reinforcement of bulkheads by shoring or the repair of leaks by wedges, boxes and concrete and the slower time repair of damage and restoration of supplies as far as is possible.

If A is the area of a hole and d is its distance below the waterline, the velocity of inflow of water, v, is given approximately by $v = 8\sqrt{d}$ ft/s, d being in feet, or $4{\cdot}5\sqrt{d}$ m/s, d being in metres. The volume of water admitted by the hole is KvA, where K is a coefficient of contraction at the hole. Assuming a value of 0·625 for K,

$$\text{inflow} = 5A\sqrt{d}\ \text{ft}^3/\text{s, with } A \text{ in ft}^2, \quad \text{or } \tfrac{11}{4}A\sqrt{d}\ \text{m}^3/\text{s, with } A \text{ in m}^2$$

Thus, a 15 cm diameter hole, 16 m below the waterline would admit

$$\frac{11}{4} \times \frac{\pi}{4} \times 0{\cdot}0225 \times 4 \times 1{\cdot}025 \times 60 = 12 \text{ tonnes of salt water per minute.}$$

UNCOMFORTABLE CARGOES

Grain, oil, explosive, hygroscopic materials, liquid gas and radioactive materials are some of the cargoes which may provide an intrinsic hazard to a ship and for which a large network of rules exists. These cannot be examined in a book of this size beyond a brief mention of their existence. See references.

In the nineteenth century, when loading was not supervised by port surveyors, grain ships were lost because voids were left under the decks. Grain and similar cargoes may shift when the ship rolls at sea, providing an upsetting moment which, if greater than the ship's maximum righting moment, causes it to capsize; if it is less, the ship heels, which may cause a further shift of the grain, causing a larger angle of heel, which may immerse openings or capsize the ship. Protection for deep sea ships is now provided as follows:

(*a*) In special bulk carriers, either upper wing tanks are fitted thus making the holds 'self-trimming' and reducing the free surface in them when full, or at least two longitudinal curtain bulkheads are used to break up the free surface.

(*b*) In general cargo carriers (mostly two deckers), shifting boards are fitted in the upper parts of the ship, usually on the centre line in conjunction with large feeder boxes. These feeders supply grain for settlement to those voids which fill in bad weather while limiting the slack surface produced higher up. When the ship's stability is adequate, feeders may be replaced by deep saucers of bagged cargo plugging the hatches.

(In both types of ship, when it is necessary for compartments to be partly filled, the shift of grain can be restricted by overstowing with bagged grain, or by strapping down with dunnage platforms and wires, or by carrying the shifting boards down through the grain.)

(*c*) In both classes of ship, it is now customary to carry grain stability information so that for bulk carriers the angle of heel after an assumed shift of grain can be found before loading and kept under 12 degrees, and in a general cargo ship, the stability required for the method of loading can be predetermined. In both cases, the loading is then planned to meet the stability requirements. To provide what is probably the most effective protection, the principal grain exporting countries appoint surveyors to check fittings and stability before loading. Later, they supervise the stowing of the grain so that voids are reduced as far as is practicable.

Oils and petrols are hazardous if they are volatile, i.e. if, at normal temperatures they emit a vapour that can accumulate until an explosive mixture with air is created. The vapour may be ignited at any mixture if it is raised to a critical and well defined temperature known as the flashpoint. For heavy and crude oils, this temperature is quite high, often above the temperature of a cigarette. Cleaning of empty crude oil tanks by steam hoses has led to several disastrous explosions due to the discharge of electrostatic build-up in the swirling fluid. Reference 8 covers the use of inert gas systems and precautions against electrostatic generation in tankers. During crude oil washing oxygen content must not exceed 8 per cent by volume.

Regulations, in general, provide for:

(*a*) total isolation of a dangerous oil cargo where possible;

(*b*) the removal of any possibility of creating ignition or overheating;

(*c*) adequate ventilation to prevent a build up of vapour to an explosive mixture;
(*d*) a comprehensive fire warning and fighting system.

Cofferdams to create an air boundary to oil spaces are a common requirement.

Explosives may be carried by ships which comply with the Merchant Shipping (Dangerous Goods) Rules. Because they must carry explosives where there is, in wartime, considerable risk of ignition by an enemy, the carriage of explosives in warships requires especial study. General regulations are contained in Naval Magazines and Explosive Regulations but these must be reviewed in the light of every new weapon development. The carriage of cordite in silk bags which were ignited by flash in the *Queen Mary* at Jutland and, probably, *Hood* in the second world war has been long discontinued; fixed robust ammunition is now carried alongside delicate guided weapons and torpedoes, some with exotic fuels. No ship construction can prevent the ingress of all offensive weapons, although the probability of causing catastrophic damage can be reduced. Structure can be arranged to minimize the effects of any incident and to this end the magazines are:

(*a*) isolated and provided with flashtight boundaries;
(*b*) kept cool by adequate insulation, ventilation and the prohibition of any electrical and mechanical devices which can be a source of unacceptable heat;
(*c*) provided with vent plates which permit an excessive pressure caused by burning to be vented to atmosphere;
(*d*) provided with fire detection and fighting facilities, often automatic.

Each dangerous or uncomfortable cargo must be examined on its own merits. Hygroscopic materials such as sisal may absorb much moisture and become so heavy that safe limits of deadweight or stability are broken; they may also swell and cause structural damage. Liquid gas is carried at very low temperatures and, if it leaks, may render the steel with which it comes in contact so brittle that it fractures instantly. Radioactive materials must, of course, have their radiations suitably insulated. Hydrogen peroxide combusts in contact with grease. Even some manure gives off dangerous organic gases.

NUCLEAR MACHINERY

A nuclear reactor is simply a type of boiler. Instead of producing heat by burning oil, a reactor produces heat by controlled decay of radioactive material. The heat is removed by circulation in a primary circuit, whence it is transferred at a heat exchanger to a secondary circuit to turn water into steam for driving the turbines. Whatever else can go wrong with the reactor there is no possibility whatever of a nuclear explosion, i.e. an explosion involving nuclear fission. There are, nevertheless, some unpleasant hazards associated with a reactor, viz.:

(*a*) under normal operation, dangerous gamma-rays are emitted from materials, activated by neutron bombardment. All personnel serving in the ship, who would be damaged by such rays must be protected by shielding of lead or other dense material. In the first nuclear merchant ship, the *Savannah*, there were nearly 2000 tons of shielding and there is similar shielding in nuclear submarines. Such heavy concentrated weight gives strength problems. Embrittlement of some materials by this irradiation is a further hazard;

(*b*) by mishandling or by accident, precise control of decay may be lost and radioactive fission products may be released. To contain such products, the reactor must be surrounded by a containment vessel which is made with scrupulous care to withstand the pressure build up due to the most serious credible accident. Safety devices are also built into the automatic control equipment which regulates the behaviour of the reactor. The subdivision of the ship referred to earlier is aimed at reducing the chances of loss of control by collision or grounding damage.

The SOLAS Conferences have made a series of recommendations for the safety of ships with nuclear reactors and these have now been incorporated into national law and classification society rules. In addition to the cofferdam protection discussed earlier, two-compartment standards are required. The Merchant Shipping Acts and the classification society rules give further details.

Other hazards

VULNERABILITY OF WARSHIPS

It is not enough to design a ship for normal operations. All ships can have accidents and warships (and sometimes merchant ships) must withstand action damage. The ability of a ship to survive in battle depends on its *susceptibility* to being hit and its *vulnerability* to the effects of a delivered weapon. Susceptibility can be reduced by reducing the various ship signatures to make it harder to detect, using jammers to defeat an enemy's detection systems, using decoys to seduce weapons that get through these defences and by hard kill weapons to destroy the incoming projectile or weapon carrier. The details of these defence systems must be suited to the weapons' characteristics. Various conventional weapons are illustrated in Fig. 5.16. They can cause structural failure, flooding, fire, blast, shock and fragment damage. To combat these requires efficient structural design, separation, zoning, redundancy, protection and containment.

An underwater explosion is likely to provide the most serious damage. Its effects are shown in Fig. 5.17. The pulsating bubble of gaseous explosion products contains about half the energy of the explosion and causes pressure waves which impact upon the hull. The frequency of these waves is close to the fundamental hull modes of vibration of small ships and the effects are most severe when the explosion occurs beneath the hull and the ship whips (Ref. 15). These motions can lead to buckling and loss of girder strength.

The other major feature of an underwater explosion is the shock wave which typically contains about a third of the energy of the explosion. It is transmitted

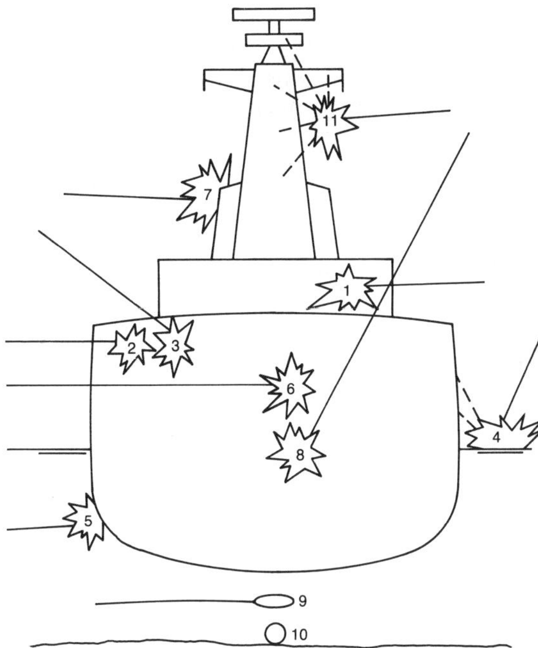

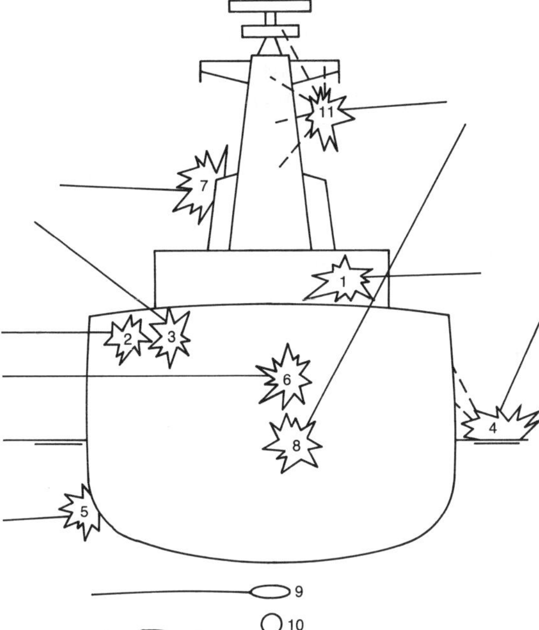

Fig. 5.16 Conventional weapon attack (Ref. 13)

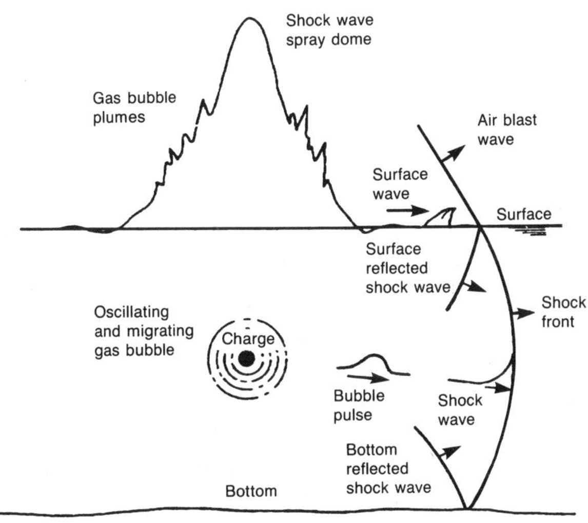

Fig. 5.17 Effects of an underwater explosion (Ref. 13)

through the water, into and through the ship's structure causing shock and possibly hull rupture. The intensity of shock experienced by the ship will depend upon the size, distance and orientation of the explosion relative to the ship. These factors are combined in various ways to produce a shock factor (Ref. 14). Since several formulations exist care is needed in their use. The intensity of shock experienced by an item of equipment will depend additionally upon its weight, rigidity, position in the ship and method of mounting. For critical systems (e.g. a nuclear power installation) it may be necessary to deduce the shock likely at a specific position in a given design. This can be done by calculation and/or model experiment using methods validated by full scale trials. More usually equipments will be fitted to several designs and in different positions, so they must be able to cope with a range of conditions. The approach is to design to a generalized shock grade curve. The overall design can be made more robust by providing shock isolation mounts for sensitive items and by siting system elements in positions where the structure offers more shock attenuation. This has the advantages that the item itself does not have to be so strong and the mounts can attenuate any noise the equipment produces, thus reducing its contribution to the underwater noise signature. The dynamic response of an item to shock will depend upon its flexibility and this must be allowed for in calculating its ability to survive and function.

As a guide, designers should avoid cantilevered components, avoid brittle materials, mount flexibly and ensure that resultant movements are not impeded by pipe or cable connections and do not cause impact with hard structure. It is important to allow for the behaviour of materials used when subject to high rates of strain. Some, notably mild steel, exhibit an increase in yield point by a factor up to two under these conditions. In warships essential equipment is designed to remain operable up to a level of shock at which the ship is likely to be lost by hull rupture. The overall design is tested by exploding large charges (up to 500 kg) fairly close to the hull of the first ship of a class.

In order to assess a design's vulnerability, and to highlight any weak elements for rectification, each new design of warship is the subject of a vulnerability assessment. This assesses the probability of various types of attack on the ship, allowing for its susceptibility. The ability of a ship to withstand each attack and retain various degrees of fighting capability, and finally survive, is computed. Essentially, the contribution of each element of the ship and its systems to each fighting capability (e.g. to detect and destroy an enemy submarine), is established. For each form of attack the probability of the individual elements being rendered non-operative is assessed using a blend of calculation, model and full scale data. If one element is particularly sensitive, or especially important, it can be duplicated (or perhaps given special protection) to reduce the overall vulnerability. The modelling for these calculations is very similar to that adopted for reliability assessments. Having made the assessments for each form of attack these can be combined, allowing for the probability of each form, to give an overall vulnerability for the design. The computations can become quite lengthy. There are also a number of difficulties which mean that any results must be carefully interpreted. These include the fact that reduced general services (electricity or chilled water, say) may be adequate to support some but not all fighting capabilities. What then happens in battle will depend upon which capabilities the command needs to deploy. This can be overcome by setting the vulnerability results in the context of various engagement scenarios. Also, in many cases, the full consequences of an attack will depend upon the actions taken by the crew in damage limitation. For instance, how effectively will they deal with fire, how rapidly will they close doors and valves to limit flooding? Recourse must be made to exercise data and statistical allowances made.

Offensive agencies may not cause damage to material but may be none the less lethal. Gas, bacteria and radioactive particles must be excluded by the crash closing of an airtight boundary. Deposits are washed away by a pre-wetting system which drenches the whole ship in salt water spray. A large air burst explosion such as that from a nuclear weapon causes a blast effect followed by a slight suction, each of which may cause structural damage against which the ship may be designed.

The effects of a ship's weapons on the ship itself must be catered for; blast, shock, heating, scouring, chemical deposit, recoil loading, noise and physical obstruction. High intensity transmissions by modern radio and radar can be hazardous to personnel and can trigger weapons which have not been properly shielded. This type of radiation hazard is called radhaz.

A good example of protection by dispersion is afforded by the unitization of machinery in large warships. Each unit, which may comprise more than one watertight compartment, is self-contained, e.g. in a steam turbine ship it would include boiler, steam turbine, gearing and auxiliaries. The total electrical generating capacity of the ship would also be divided amongst the various units. A ship designed for about 30 knots on two units could attain more than 25 knots on one unit. Thus, keeping one unit available after damage, affords considerable mobility to the ship. A ship unable to move would be a sitting duck for further attack.

SHIP SIGNATURES

As discussed in the previous section, the ability of a ship to survive depends, in part, on its susceptibility to detection and attack. Many ways of reducing the susceptibility lie in the province of the weapons engineers although the naval architect will be directly concerned with integrating them into the design. However, the designer will have a direct responsibility for, and influence on signatures. If these can be kept low, the enemy will have difficulty in detecting and classifying his prey and decoys are more likely to be effective. Each signature brings its own problems for the designer and they must all be considered. To reduce all but one would leave the ship susceptible in that respect. The signatures of most concern are:

(*a*) the ship has a natural magnetism which can be used to trip the fuse of a magnetic mine; such a signature can be counteracted by passing a current through electric cables wound round the inside of the ship. This is known as degaussing. Such magnetism can also be caused by induced eddy currents in a ship, particularly an aluminium alloy ship, rolling and thereby cutting the earth's magnetic lines of force;

(*b*) a point at the bottom of the sea experiences a pressure variation as a ship passes over it which can be used to trip a fuse in a pressure mine;

(*c*) the heat given off by a ship produces an infra-red radiation which can be used to home a guided weapon;
(*d*) radio and radar emissions can be used to detect the position of a ship and to home a guided weapon;
(*e*) underwater noise gives away a ship's position and can be used for homing;
(*f*) detritis and hydrodynamic disturbance allows the ship's track to be detected many hours after its passage.

GENERAL VULNERABILITY OF SHIPS

Ships are vulnerable to accident at any time but especially in storm conditions. The statistics of ship casualties over many years enables us to depict the frequency of events which has occurred in the past and to use that pattern to predict the likelihood of a particular event in the future. For example, it would not be surprising to discover that all of the lengths of damage due to collision and grounding which had occurred in the past, showed a frequency density as in Fig. 5.18. The area of any strip δx represents the fraction of the total number of ships to have suffered a damage length lying between x_1 and $x_1 + \delta x$ of ship length.

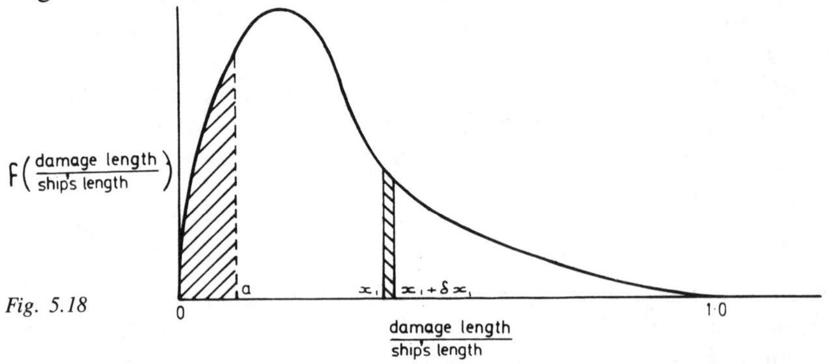

Fig. 5.18

This information could be used to predict that the probability of a damage length equal to a fraction *a* of ship length or less occurring to any ship in the future is given by the area shaded up to *a*. This is the basis for a new approach to ship vulnerability promulgated for merchant ships by IMO in 1973. The probability of two or more independent events is found by multiplying together their separate probabilities. Thus, the probability that a ship in collision will suffer a damage length *l* and a penetration from the side *d* and at the same time be in a condition with a permeability μ is given by

$$p\,(l) \times p\,(d) \times p\,(\mu)$$

Moreover, this must be compounded with the probability of collision at a certain position along the length of the ship. Probability of collision or grounding occurring at all is clearly dependent upon the routeing of the ship which is affected by the weather, sea state and traffic density met; it will be high in the English Channel or St Lawrence and low in the Indian Ocean. It must also depend upon the number and competence of the crew.

The effects upon the ship of a stated set of damage can be found in terms of trim, flotation and stability for which permissible values have already been declared. Thus, the vulnerability of the ship—that is to say the likelihood of unacceptable damage occurring—can be determined from the architecture of the ship and the probability density functions for: position of damage; length of damage; penetration of damage; permeability at the time of damage; occurrence of an incident; crew competence.

'Standard' shapes for the first four of these have been proposed by IMO who have also described how they should be applied. There are so many variables that computer programs have been written for the calculation. Similar standard functions have been in use by various navy departments for some while to determine the vulnerabilities of warships which are subjected to weapons with different homing devices and explosive content.

A steady change to probabilistic standards, begun in Code A265, is IMO's aim. Both passenger ship vulnerability and, from 1992, cargo ship safety are examined in this way. These standards should not be applied blindly or to the exclusion of an examination of compartment standard or, indeed, from a complete safety survey to be considered in Chapter 15.

The probabilistic standards are formulated upon a required subdivision index, R, such that:

$R = (0{\cdot}002 + 0{\cdot}009L_s)^{1/3}$ to be compared with an achieved subdivision index, A,

$A = \Sigma p_i S_i$ where p = the probability of flooding a compartment i and S = the probability of survival after flooding compartment i,

$S = C\sqrt{0{\cdot}5GZ_{max}(\text{range})}$ and $C\sqrt{(30-\theta_e)/5}$ or $= 1$ if $\theta_e \leq 25°$ or $= 0$ if $\theta_e \geq 30°$

θ_e = equilibrium angle in degrees.

There are many caveats to these basic requirements known commonly (but incorrectly) as SOLAS 90. Note that there is no direct relationship between the value of A obtained from these crude formulae and the actual survivability standards of a ship which owners would wish to know as part of their safety survey.

It is not easy, once it is acknowledged that nothing is absolutely safe or absolutely reliable, to declare a probability of catastrophe which, however tiny, is nevertheless acceptable. Actuaries have had to make such judgments for many years of course in settling the levels of premium for insurance involving human life. Such a cold declaration in monetary terms does not commend itself to the public, whose outrage at a particular catastrophe may be whipped up by the press. Nevertheless, a standard of acceptable probability must be declared and this will fall to national authorities and insurers, generally on the basis of what has been found acceptable in the past.

ABNORMAL WAVES

Abnormal waves can be created by a combination of winds, currents and seabed topography. In one case a ship may be heading into waves 8 m high when suddenly the bow falls into a long, sloping trough so that in effect it is steaming downhill. At the bottom it may meet a steep wall of water, perhaps 18 m high and about to break, bearing down on it at 30 knots.

Vessels can suffer severe damage. In 1973 the 12,000 grt *Bencruachan* had the whole bow section forward of the break of the 120 ft long forecastle bent downwards at 7 degrees. A 15,000 dwt freighter was broken in two by a freak wave on its maiden voyage.

Environmental pollution

Besides ensuring that the ship can meet the hazards imposed on it by the environment, it is important to ensure that the ship itself does not hazard the environment. Under one of the fundamental conventions of IMO, MARPOL (Ref. 2); requirements are laid down to restrict or totally prevent pollution in certain areas of sea particularly those close to land and in land-locked seas such as the Mediterranean. These deal with oil, sewage, food disposal and packaging materials. Ships must either store waste products or process them on board. Sewage treatment plants can be fitted to turn raw sewage into an inoffensive effluent that can be pumped overboard; incinerators can burn many packaging materials; crushers can compact waste so that it can be stored more conveniently.

Problems

1. A water lighter is in the form of a rectangular barge, 42·7 m long by 5·5 m beam, floating at an even draught of 2·44 m in sea water. Its centre of gravity is 1·83 m above the keel. What will be the new draughts at the four corners if 20 m^3 of fresh water are pumped out from an amidships tank whose centre of volume is 0·3 m from the port side and 0·91 m above the keel? Ignore free surface effects.
2. A box-form vessel, 150 ft long, 50 ft wide and 20 ft deep floats on an even keel with a draught of 6 ft in salt water. A transverse watertight bulkhead is fitted 25 ft from the forward end. If the compartment thus formed is open to the sea, estimate the new draughts forward and aft to the nearest inch.
3. A methane carrier, displacing 6000 m^3 floats at draughts of 5·1 m forward and 5·8 m aft, measured at marks which are 80·5 m forward and 67·5 m aft amidships. The vessel grounds, on a falling tide, on a rock at a position 28 m forward of amidships. Calculate the force on the ship's bottom and the new draughts after the tide has fallen 25 cm.

The rock then ruptures the bottom and opens to the sea, a compartment between two transverse bulkheads 14·5 m and 28·5 m forward of amidships and between two longitudinal bulkheads 5·2 m each side of the centre line. Estimate the draughts of the ship when the tide has risen to float the ship free.

The hydrostatic curves show for about these draughts:

WP area = 1850 m² Length BP = 166 m
CF abaft amidships = 6 m
$\overline{GM_L}$ = 550 m

4. A catamaran is made up of two hulls, each 80 m long and with centres 30 m apart. The constant cross-section of each hull has the form of an equilateral triangle, each side being of 10 m length. The draught of the craft is 4 m and its $\overline{KG}$ is 10 m.

Calculate the heeling couple and the angle of heel necessary to bring one hull just clear of the water, assuming that the broadside wind force does not effectively increase the displacement.

Each of the hulls has main bulkheads 10 m each side of amidships. If these central compartments were both open to the sea, calculate the virtual $\overline{GM}$ of the craft in the bilged condition, comparing the added weight and lost buoyancy methods.

5. A ship displacing 2200 tonf and 360 ft in length has grounded on a rock at a point 110 ft forward of amidships. At low water the draughts forward and aft are 8·92 and 13·24 ft respectively. Given the following hydrostatic particulars of the ship in level trim and taking the $\overline{KG}$ as 16·53 ft, calculate; (i) the force on the rock at low tide; (ii) the virtual $\overline{GM}$ at low tide; (iii) the rise in tide necessary to refloat the ship.

Mean draught (ft)	Displacement (tonf)	LCB (ft aft)	VCB (ft)	LCF (ft aft)	$\overline{BM_L}$ (ft)	$\overline{BM_T}$ (ft)
12·0	2220	7·62	7·42	22·88	943	12·16
11·0	1939	5·64	6·83	20·76	1026	13·57

6. A ship, for which the hydrostatic particulars of the preceding question apply at level trim, has a length of 360 ft and displaces 2035 tonf at level trim and with a $\overline{KG}$ of 16·67 ft. A collision causes uncontrollable flooding of the auxiliary machinery space. Calculate (*a*) the final draughts at the forward and aft perpendiculars; (*b*) the virtual $\overline{GM}$ in the bilged condition. Particulars of the flooded compartment are as follows: after bulkhead, 20 ft forward of amidships, length of compartment, 20 ft, beam above 10 ft waterline assumed constant at 38 ft, area of section assumed constant at 320 ft^2 up to 10 ft waterline, centroid of section above keel, 6·6 ft. Hydrostatic data may be linearly extrapolated when required.

7. A rectangular pontoon, 300 ft × 12 ft × 4 ft draught, is divided by a longitudinal bulkhead at the middle line and by four equally spaced transverse bulkheads. The metacentric height is 2 ft. Find the draughts at the four corners when one corner compartment is bilged. Ignore rotation of the principal axes of the waterplane.

8. A box-shaped vessel 100 m long and 20 m broad, floats at a draught of 6 m forward and 10 m aft, the metacentric height being 2·25 m. Find the virtual metacentric height when the keel just touches level blocks throughout its length.

9. A vessel of box form, 150 ft long and 25 ft broad, floats at an even draught of

8 ft, and has a watertight deck $8\frac{1}{2}$ ft above the keel. If a central compartment, 30 ft long, bounded by two transverse bulkheads extending up to the deck, is bilged, what will be (*a*) the new draught of the vessel, (*b*) the alteration in metacentric height if the water admitted is regarded as lost buoyancy

10. A ship 120 m long and floating at a level draught of 4·0 m has a displacement vol = 3140 m³. Its centre of gravity is 5·1 m above the keel and, at the 4 m waterline, the TPM = 1200, MCT = 8000 tonnef m/m, $\overline{KB}$ = 2·2 m and the centre of flotation is 5·1 m abaft amidships.

After striking a rock at a point 2 m abaft the fore perpendicular, the foremost 7 m are flooded up to the waterline. The displacement lost up to the 4 m waterline is 40 m³ with its centre of buoyancy, 4·5 m abaft the FP and 2 m above the keel. Twenty square metres of waterplane are lost, with the centroid 4·5 m abaft the FP

Making reasonable assumptions, estimate: (*a*) the force on the rock immediately after the accident; (*b*) the force on the rock when the tide has fallen 30 cm; (*c*) the rise of tide necessary to lift the bow just clear of the rock. rock.

11. The following particulars refer to a 30,000 tonf aircraft carrier:

length, 700 ft	$\overline{KB}$, 16 ft
beam, 90 ft	$\overline{KG}$, 30 ft
draught, 28 ft	TPI, 100
$\overline{GM_T}$, 6 ft	

The design is modified by increasing all athwartships dimensions by 3 per cent, and all longitudinal dimensions by 2 per cent. The weight of the flight deck armour is adjusted to limit the increase in displacement to 900 tonf. It should be assumed that the height of the flight deck above keel remains constant at 70 ft and that the weight of the ship varies directly as the length and beam.

In compiling the flooding board for the ship, a compartment 60 ft long by 20 ft wide is assumed flooded, and open to the sea. If the compartment extends the full depth of the ship, with its centre 40 ft to starboard of the middle line, estimate the resulting angle of heel. The effects of changing trim should be ignored.

12. A vessel of length 120 m between perpendiculars floats at a uniform draught of 5·0 m with the following particulars:

Displacement, 3000 tonnef
WP area, 1300 m²
CF abaft amidships, 3 m
CB above keel, 2·70 m
CG above keel, 4·52 m
Transverse metacentre above keel, 6·62 m
MCT BP, 7300 tonnef m/m

It may be assumed to be wall-sided in the region of the waterplane.

The vessel is involved in an accident in which 60 tonnef outer bottom (with centre of gravity at keel level and 3·2 m from the middle line) are torn away and a compartment 3·1 m wide and 7·9 m long is opened to the sea.

Estimate the resulting angle of heel, assuming that the compartment extends from keel level to above the new waterplane and neglecting any change of trim. The centroids of volume and plan area of the flooded compartment are 3·0 m from the middle line.

13. A ship 120 m BP for which hydrostatic and other data are given in question 12, floats at a uniform draught of 5·0 m.

The vessel grounds on a rock at keel level, 20 m forward of amidships and 3·0 m from the middle line. Making reasonable assumptions, calculate the force on the rock and the angle of heel when the tide has fallen 15 cm.

14. A rectangular pontoon 12 ft deep, 36 ft wide and 105 ft long is divided into three equal sections by two transverse bulkheads. The centre compartment is further subdivided into three equal compartments by two longitudinal bulkheads. All compartments extend the full depth of the pontoon. The centre of gravity is 6 ft above the keel.

The pontoon is damaged in such a way that the three central compartments are open to the sea at keel level. Of the three compartments, the centre one is vented at deck level and the wing compartments are airtight above keel level. The draught after damage is 7·18 ft and the depth of water in the wing tanks is 1·71 ft. The position of the centre of gravity of the pontoon may be assumed unaffected.

Calculate the virtual metacentric height in the damaged condition, assuming that the height of the sea water barometer is 33 ft.

15. A ship floating at a draught of 2 m has a hole of area 0·3 m² in the bottom at the keel, giving access to a rectangular compartment right across the ship. The compartment is 10 m long and has an even permeability of 0·80 throughout. This ship has initially WP area = 300 m². Its length is 70 m, beam 8 m and it does not trim due to the damage. Calculate how long it will be before the maximum flooded draught occurs.

16. Calculate the probability of losing half and total power of worked example 1, assuming that there was no 15 m separation of the machinery units.

References

1. *Merchant Shipping (Passenger Ship Construction and Survey) Regulations* 1984 as amended 1990 (No 892) and 1992 (No 2358), HMSO.
2. MARPOL 73/78 Regulations and Guidelines, IMO.
3. Sarchin, T. H. and Goldberg, L. L. Stability and Buoyancy criteria for US naval surface ships, *TSNAME*, 1962.
4. *Merchant Shipping (Fire Appliances) Regulations*, HMSO.
5. *Merchant Shipping (Life Saving Appliances) Regulations and Supplement*, HMSO.
6. *Merchant Shipping (Grain) Regulations*, HMSO.
7. *Merchant Shipping (Dangerous Goods) Regulations*, HMSO.
8. *International Maritime Dangerous Goods Code*, IMO.
9. Holstead, R. Life Saving appliances — the changing scene, *TRINA*, 1985.
10. Tankers and bulk carriers — the way ahead, *RINA*, 1992.
11. Harvey, R. C. A brief history of the anchor with some thoughts on future developments, *NA*, July 1980.
12. The Kummerman International Conferences on RoRo safety and vulnerability, *RINA*, 1987 and 1991.
13. Brown, D. K. and Tupper, E. C. The naval architecture of surface warships, *TRINA*, 1989.
14. Greenhorn, J. The assessment of surface ship vulnerability to underwater attack. *TRINA*, 1989.
15. Hicks, A. N. *Explosure induced hull whipping*. Advances in Marine Structure, Elsevier, 1986.
16. Francescutto, A. Is it really impossible to design safe ships? *TRINA*, 1992.

6 The ship girder

Few who have been to sea in rough weather can doubt that the structure of a ship is subject to strain. Water surges and crashes against the vessel which responds with groans and shudders and creaks; the bow is one moment surging skywards, the next buried beneath green seas; the fat middle of the ship is one moment comfortably supported by a wave and the next moment abandoned to a hollow. The whole constitutes probably the most formidable and complex of all structural engineering problems in both the following aspects:

(*a*) the determination of the loading
(*b*) the response of the structure.

As with most complex problems, it is necessary to reduce it to a series of unit problems which can be dealt with individually and superimposed. The smallest units of structure which have to be considered are the panels of plating and single stiffeners which are supported at their extremities by items which are very stiff in comparison; they are subject to normal and edge loads under the action of which their dishing, bowing and buckling behaviour relative to the supports may be assessed. Many of these small units together constitute large flat or curved surfaces of plating and sets of stiffeners called grillages, supported at their edges by bulkheads or deck edges which are very stiff in comparison; they are subject to normal and edge loading and their dishing and buckling behaviour as a unit relative to their supports may be assessed. Finally, many bulkheads, grillages and decks, together constitute a complete hollow box whose behaviour as a box girder may be assessed. It is to this last unit, the whole ship girder, that this chapter is confined, leaving the smaller units for later consideration.

Excluding inertia loads due to ship motion, the loading on a ship derives from only two sources, gravity and water pressure. It is impossible to conceive a state of the sea whereby the loads due to gravity and water pressure exactly cancel out along the ship's length. Even in still water, this is exceedingly unlikely

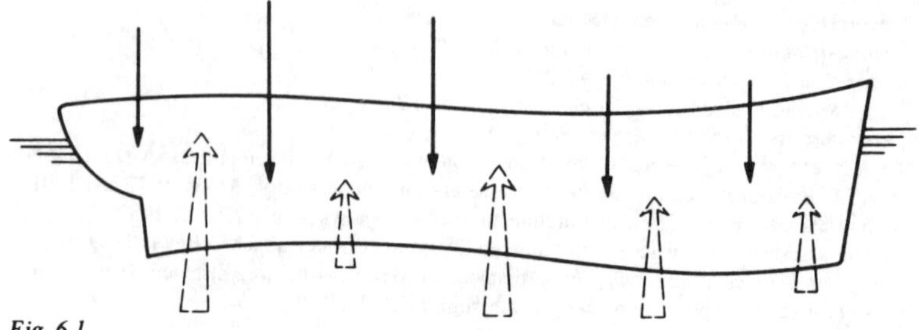

Fig. 6.1

but in a seaway where the loading is changing continuously, it is inconceivable. There is therefore an uneven loading along the ship and, because it is an elastic structure, it bends. It bends as a whole unit, like a girder on an elastic foundation and is called the *ship girder*. The ship will be examined as a floating beam subject to the laws deduced in other textbooks for the behaviour of beams.

In still water, the loading due to gravity and water pressure are, of course, weight and buoyancy. The distribution of buoyancy along the length follows the curve of areas while the weight is conveniently assessed in unit lengths and might, typically, result in the block diagram of Fig. 6.2. (Clearly, the areas representing total weight and total buoyancy must be equal.) This figure

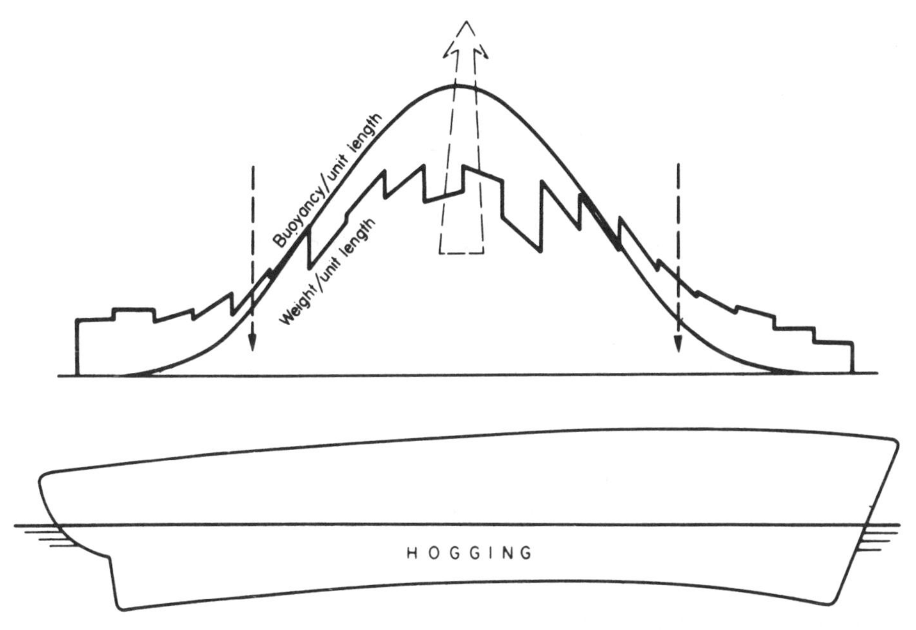

Fig. 6.2 Still water hogging

would give the resultants dotted which would make the ship bend concave downwards or *hog*. The reverse condition is known as *sagging*. Because it is not difficult to make some of the longer cargo ships break their backs when badly loaded, consideration of the still water hogging or sagging is vital in assessing a suitable cargo disposition. It is the first mate's yardstick of structural strength.

It is not difficult to imagine that the hog or sag of a ship could be much increased by waves. A long wave with a crest amidships would increase the upward force there at the expense of the ends and the hogging of the ship would be increased. If there were a hollow amidships and crests towards the ends sagging would be increased (Fig. 6.3). The loads to which the complete hull girder is subject are, in fact:

(*a*) those due to the differing longitudinal distribution of the downward forces of weight and the upward forces of buoyancy, the ship considered at rest in still water;

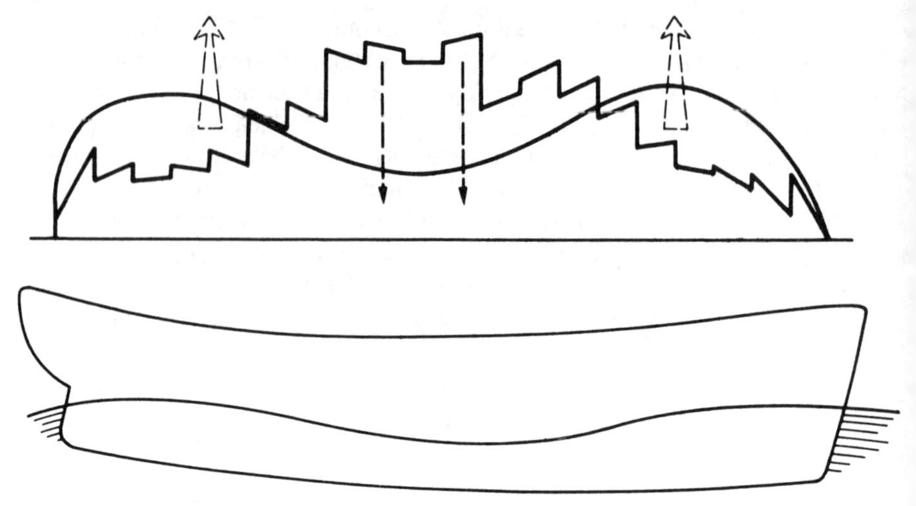

Fig. 6.3 Sagging on a wave

(*b*) the additional loads due to the passage of a train of waves, the ship remaining at rest;
(*c*) loads due to the superposition on the train of waves of the waves caused by the motion of the ship itself through still water;
(*d*) the variations of the weight distribution due to the accelerations caused by ship motion.

Consideration of the worst likely loading effected by (*a*) and (*b*) is the basis of the standard calculation. The effects of (*c*) and (*d*) are smaller and are not usually taken into account except partly by the statistical approach outlined later.

The standard calculation

This is a simple approach but one that has stood the test of time. It relies on a comparison of a new design with previous successful design. The calculated stresses are purely notional and based on those caused by a single wave of length equal to the ship's length, crest normal to the middle line plane and with

(*a*) a crest amidships and a hollow at each end causing maximum hogging, and
(*b*) a hollow amidships and a crest at each end causing maximum sagging.

The ship is assumed to be momentarily still, balanced on the wave with zero velocity and acceleration and the response of the sea is assumed to be that appropriate to static water. In this condition, the curves of weight and buoyancy are deduced. Subtracted one from the other, the curves give a curve of net loading p'. Now, a fundamental relationship at a point in an elastic beam is

$$p' = \frac{\mathrm{d}S}{\mathrm{d}x} = \frac{\mathrm{d}^2M}{\mathrm{d}x^2}$$

where p' is the load per unit length, S is the shearing force, M is the bending moment, and x defines the position along the beam.

$$\therefore \quad S = \int p' \, dx \text{ and } M = \int S \, dx$$

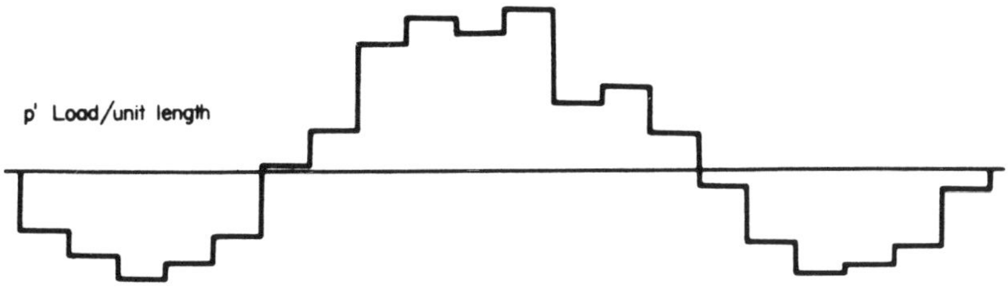

Fig. 6.4 Loading curve, p'

Thus, the curve of loading p' must be integrated along its length to give a curve of shearing force and the curve of shearing force must be integrated along its length to give the bending moment curve. From the maximum bending moment, a figure of stress can be obtained,

$$\text{stress } \sigma = \frac{M}{I} y$$

I/y being the modulus of the effective structural section.

A closer look at each of the constituents of this brief summary is now needed.

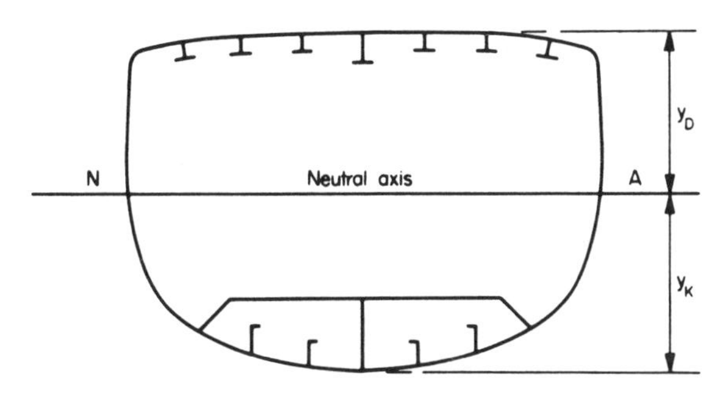

Fig. 6.5 Simplified structural section

THE WAVE

Whole books have been written about ocean waves. Nevertheless, there is no universally accepted standard ocean wave which may be assumed for the standard longitudinal strength calculation. While the shape is agreed to be trochoidal, the observed ratios of length to height are so scattered that many 'standard' lines can be drawn through them. Fortunately, this is not of primary importance; while the calculation is to be regarded as comparative, provided that the same type of wave is assumed throughout for design and type ships, the comparison is valid.

A trochoid is a curve produced by a point at radius r within a circle of radius R rolling on a flat base. The equation to a trochoid with respect to the axes shown in Fig. 6.6, is

$$x = R\theta - r\sin\theta$$

$$z = r(1-\cos\theta)$$

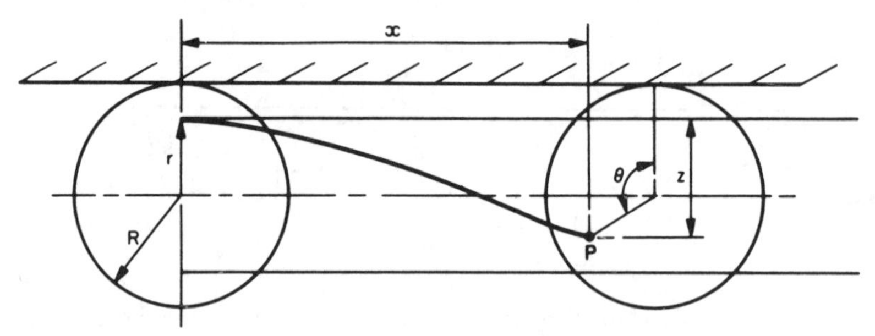

Fig. 6.6 Construction of a trochoid

One accepted standard wave is that having a height from trough to crest of one twentieth of its length from crest to crest. In this case, $L = 2\pi R$ and $r = h/2 = L/40$ and the equation to the wave is

$$x = \frac{L}{2\pi}\theta - \frac{L}{40}\sin\theta$$

$$z = \frac{L}{40}(1-\cos\theta)$$

The co-ordinates of this wave at equal intervals of x are:

$\frac{x}{L/20}$	0	1	2	3	4	5	6	7	8	9	10
$\frac{z}{L/20}$	0	0·034	0·128	0·266	0·421	0·577	0·720	0·839	0·927	0·982	1·0

Research has shown that the $L/20$ wave is somewhat optimistic for wavelengths from 300 ft up to about 500 ft in length. Above 500 ft, the $L/20$ wave becomes progressively more unsatisfactory and at 1000 ft is probably so exaggerated in height that it is no longer a satisfactory criterion of comparison. This has resulted in the adoption of a trochoidal wave of height $1{\cdot}1\sqrt{L}$ as a standard wave in the comparative longitudinal strength calculation. This wave has the equation

$$x = \frac{L}{2\pi}\theta - \frac{1{\cdot}1\sqrt{L}}{2}\sin\theta \qquad x, z \text{ and } L \text{ in feet}$$

$$z = \frac{1{\cdot}1\sqrt{L}}{2}(1-\cos\theta)$$

The $1{\cdot}1\sqrt{L}$ wave has the slight disadvantage that it is not non-dimensional, and units must be checked with care when using this wave and the formulae derived from it. Co-ordinates for plotting are conveniently calculated from the equations above for equal intervals of θ. In metric units $1{\cdot}1\sqrt{L} = 0{\cdot}607\sqrt{L}$.

The length of the wave is, strictly, taken to be the length of the ship on the load waterline; in practice, because data is more readily available for the displacement stations, the length is often taken between perpendiculars (if this is different) without making an appreciable difference to the bending moment. Waves of length slightly greater than the ship's length can, in fact, produce theoretical bending moments slightly in excess of those for waves equal to the ship's length, but this has not been an important factor while the calculation continued to be comparative.

Waves steeper than $L/7$ cannot remain stable. Standard waves of size $L/9$ are used not uncommonly for the smaller coastal vessels. It is then a somewhat more realistic basis of comparison.

WEIGHT DISTRIBUTION

Consumable weights are assumed removed from those parts of the ship where this aggravates the particular condition under investigation; in the sagging condition, they are removed from positions near the ends and, in the hogging condition, they are removed amidships. The *influence lines* of a similar design should be consulted before deciding where weights should be removed, and the decision should be verified when the influence lines for the design have been calculated (see p. 191), provided that the weights are small enough.

The longitudinal distribution of the weight is assessed by dividing the ship into a large number of intervals. Twenty displacement intervals are usually adequate. The weight falling within each interval is assessed for each item or group in the schedule of weights and tabulated. Totals for each interval divided by the length give mean weights per unit length. It is important that the centre of gravity of the ship divided up in this way should be in the correct position. To ensure this, the centre of gravity of each individual item should be checked after it has been distributed.

One of the major items of weight requiring distribution is the hull itself, and this will sometimes be required before detailed structural design of the hull has been completed. A useful first approximation to the hull weight distribution is obtained by assuming that two-thirds of its weight follows the still water buoyancy curve and the remaining one-third is distributed in the form of a trapezium, so arranged, that the centre of gravity of the whole hull is in its correct position (Fig. 6.7).

Having obtained the mean weight per unit length for each interval, it is plotted as the mid-ordinate of the interval and a straight line is drawn through it, parallel to the chord of the buoyancy curve. This is a device for simplifying later stages of the calculation which introduces usually insignificant inaccuracies. A sawtooth distribution of weight per unit length, as shown in Fig. 6.3, results.

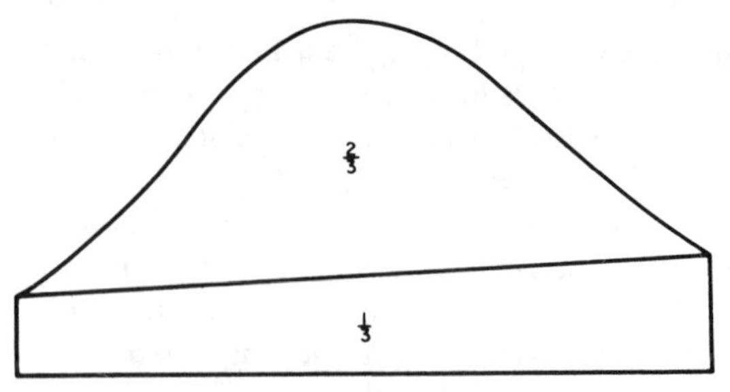

Fig. 6.7 Approximate hull weight distribution

BUOYANCY AND BALANCE

If the standard calculation is performed by hand, the wave is drawn on tracing paper and placed over the contracted profile of the ship on which the Bonjean curve has been drawn at each ordinate. Figure 6.8 shows one such ordinate. It is necessary for equilibrium to place the wave at a draught and trim such that

(*a*) the displacement equals the weight and
(*b*) the centre of buoyancy lies in the same vertical plane as the centre of gravity.

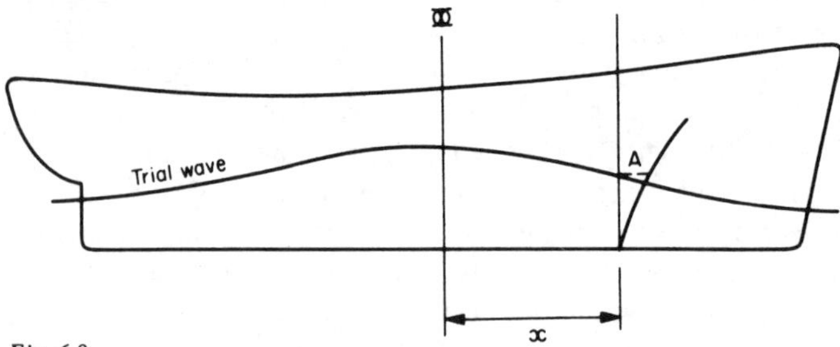

Fig. 6.8

Simple hydrostatic pressure is assumed; the areas immersed at each ordinate can be read from the Bonjean curves where the wave profile cuts the ordinate, and these areas are subjected to normal approximate integration to give displacement and LCB position. The position of the wave to meet the two conditions can be found by trial and error. It is usual to begin at a draught to the midships point of the trochoid about 85 per cent of still water draught for the hogging condition and 120 per cent for the sagging wave. A more positive way of achieving balance involves calculating certain new tools. Consider a typical transverse section of the ship, x m forward of amidships, at which the Bonjean curve shows the area immersed by the first trial wave surface to be A m^2.

If ∇ m^3 and M m^4 are the volume of displacement and the moment of this volume before amidships for the trial wave surface, then, over the immersed length,

$$\nabla = \int A \, dx \text{ and } M = \int Ax \, dx$$

Suppose that ∇_0 m³ and M_0 m⁴ are the volume of displacement and moment figures for equilibrium in still water and that ∇ is less than ∇_0 and M is less than M_0. The adjustment to be made to the trial wave waterline is therefore a parallel sinkage and a trim by the bow in order to make $\nabla = \nabla_0$ and $M = M_0$. Let this parallel sinkage at amidships be z_0 and the change of trim be m radians, then the increased immersion at the typical section is

$$z = (z_0 + mx) \text{ m}$$

Let the slope of the Bonjean curve be $s = \mathrm{d}A/\mathrm{d}z$ m. Assuming that the Bonjean curve is straight over this distance, then

$$A + zs = A + (z_0 + mx)s \text{ m}^2$$

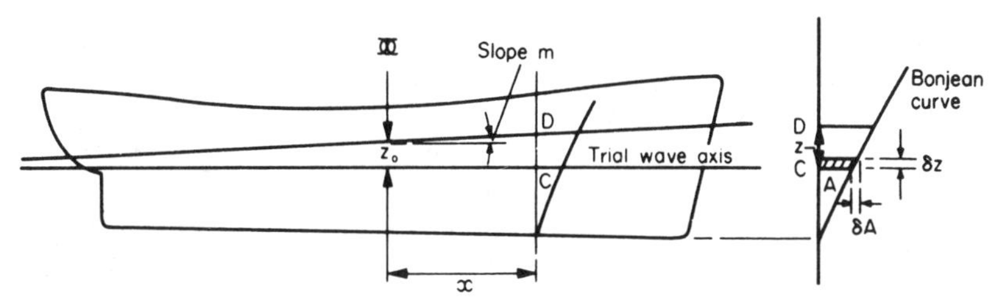

Fig. 6.9

In order to satisfy the two conditions, $\nabla = \nabla_0$ and $M = M_0$, therefore,

$$\int [A + (z_0 + mx)s] \, \mathrm{d}x = \nabla_0$$

and

$$\int [A + (z_0 + mx)s] x \, \mathrm{d}x = M_0$$

Now put

$$\int s \, \mathrm{d}x = \delta\nabla \text{ m}^2$$

$$\int sx \, \mathrm{d}x = \delta M \text{ m}^3$$

$$\int sx^2 \, \mathrm{d}x = \delta I \text{ m}^4$$

Then

$$z_0 \,.\, \delta\nabla + m\delta M = \nabla_0 - \nabla$$

and

$$z_0 \,.\, \delta M + m\delta I = M_0 - M$$

$\delta\nabla$, δM and δI can all be calculated by approximate integration from measurement of the Bonjean curves. From the pair of simultaneous equations above the required sinkage z_0 and trim m can be calculated. A check that this wave position in fact gives the required equilibrium should now be made. If the trial waterplane was a poor choice, the process may have to be repeated. Negative values of z_0 and m, mean that a parallel rise and a trim by the stern are required.

Having achieved a balance, the curve of buoyancy per metre can be drawn as a smooth curve. It will have the form shown in Fig. 6.2 for the hogging calculation and that of Fig. 6.3 for the sagging condition.

LOADING, SHEARING FORCE AND BENDING MOMENT

Because the weight curve has been drawn parallel to the buoyancy curve, the difference between the two, which represents the net loading p' for each interval, will comprise a series of rectangular blocks. Using now the relationship $S = \int p' \, dx$, the loading curve is integrated to obtain the distribution of shearing force along the length of the ship. The integration is a simple cumulative addition starting from one end and the shearing force at the finishing end should, of course, be zero; in practice, due to the small inaccuracies of the preceding steps, it will probably have a small value. This is usually corrected by canting the base line, i.e. applying a correction at each section in proportion to its distance from the starting point.

The curve of shearing force obtained is a series of straight lines. This curve is now integrated in accordance with the relationship $M = \int S \, dx$ to obtain the distribution of bending moment M. Integration is again a cumulative addition of the areas of each trapezium and the inevitable final error, which should be small, is distributed in the same way as is the shearing force error. If the error is large, the calculations must be repeated using smaller intervals for the weight distribution.

The integrations are performed in a methodical, tabular fashion as shown in the example below. In plotting the curves, there are several important features which arise from the expression

$$p' = \frac{dS}{dx} = \frac{d^2M}{dx^2}$$

which will assist and act as checks. These are

(*a*) when p' is zero, S is a maximum or a minimum and a point of inflexion occurs in the M curve,
(*b*) when p' is a maximum, a point of inflexion occurs in the S curve,
(*c*) when S is zero, M is a maximum or minimum.

A typical set of curves is shown in Fig. 6.10.

EXAMPLE 1. The mean immersed cross-sectional areas between ordinates of a ship, 300 m in length, balanced on a hogging wave, as read from the Bonjean curves are given below, together with the mass distribution. The second moment of area of the midship section is 752 m^4 and the neutral axis is 9·30 m from the keel and 9·70 m from the deck.

Calculate the maximum direct stresses given by the comparative calculation and the maximum shearing force.

Making the assumption that the second moment of area is constant along the length, estimate the difference in slope between ordinates number 4 and 10 and the deflection of the middle relative to the ends.

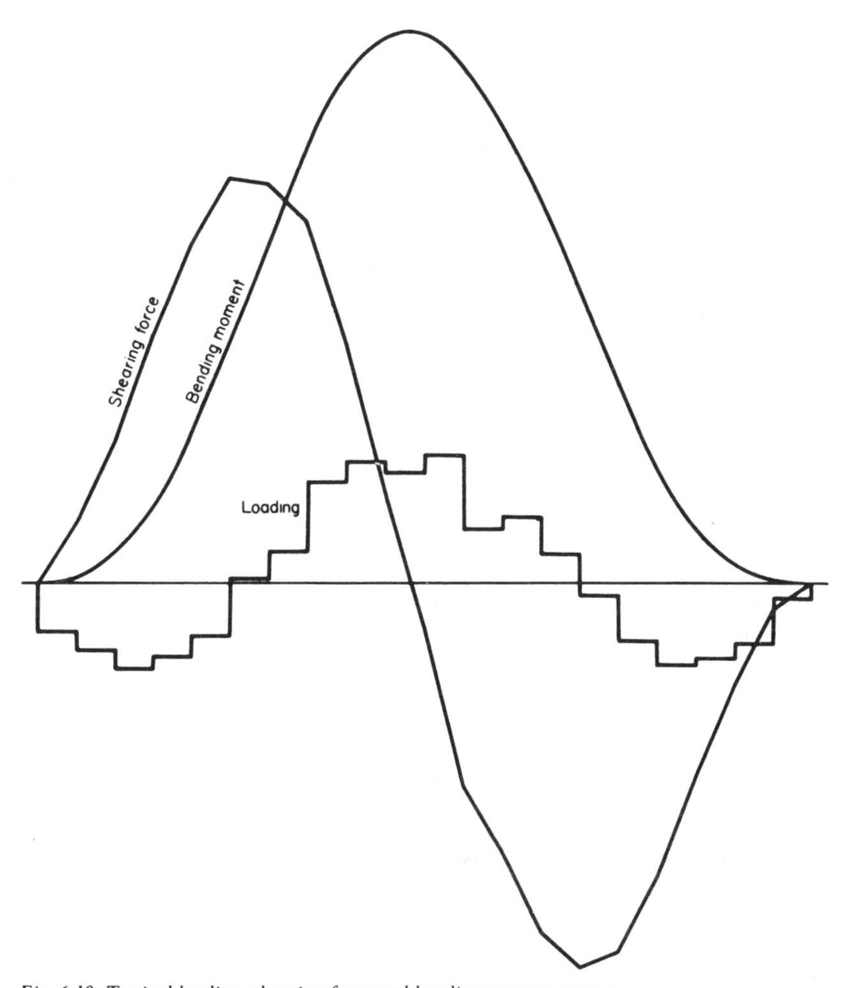

Fig. 6.10 Typical loading, shearing force and bending moment curves

Ordinate	11		10		9		8		7		6		5		4		3		2		1
Area A, m²		10·5		49·5		243		490		633		565		302		71		2·5		0·5	
Mass M, Mg		5030		6790		10,670		9900		9370		7830		9260		6400		4920		2690	

Solution: Integrate in tabular form using trapezoidal rule

Ordinate spacing = 30 m

$$\text{Weight per metre} = \frac{M}{30}\frac{\text{Mg}}{\text{m}} \cdot \frac{9{\cdot}81\ \text{m}}{s^2} \cdot \frac{1000\ \text{kg}}{\text{Mg}} = 327M, \text{N/m}$$

$$\text{Buoyancy per metre} = A\ \text{m}^3 \cdot \frac{1{\cdot}026\ \text{Mg}}{\text{m}^3} \cdot \frac{9{\cdot}81\ \text{m}}{\text{s}^2} \cdot \frac{1000\ \text{kg}}{\text{Mg}} = 10{,}060A, \text{N/m}$$

a Ord.	b wt./m $\frac{327M}{10^6}$ (MN)	c buoy./m $\frac{10{,}060A}{10^6}$ (MN)	d load/m b − c (MN)	e δSF 30d (MN)	f SF Σe (MN)	g mid SF (MN)	h δBM 30g (MN m)	i BM Σh (MN m)	j corrn.	k BM (MN m)
1					0			0		0
	0·880	0·005	0·875	26·25		13·13	394			
2					26·25			394	−4	390
	1·609	0·025	1·584	47·52		50·01	1500			
3					73·77			1894	−8	1886
	2·093	0·714	1·379	41·37		94·46	2834			
4					115·14			4728	−12	4716
	3·028	3·038	−0·010	−0·30		114·99	3450			
5					114·84			8178	−16	8162
	2·560	5·684	−3·124	−93·72		67·98	2039			
6					21·12			10,217	−20	10,197
	3·064	6·368	−3·304	−99·12		−28·44	−853			
7					−78·00			9364	−24	9340
	3·237	4·929	−1·692	−50·76		−103·38	−3101			
					−128·76			6263	−28	6235
	3·489	2·445	1·044	31·32		−113·10	−3393			
9					−97·44			2870	−32	2838
	2·220	0·498	1·722	51·66		−71·61	−2148			
10					−45·78			722	−36	686
	1·645	0·106	1·539	46·17		−22·70	−681			
11					0·39			41	−40	1

Maximum shearing force = 128·76 MN
Maximum bending moment = 10,220 MN m (by plotting column k)

$$\text{Keel stress} = \frac{10{,}220\ \text{MN m}}{752\ \text{m}^4} \times 9{\cdot}3\ \text{m} = 126{\cdot}5\ \text{N/mm}^2\ (8{\cdot}2\ \text{tonf/in}^2)$$

$$\text{Deck stress} = \frac{10{,}220}{752} \times 9{\cdot}7 = 132\ \text{N/mm}^2\ (8{\cdot}6\ \text{tonf/in}^2)$$

$$\text{Slope} = \frac{1}{EI}\int M\,\mathrm{d}x = \frac{\mathrm{d}y}{\mathrm{d}x} = \frac{10^{-6}}{0{\cdot}208 \times 752}\int M\,\mathrm{d}x \quad \text{and} \quad y = \int \frac{\mathrm{d}y}{\mathrm{d}x}\cdot \mathrm{d}x$$

Columns e and h are not strictly necessary to this table since the multiplier 30 could have been applied at the end; they have been included merely to give the student an idea of the values expressed in meganewtons.

a Ord	b BM (MN m)	c mid BM	d Σc (MN m^2)	e mid d	f Σe	g $y = 5{\cdot}75$f ($\times 10^{-6}$)	h chord	i h − g
1	0		0		0	0	0	
		195		98				
2	390		195		98			
		1138		764				
3	1886		1333		862			
		3301		2984				
4	4716		4634		3846			
		6439		7854				
5	8162		11,073		11,700	0·066	0·484	0·418
		9180		15,663				
6	10,197		20,253		27,363	0·154	0·605	0·451
		9769		25,138				

(*continued on p. 185*)

(*continued from p. 184*)

a Ord	b BM (MN m)	c mid BM	d Σc (MN m²)	e mid d	f Σe	g $y = 5{\cdot}75$f ($\times 10^{-6}$)	h chord	i h−g
7	9340		30,022		52,501	0·296	0·726	0·430
		7788		33,916				
8	6235		37,810		86,417			
		4537		40,079				
9	2838		42,347		126,496			
		1762		43,228				
10	686		44,109		169,724			
		343		44,280				
11	1		44,452		214,004	1·209	1·209	

$$\text{Difference in slope 4–10 Ords.} = \frac{10^{-6}}{0{\cdot}208 \times 752}(44{,}109 - 4634) \times 30 = 0{\cdot}0076 \text{ radians}$$

$$= 0{\cdot}0076 \times \frac{180}{\pi} \times 60 = 26 \text{ minutes of arc}$$

$$\text{Deflection } y = \frac{30 \times 30 \times 10^{-6}}{0{\cdot}208 \times 752}\text{f} = 5{\cdot}75\text{f} \times 10^{-6}$$

Central deflection = 0·451 m

Column h is necessary because, as shown in Fig. 6.11 by integrating from no. 1 ord., zero slope has been assumed there; to obtain deflection relative to the ends, the chord deflection must be subtracted.

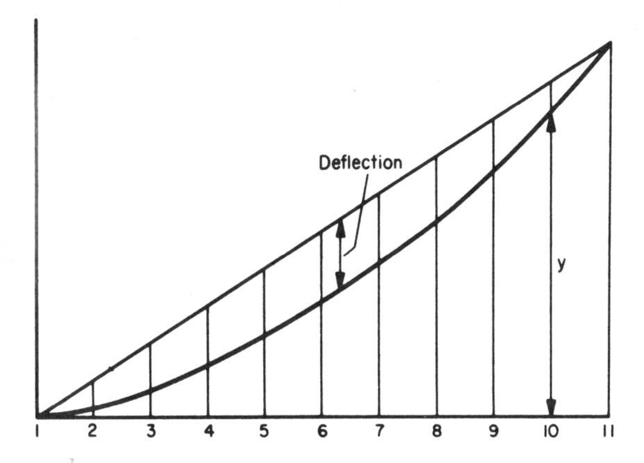

Fig. 6.11

SECOND MOMENT OF AREA

It is now necessary to calculate the second moment of area, I, of the section of the ship girder. For a simple girder this is a straightforward matter. For a ship girder composed of plates and sections which may not extend far longitudinally or which might buckle under compressive load or which may be composed of differing materials, a certain amount of adjustment may be judged necessary. Unless there is some good reason to the contrary (such as inadequate jointing in a wood sheathed or armoured deck), it is assumed here that local structural design has been sufficient to prevent buckling or tripping or other forms of load shirking. In fact, load shirking by panels having a width/thickness ratio in excess of seventy is likely and plating contribution should be limited to seventy times the thickness (see Chapter 7). No material is assumed to contribute to the section modulus unless it is structurally continuous for at least half the length of the ship amidships. Differing materials are allowed for in the manner described on the next page. In deciding finally on which parts of the cross section to include, reference should be made to the corresponding assumptions made for the designs with which final stress comparisons are made. Similar comparisons are made also about members which shirk their load.

Having decided which material to include, the section modulus is calculated in a methodical, tabular form. An *assumed neutral axis* (ANA) is first taken near the mid-depth. Positions and dimensions of each item forming the structural mid-section are then measured and inserted into a table of the following type:

Table 6.1
Modulus calculation

1	2	3	4	5	6	7
Item	A (in^2 or cm^2)	h (ft or m)	Ah (in^2 ft or cm^2 m)	Ah^2 (in^2 ft^2 or cm^2 m^2)	k^2 (ft^2 or m^2)	Ak^2 (in^2 ft^2 or cm^2 m^2)
Each item above ANA						
Totals above ANA	ΣA_1		$\Sigma A_1 h_1$	$\Sigma A_1 h_1^2$		$\Sigma A_1 k_1^2$
Each item below ANA						
Totals below ANA	ΣA_2		$\Sigma A_2 h_2$	$\Sigma A_2 h_2^2$		$\Sigma A_2 k_2^2$

where A = cross-sectional area of item, h = distance from ANA, k = radius of gyration of the structural element about its own NA.

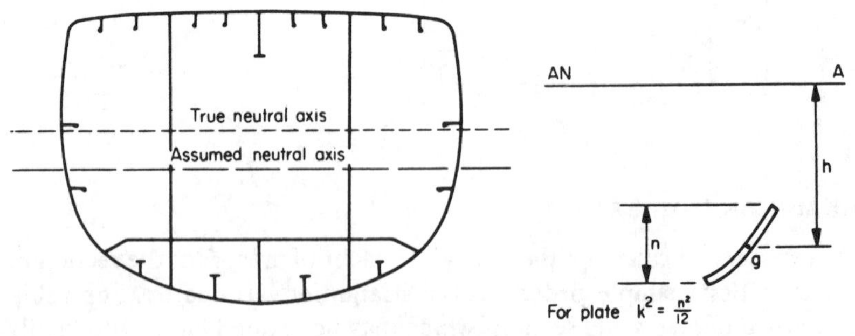

Fig. 6.12

Subscript 1 is used to denote material above the ANA and subscript 2 for material below.

$$\text{Distance of true NA above ANA} = \frac{\Sigma A_1 h_1 - \Sigma A_2 h_2}{\Sigma A_1 + \Sigma A_2} = d$$

Second moment of area about true NA

$= \Sigma A_1 h_1^2 + \Sigma A_2 h_2^2 + \Sigma A_1 k_1^2 + \Sigma A_2 k_2^2 - (\Sigma A_1 + \Sigma A_2) d^2 = I$

Lever above true neutral axis from NA to deck at centre $= y_D$

Lever below true neutral axis from NA to keel $= y_K$

The strength section of some ships is composed of different materials, steel, light alloy, wood or plastic. How is the second moment of area calculated for these composite sections? Consider a simple beam composed of two materials, suffixes a and b. From the theory of beams, it is known that the stress is directly proportional to the distance from the neutral axis and that, if R is the radius of curvature of the neutral axis and E is the elastic modulus,

$$\text{stress } \sigma = \frac{E}{R} h$$

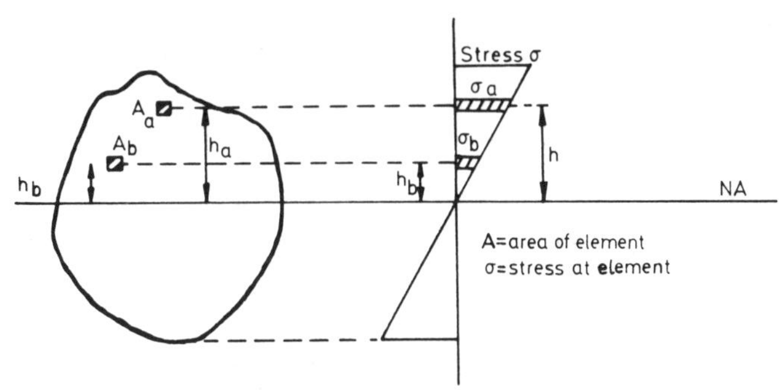

Fig. 6.13 Composite section

Consider the typical element of area A of Fig. 6.13. For equilibrium of the cross-section, the net force must be zero, therefore

$$\Sigma(\sigma_a A_a + \sigma_b A_b) = 0$$

$$\Sigma\left(\frac{E_a}{R} A_a h_a + \frac{E_b}{R} A_b h_b\right) = 0$$

i.e.

$$\Sigma\left(A_a h_a + \frac{E_b}{E_a} A_b h_b\right) = 0$$

Now the contribution to the bending moment by the element A is the force σA multiplied by the distance from the neutral axis, h. For the whole section,

$$M = \Sigma(\sigma_a A_a h_a + \sigma_b A_b h_b)$$

$$= \frac{E_a}{R} \Sigma\left(A_a h_a^2 + \frac{E_b}{E_a} A_b h_b^2\right)$$

$$= \frac{E_a}{R} \cdot I_{eff} \cdot$$

where I_{eff} is the effective second moment of area.

It follows from the equations above that the composite section may be assumed composed wholly of material a, provided that an effective area of material b, $(E_b/E_a)A_b$, be used instead of its actual area. The areas of material b used in columns 2, 4, 5 and 7 of Table 6.1 must then be the actual areas multiplied by the ratio of the elastic moduli. The ratio E_b/E_a for different steels is approximately unity; for wood/steel it is about $\frac{1}{16}$ in compression and $\frac{1}{25}$ in tension. For aluminium alloy/steel it is about $\frac{1}{3}$ and for glass reinforced plastic/steel it is between $\frac{1}{15}$ and $\frac{1}{30}$. The figures vary with the precise alloys or mixtures and should be checked.

BENDING STRESSES

Each of the constituents of the equation, $\sigma = (M/I)y$, has now been calculated, maximum bending moment M, second moment of area, I, and the levers y for two separate conditions, hogging and sagging. Values of the maximum direct stress at deck and keel arising from the standard comparative longitudinal strength calculation can thus be found.

Direct stresses occurring in composite sections are, following the work of the previous section

$$\text{in material a,} \quad \sigma_a = \frac{E_a y}{R} = \frac{M}{I_{\text{eff}}} y$$

$$\text{in material b,} \quad \sigma_b = \frac{E_b y}{R} = \frac{My}{I_{\text{eff}}} \frac{E_b}{E_a}$$

It is at this stage that the comparative nature of the calculation is apparent, since it is now necessary to decide whether the stresses obtained are acceptable or whether the strength section needs to be modified. It would, in fact, be rare good fortune if this was unnecessary, although it is not often that the balance has to be repeated in consequence. The judgment on the acceptable level of stress is based on a comparison with similar ships in similar service for which a similar calculation has previously been performed. This last point needs careful checking to ensure that the same wave has been used and that the same assumptions regarding inclusion of material in weight and modulus calculations are made. For example, a device was adopted many years ago to allow for the presence of rivet holes, either by decreasing the effective area of section on the tension side by $\frac{2}{11}$ or by increasing the tensile stress in the ratio $\frac{11}{9}$. Although this had long been known to be an erroneous procedure, some authorities continued to use it in order to maintain the basis for comparison for many years.

Stresses different from those found acceptable for the type ships may be considered on the following bases:

(*a*) Length of ship. An increase of acceptable stress with length is customary on the grounds that standard waves at greater lengths are less likely to be met. This is more necessary with the $L/20$ wave than the $1{\cdot}1\sqrt{L}$ wave for which the probability varies less with length. If a standard thickness is allowed for corrosion, it will constitute a smaller proportion of the modulus for larger ships.

(*b*) Life of ship. Corrosion allowance is an important and hidden factor. Classification societies demand extensive renewals when survey shows the plating thicknesses and modulus of section to be appreciably reduced. It could well be economical to accept low initial stresses to postpone the likely time of renewal. Comparison with type ship ought to be made before corrosion allowance is added and the latter assessed by examining the performance of modern paints and anti-corrosive systems (see Chapter 14).
(*c*) Conditions of service. Classification societies permit a reduced modulus of section for service in the Great Lakes or in coastal waters. Warship authorities must consider the likelihood of action damage by future weapons and the allowance to be made in consequence. Warships are not, of course, restricted to an owner's route.
(*d*) Local structural design. An improvement in the buckling behaviour or design at discontinuities may enable higher overall stress to be accepted.
(*e*) Material. Modern high grade steels permit higher working stresses and the classification societies encourage their use, subject to certain provisos.
(*f*) Progress. A designer is never satisfied; a structural design which is entirely successful suggests that it was not entirely efficient in the use of materials and he is tempted to permit higher stresses next time. Such progress must necessarily be cautiously slow.

A fuller discussion of the nature of failure and the aim of the designer of the future occurs later in the chapter. Approximate values of total stress which have been found satisfactory in the past are given below:

Ships	Wave	Design stresses Deck		Keel	
		tonf/in^2	N/mm^2	tonf/in^2	N/mm^2
100 m frigate	$L/20$	7	110	6	90
150 m destroyer	$L/20$	8	125	7	110
200 m general cargo vessel	$1{\cdot}1\sqrt{L}$	7	110	6	90
250 m aircraft carrier	$L/20$	9	140	8	125
300 m oil tanker	$1{\cdot}1\sqrt{L}$	9	140	8	125

SHEAR STRESSES

The shearing force at any position of the ship's length is that force which tends to move one part of the ship vertically relative to the adjacent portion. It tends to distort square areas of the sides into rhomboids. The force is distributed over the section, each piece of material contributing to the total. It is convenient to consider shear stress, the force divided by the area and this is divided over the cross-section of a simple beam according to the expression

$$\tau = \frac{SA\bar{y}}{Ib}$$

$A\bar{y}$ is the moment about the NA of that part of the cross-section above section PP where the shear stress τ is required. I is the second moment of area of the

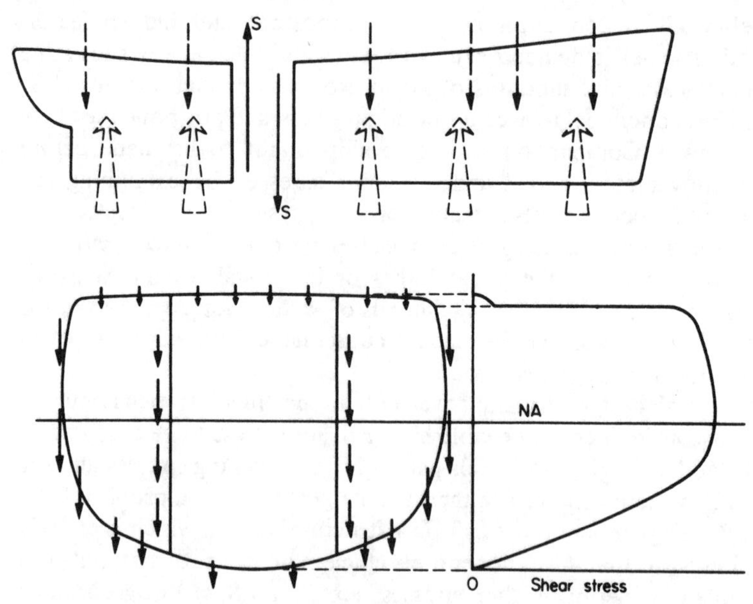

Fig. 6.14 Shear loading on a ship girder

whole cross-section and b is the total width of material at section P (Fig. 6.15). The distribution of shear stress over a typical cross-section of a ship is shown in Fig. 6.14. The maximum shear stress occurs at the neutral axis at those points along the length where the shearing force is a maximum. A more accurate distribution would be given by shear flow theory applied to a hollow box girder, discussed presently.

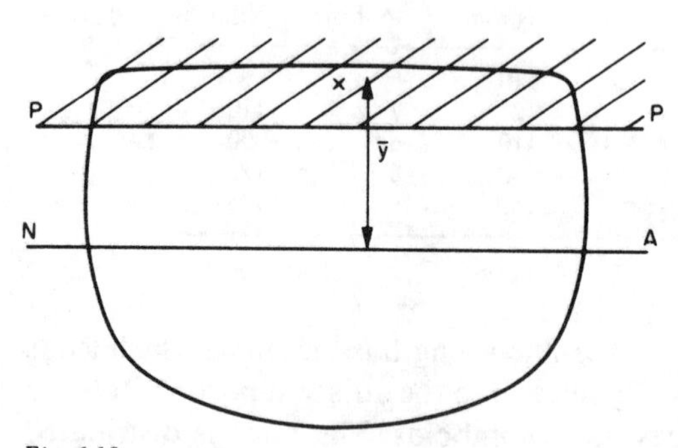

Fig. 6.15

Acceptable values of shear stress depend on the particular type of side construction. Failure under the action of shear stress would normally comprise wrinkling of panels of plating diagonally. The stress at which this occurs depends on the panel dimensions, so while the shear stress arising from the standard longitudinal strength calculation clearly affects side plating thickness and stiffener spacings, it does not have a profound effect on the structural cross-section of the ship.

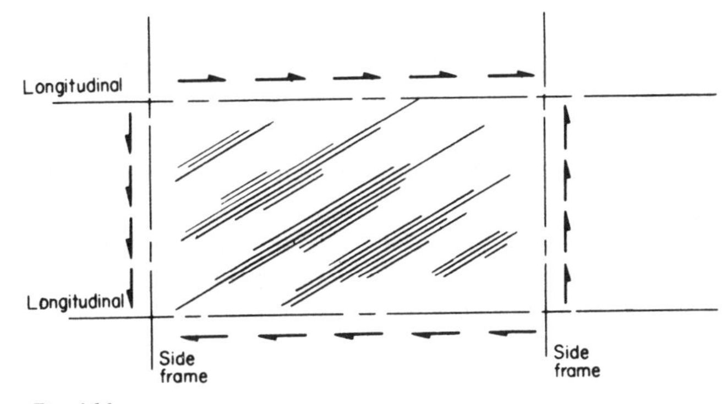

Fig. 6.16

Already, there is a tendency to assume that the stresses obtained from the standard comparative calculation actually occur in practice, so that the design of local structure may be effected using these stresses. Because local structure affected by the ship girder stresses cannot otherwise be designed, there is no choice. Strictly, this makes the local structure comparative. In fact, the stresses obtained by the comparative calculation for a given wave profile have been shown by full scale measurements to err on the safe side so that their use involves a small safety factor. Often, other local loading is more critical. This is discussed more fully under 'Criterion of failure.'

INFLUENCE LINES

The ship will not often, even approximately, be in the condition assumed for the standard calculation. It is important for designers and operators to know at a glance, the effect of the addition or removal of weight on the longitudinal strength. Having completed the standard calculation, the effects of small additions of weight are plotted as influence lines in much the same way as for bridges and buildings. (See Fig. 6.17.)

An influence line shows the effect on the *maximum* bending moment of the addition of a unit weight at any position along the length. The height of the line at P represents the effect on the maximum bending moment at X of the addition

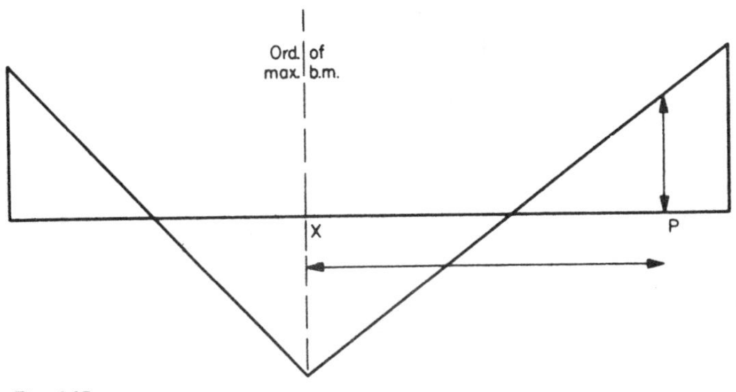

Fig. 6.17

of a unit weight at P. Two influence lines are normally drawn, one for the hogging and one for the sagging condition. Influence lines could, of course, be drawn to show the effect of additions on sections other than that of the maximum BM but are not generally of much interest.

Consider the addition of a weight P, x aft of amidships, to a ship for which the hogging calculation has been performed as shown in Fig. 6.18. It will cause a parallel sinkage s and a change of trim t over the total length L. If A and I are the area and least longitudinal second moment of area of the curved waterplane and the ship is in salt water of reciprocal weight density u, then, approximately,

$$\text{sinkage } s = \frac{uP}{A} \quad \text{and} \quad \text{trim } t = \frac{P(x-f)uL}{I}$$

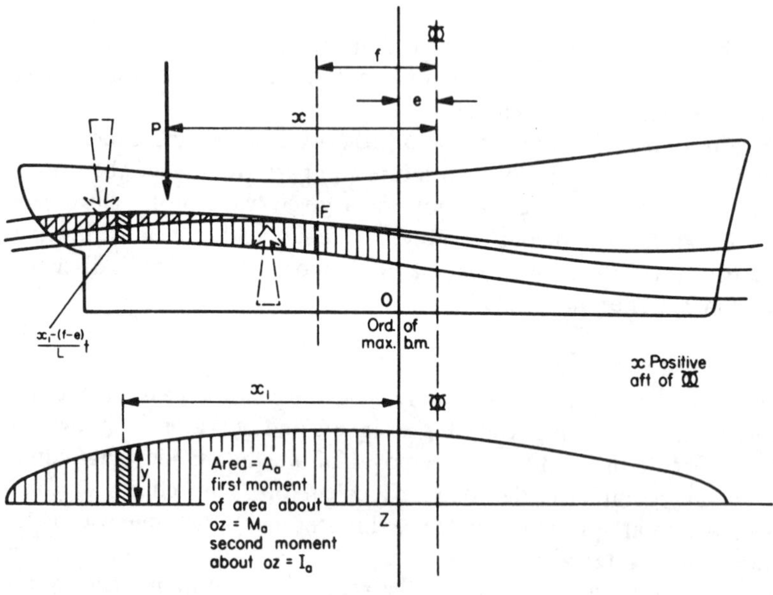

Fig. 6.18

The increase in bending moment at OZ is due to the moment of added weight less the moment of buoyancy of the wedges aft of OZ less the moment of buoyancy of the parallel sinkage aft of OZ.

$$\text{Moment of buoyancy of parallel sinkage about OZ} = \frac{sM_a}{u} = \frac{PM_a}{A}$$

Moment of buoyancy of wedges

$$= \frac{1}{u}\int 2yx_1 \times \frac{(x_1 - \overline{f-e})}{L} t \, dx_1$$

$$= \frac{I_a t}{uL} - \frac{M_a(f-e)t}{uL} = \{I_a - M_a(f-e)\}\frac{P(x-f)}{I}$$

Moment of weight which is included only if the weight is aft of OZ and therefore only if positive

$= P[x-e]$, positive values only.

Increase in BM for the addition of a unit weight,

$$\frac{\delta M}{P} = -\{I_a - M_a(f-e)\}\frac{(x-f)}{I} - \frac{M_a}{A} + [x-e]$$

Note that this expression is suitable for negative values of x (i.e. for P forward of amidships), provided that the expression in square brackets, [ ], is discarded if negative. A discontinuity occurs at OZ, the ordinate of maximum bending moment. The influence lines are straight lines which cut the axis at points about 0·2–0·25 of the length from amidships. It is within this length, therefore, that weights should be removed to aggravate the hogging condition and outside this length that they should be removed to aggravate the sagging condition.

Certain simplifying assumptions are sometimes made to produce a simpler form of this equation. There is little point in doing this because no less work results. However, if $(f-e)$ is assumed negligible, n put equal to I_a/I and F put equal to M_a/A

$$\frac{\delta M}{P} = -n(x-f) - F + [x-e]$$

This gives the results shown in Fig. 6.19.

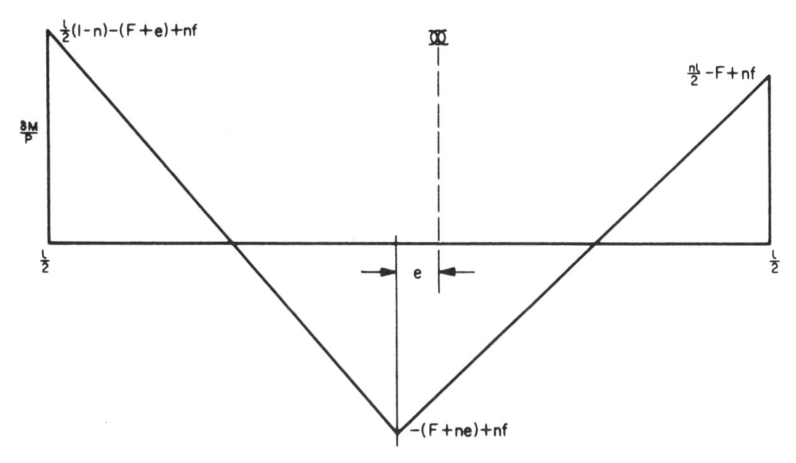

Fig. 6.19

A note of caution is necessary in the use of influence lines. They are intended for small weight changes and are quite unsuitable for large changes such as might occur with cargo, for example. In general, weight changes large enough to make a substantial change in stress are too large for the accuracy of influence lines.

CHANGES TO SECTION MODULUS

Being a trial and error process, the standard calculation will rarely yield a suitable solution first time. It will almost always be necessary to return to the structural section of the ship to add or subtract material in order to adjust the resulting stress. The effect of such an addition, on the modulus I/y, is by no means obvious. Moreover, it is not obvious in the early stages, whether the superstructure should be included in the strength section or whether the construction should be such as to discourage a contribution from the superstructure. Some measure of the effect of adding material to the strength section is needed if we are to avoid calculating a new modulus for each trial addition.

In order to provide such a measure, consider (Fig. 6.20) the addition of an area a at a height z above the neutral axis of a structural section whose second moment of area has been calculated to be Ak^2, area A, radius of gyration k and levers y_1 to the keel and y_2 to the deck. Now

$$\text{stress } \sigma = \frac{M}{I/y}$$

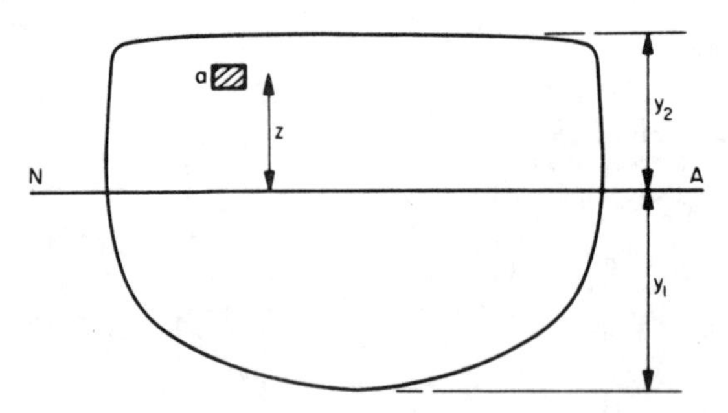

Fig. 6.20

By considering changes to I and y it can be shown that for a given bending moment, M, stress σ will obviously be reduced at the deck and will be reduced at the keel if $z > k^2/y_1$, when material is added within the section, $z < y_2$.

If the material is added above the deck, $z > y_2$ then the maximum stress occurs in the new material and there will be a reduction in stress above the neutral axis (Fig. 6.21) if

$$a > \frac{A(z/y_2 - 1)}{(z^2/k^2 + 1)}$$

The position is not quite so simple if the added material is a superstructure and attention is drawn to p. 224.

EXAMPLE 2. In converting a steel survey ship, it is proposed to extend the short forecastle for the whole length of the ship and to arrange the structure so that it

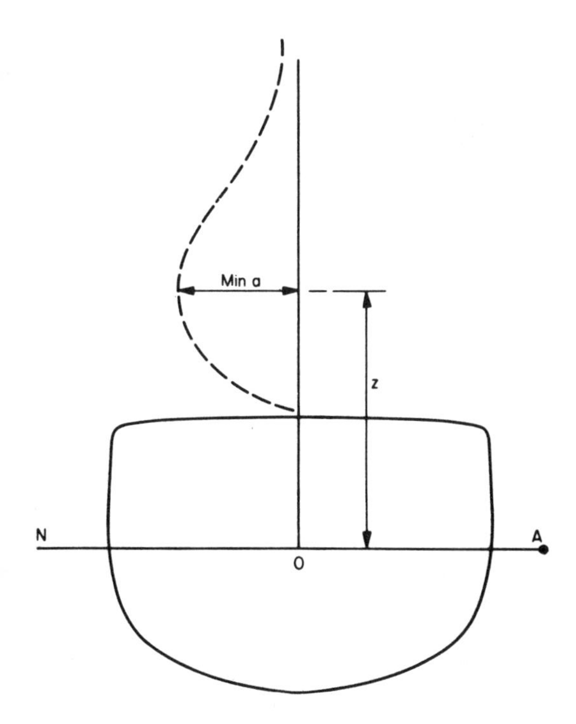

Fig. 6.21

contributes 100 per cent to the hull girder. The new structure is wholly of light alloy. Estimate the new nominal stresses due to the change in section modulus, assuming that the bending moment remains unchanged.

Before conversion: BM = 25,000 tonf ft, I = 40,000 in^2 ft^2, A = 700 in^2, y, deck = 9·5 ft; y, keel = 10·0 ft.

Added structure: side plating 7·5 ft × $\frac{3}{8}$ in., stiffened by one 4·0 in^2 girder at mid-height; deck plating 36 ft × $\frac{3}{8}$ in. stiffened by five 4·0 in^2 girders with centre of area 3 in. below the deck.

E, light alloy: 10×10^6 lbf/in^2
E, steel: 30×10^6 lbf/in^2

Above upper deck

Item	A (in^2)	h (ft)	Ah (in^2 ft)	Ah^2 (in^2 ft^2)	k^2 (ft^2)	Ak^2 (in^2 ft^2)
two sides	67·5	3·75	253	949	4·7	318
two side girders	8·0	3·75	30	113	0	0
deck	162·0	7·5	1215	9113	0	0
five deck girders	20·0	7·25	145	1051	0	0
	257·5	6·38	1643	11,226		318

Solution:

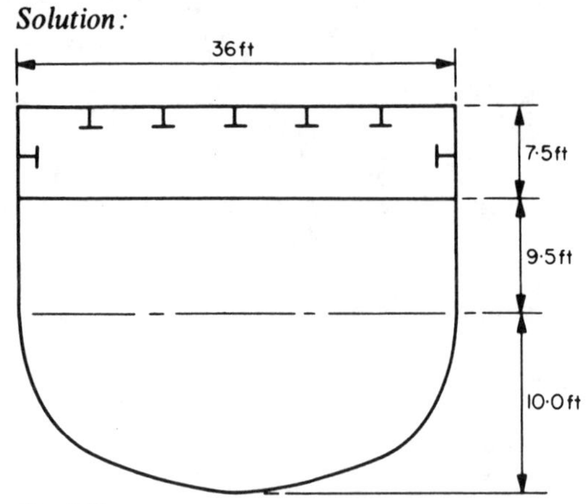

Fig. 6.22

$$\text{New effective area} = 700 + \tfrac{1}{3} \times 257{\cdot}5 = 785{\cdot}8 \text{ in}^2$$

$$\text{Movement of neutral axis} = \frac{\frac{1}{3} \times 257{\cdot}5(9{\cdot}5 + 6{\cdot}38)}{785{\cdot}8} = 1{\cdot}73 \text{ ft}$$

$$\begin{aligned}
I \text{ about old NA} &= 40{,}000 + \tfrac{1}{3}\{11{,}226 + 318 - 257{\cdot}5(6{\cdot}38)^2 + 257{\cdot}5(6{\cdot}38 + 9{\cdot}5)^2\} \\
&= 62{,}009 \text{ in}^2 \text{ ft}^2 \\
I \text{ about new NA} &= 62{,}009 - 785{\cdot}8(1{\cdot}73)^2 = 59{,}652 \text{ in}^2 \text{ ft}^2 \\
y, \text{ new deck} &= 15{\cdot}27 \text{ ft} \\
y, \text{ keel} &= 11{\cdot}73 \text{ ft}
\end{aligned}$$

$$\text{Old stresses were, } \sigma, \text{ keel} = \frac{25{,}000}{40{,}000} \times 10{\cdot}0 = 6{\cdot}25 \text{ tonf/in}^2 \text{ (steel)}$$

$$\sigma, \text{ deck} = \frac{25{,}000}{40{,}000} \times 9{\cdot}5 = 5{\cdot}94 \text{ tonf/in}^2 \text{ (steel)}$$

$$\text{New stresses are, } \sigma, \text{ keel} = \frac{25{,}000}{59{,}650} \times 11{\cdot}73 = 4{\cdot}92 \text{ tonf/in}^2 \text{ (steel)}$$

$$\sigma, \text{ deck} = \frac{1}{3} \times \frac{25{,}000}{59{,}650} \times 15{\cdot}27 = 2{\cdot}13 \text{ tonf/in}^2 \text{ (alloy)}$$

$$\sigma, \text{ deck} = \frac{25{,}000}{59{,}650} \times 7{\cdot}77 = 3{\cdot}26 \text{ tonf/in}^2 \text{ (steel)}$$

SLOPES AND DEFLECTIONS

The bending moment on the ship girder has been found by integrating first the loading p' with respect to length to give shearing force S, and then by integrating the shearing force S with respect to length to give M. There is a further relationship with which students will already be familiar.

$$M = EI\frac{d^2y}{dx^2}$$

or

$$\frac{dy}{dx} = \frac{1}{E}\int\frac{M}{I}dx \quad \text{and} \quad y = \int\frac{dy}{dx}dx$$

Thus, the slope at all points along the hull can be found by integrating the M/EI curve and the bent shape of the hull profile can be found by integrating the slope curve. It is not surprising that the errors involved in integrating approximate data four times can be quite large. Moreover, the calculation of the second moment of area of all sections throughout the length is laborious and the standard calculation is extended this far only rarely—it might be thus extended, for example, to give a first estimate of the distortion of the hull between a master datum level and a radar aerial some distance away. It may also be done to obtain a deflected profile in vibration studies.

HORIZONTAL FLEXURE

Flexure perpendicular to the plane with which we are normally concerned may be caused by vibration, flutter, uneven lateral forces or bending while rolling. Vibrational modes are discussed in Chapter 9. Unsymmetrical bending can be resolved into bending about the two principal axes of the cross-section. The only point in the ship at which the maxima of the two effects combine is the deck edge and if the ship were to be balanced on a standard wave which gave a bending moment M, the stress at the deck edge would be

$$\sigma = \frac{Mz}{I_{yy}}\cos\theta + \frac{My}{I_{zz}}\sin\theta$$

However, it is not quite so easy as this; if the ship were heading directly into a wave train, it would not be rolling. Only in quartering or bow seas will bending

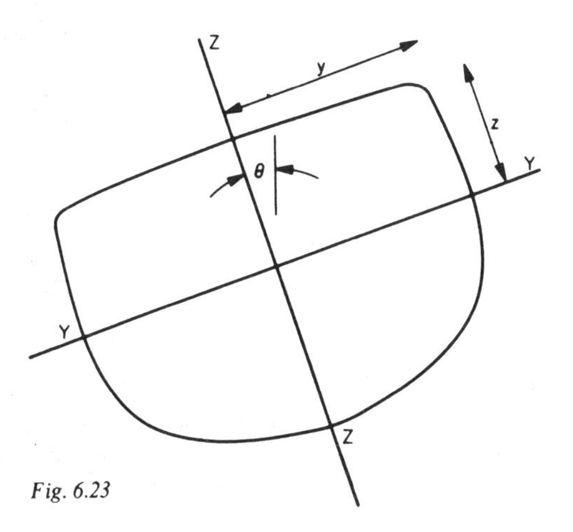

Fig. 6.23

and rolling be combined and, in such a case, the maximum bending moments in the two planes would not be in phase, so that the deck edge effect will be less than that obtained by superposition of the two maxima. Some limited research into this problem indicates that horizontal bending moment maxima are likely to be of the order of 40 per cent of the vertical bending moment maxima and the ratio of the stresses is likely to be about 35 per cent in common ship shapes. The increase in deck edge stress over that obtained with head seas is thought to be of the order of 20–25 per cent. This is an excellent reason to avoid stress concentrations in this area.

BEHAVIOUR OF A HOLLOW BOX GIRDER

The simple theory of bending of beams assumes that plane sections of the beam remain plane and that the direct stress is directly proportional to the distance from the neutral axis. A deck parallel to the neutral axis would therefore be expected to exhibit constant stress across its width. In fact, the upper flange of a hollow box girder like a ship's hull can receive its load only by shear at the deck edge. The diffusion of this shear into a plane deck to create the direct stresses is a difficult problem mathematically; the diffusion from the edge elements across the deck is such that the plane sections do not remain plane, and the mathematics shows that the direct stresses reduce towards the middle of the deck. This has been borne out by observations in practice. The effect is known as *shear lag* and its magnitude depends on the type of loading and dimensions of the ship; it is more pronounced, for example, under concentrated loads. In common ship shapes, it accounts for a difference in stress level at mid-deck of only a few per cent. It is more important in the consideration of the effects of superstructures and of the effective breadth of plating in local strength problems which will be discussed later.

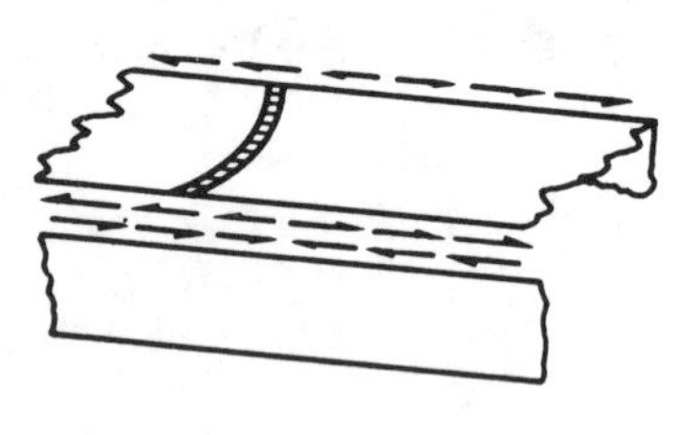

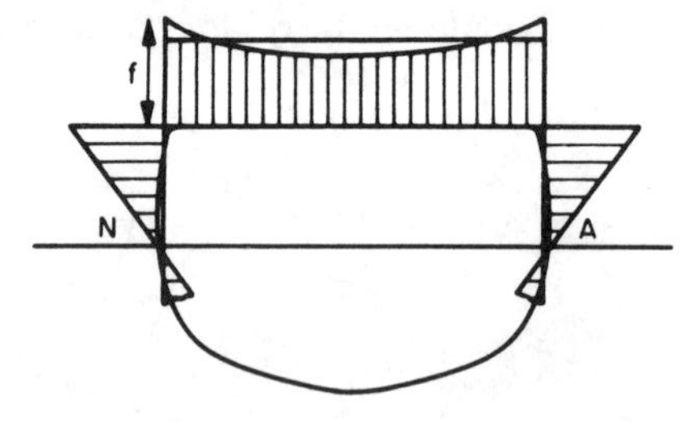

Fig. 6.24 Direct stresses in practice

WAVE PRESSURE CORRECTION

This is usually known as the *Smith Correction* (Ref. 1). In calculating the buoyancy per unit length at a section of the ship, the area of the section given by the Bonjean curve cut by the wave surface was taken. This assumes that the pressure at a point P on the section is proportional to the depth h of the point below the wave surface and the buoyancy $= w \int h \, dB = w \times$ area immersed (Fig. 6.25).

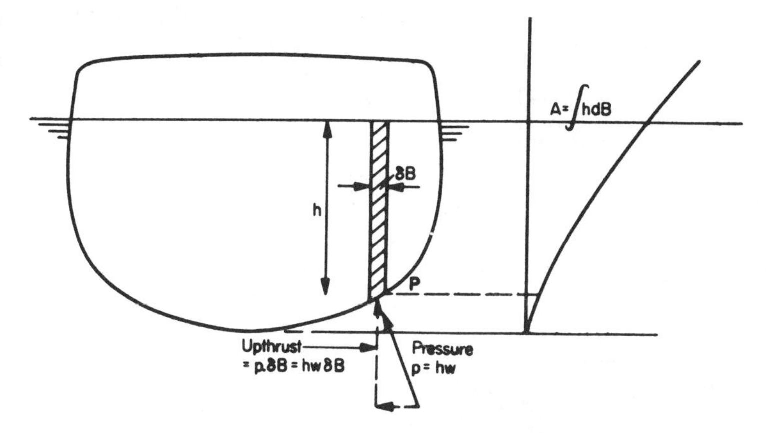

Fig. 6.25

It is shown in Chapter 9, that this simple hydrostatic law is not true in a wave because the wave is caused by the orbital motion of water particles. It can be shown that the pressure at a point P in a wave at a depth h below the wave surface is the same as the hydrostatic pressure at a depth h', where h' is the distance between the mean (or still water) axis of the surface trochoid and the subsurface trochoid through P (Fig. 6.26). Thus, the buoyancy at a section of a ship is

$$w \int h' \mathrm{d}B = w \times (\text{effective area immersed}).$$

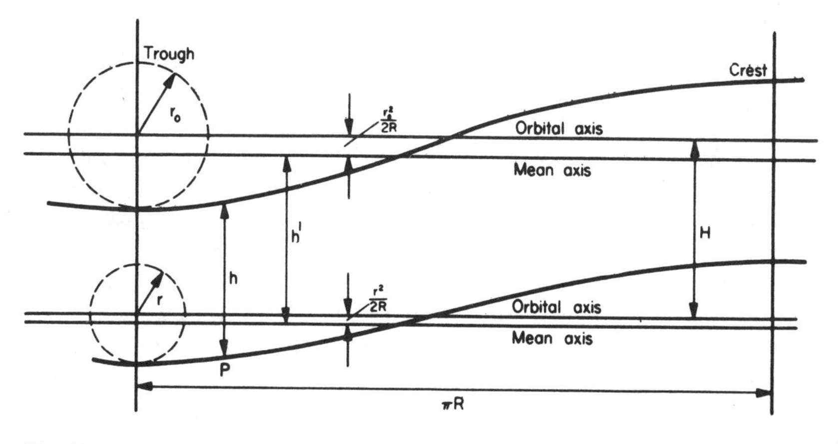

Fig. 6.26

Setting up h' from the sub-surface trochoid through P and similar heights from a series of sub-surface trochoids drawn on the wave profile, the effective immersed area up to W'L is obtained. New Bonjean curves can thus be plotted (Fig. 6.27). The formula for h' is

$$h' = H - \frac{r_0^2}{2R}\left\{1 - \exp\left(-\frac{2H}{R}\right)\right\}$$

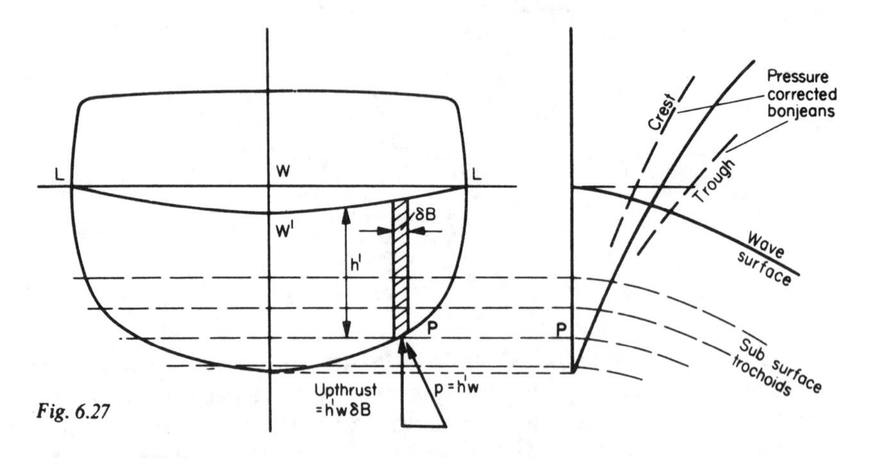

Fig. 6.27

An approximation to the Smith correction is:

$$\text{corrected wave BM} = \text{wave BM} \times \exp\left(-\frac{nT}{L}\right)$$

where T is the draught, L the length and n a constant given by:

Block coeff. C_B	n, sagging	n, hogging
0·80	5·5	6·0
0·60	5·0	5·3

LONGITUDINAL STRENGTH STANDARDS BY RULE

Until 1960 the Classification Societies prescribed the structure of merchant ships through tables of dimensions. They then changed to the definition of applied load and structural resistance by formulae. In the 1990s the major Classification Societies under the auspices of the International Association of Classification Societies (IACS) agreed a common minimum standard for the longitudinal strength of ships supported by the statistics of structural failure. There is today widespread acceptance of the principle that there is a very remote probability that load will exceed strength during the whole lifetime of a ship. This probability may be as low as 10^{-8} at which level the IACS requirement is slightly more conservative than almost every Classification Society standard.

Loading on a merchant ship is separated into two parts:

(*a*) the bending moment and shear force due to the weight of the ship and the buoyancy in still water,
(*b*) the additional effects induced by waves.

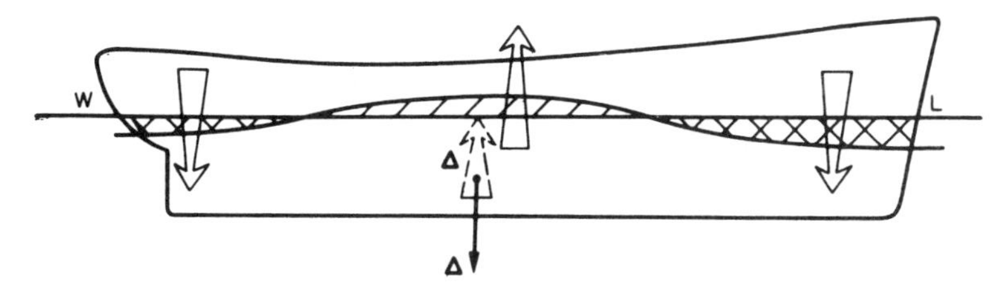

Fig. 6.28 Wave induced bending forces

Still water loading is calculated by the simple methods described at the beginning of this chapter without, of course, the wave which is replaced by the straight waterline of interest. Several such waterlines will usually be of concern. Stresses caused by such loading may be as much as 40 per cent of the total stresses allowed and incorrect acceptance of cargo or unloading of cargo has caused spectacular failures of the ship girder. With such ships as bulk carriers in particular, Masters must follow the sequences of loading and unloading recommended by their Classification Societies with scrupulous care. Not only may the ship break its back by bending but failure could be caused by the high shear forces that occur between full and empty holds. Indeed, a simple desktop computer to assess changes as they are contemplated has become highly desirable since high capacity cargo handling has evolved.

Wave induced bending moment (WIBM) is now accepted to be represented by the formulae:

$$\text{Hogging WIBM} = 0{\cdot}19MCL^2BC_b \quad \text{kN m}$$
$$\text{Sagging WIBM} = -0{\cdot}11MCL^2B(C_b + 0{\cdot}7) \quad \text{kN m}$$

where L and B are in metres and $C_b \geq 0{\cdot}6$

$$\text{and } C = 10{\cdot}75 - \left(\frac{300 - L}{100}\right)^{1{\cdot}5} \quad \text{for } 90 \leq L \leq 300 \text{ m}$$
$$= 10{\cdot}75 \quad \text{for } 300 < L < 350 \text{ m}$$
$$= 10{\cdot}75 - \left(\frac{L - 350}{150}\right)^{1{\cdot}5} \quad \text{for } 350 \leq L$$

M is a distribution factor along the length of the ship,

$M = 1{\cdot}0$ between $0{\cdot}4L$ and $0{\cdot}65L$ from the stern

$= 2{\cdot}5x/L$ at x metres from the stern up to $0{\cdot}4L$

$= 1{\cdot}0 - \dfrac{x - 0{\cdot}65L}{0{\cdot}35L}$ at x metres from the stern between $0{\cdot}65L$ and L.

Wave induced shear force is given by IACS as

$$\text{Hogging condition } S = 0{\cdot}3F_1CLB(C_b + 0{\cdot}7) \quad \text{kN}$$
$$\text{Sagging condition } S = -0{\cdot}3F_2CLB(C_b + 0{\cdot}7) \quad \text{kN}$$

where L, B and C_b are as given above and F_1 and F_2 by Fig. 6.29.

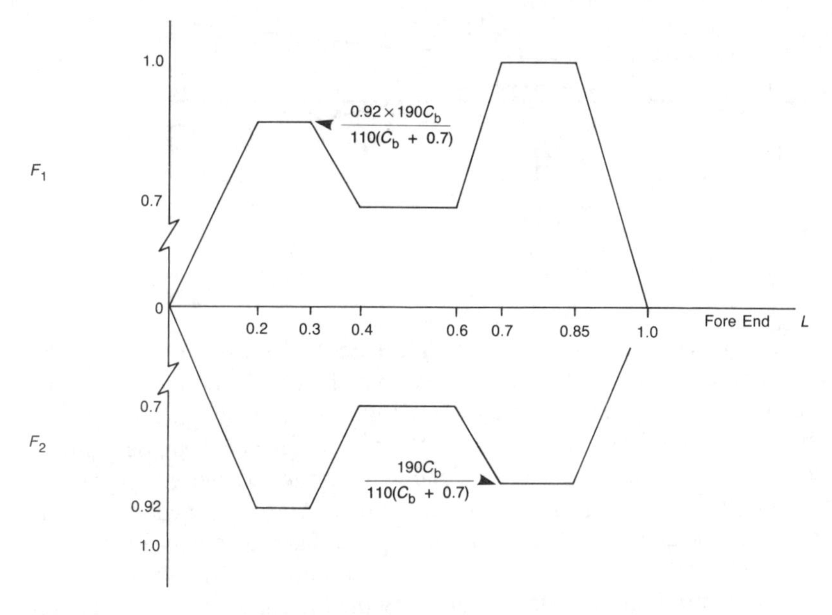

Fig. 6.29 Factors F_1 and F_2 (from Ref. 3)

The modulus of cross section amidships must be such that bending stress caused by combining still water and wave induced BM is less than 175 N/mm^2. There is the further proviso that, despite the waters in which the vessel may sail, this midship section modulus shall not be less than:

$$CL^2B(C_b + 0{\cdot}7) \quad \text{cm}^3$$

which represents a stress of 110 N/mm^2.

The maximum permissible total shear stress, wave induced plus still water is also 110 N/mm^2.

The formulae given above embrace most common ship types. Special consideration must be given to vessels with large deck openings, large flare, heated cargoes or which fall outside limits when:

$L \geq 500$ m $\quad L/B \leq 5$
$C_b < 0{\cdot}60 \quad B/D \geq 2{\cdot}5$

Classification Societies now rely upon computer programs based on the methods presently to be described in this chapter. Fundamental to this is the representation of the wave loading by wave spectra and by strip theory. Strip theory is linear and, in essence, assumes the vessel to be wallsided. Thus, wave induced loading would be the same in both hogging and sagging conditions which is known in practice not to be so. Sagging stresses are greater by about 10 per cent than hogging stresses due in part to the support at the end of the ship provided by flare. This accounts for the different coefficients given in the formulae above, providing some degree of non-linearity in the treatment.

This gratifying conformity among Classification Societies is, of course, reviewed periodically. It may be reviewed if statistics show that standards have become inadequate or if oceanographers demonstrate that world wave patterns are changing with climate. Variations remain among the Classification Societies in the more detailed aspects of structural design and in allowances made for high quality steels. Variations also occur in the stresses permitted among different ship types and with restricted service conditions.

FULL SCALE TRIALS

Full scale trials can be carried out for a variety of purposes.

(*a*) To load a hull, possibly to destruction, noting the (ultimate) load and the failure mechanisms.
(*b*) To measure the strains in a ship at sea, over a short period, in a seaway which is itself measured.
(*c*) To measure strains in ships at sea over a prolonged period of service with the crew noting any signs of failure. This is the only trial which is likely to throw light on fatigue failure.

There have been several attempts to measure the behaviour of the ship girder by loading ships and recording their behaviour at various positions by strain gauges and measurements of deflection. *Neverita* and *Newcombia*, *Clan Alpine* and *Ocean Vulcan* were subject to variations in loading by filling different tanks while they were afloat. *Wolf*, *Preston*, *Bruce* and *Albuera* were supported at the middle or the ends in dry dock and weight added until major structural failure occurred.

Short duration sea trials pose difficulty in finding rough weather and in measuring the wave system accurately. In one successful trial (Ref. 11) two frigates of significantly different design were operated in close company in waves with significant heights up to 8 metres. Electrical resistance gauges were used to record stress variations with time and the seas measured using wave buoys. Of particular interest was the slamming which excites the hull girder whipping modes. It was found that the slam transient increased the sagging bending moment much more than the hogging. In frigates whipping oscillations are damped out quite quickly but they can persist longer in some vessels.

Long term trials are of great value. They provide data that can be used in the more representative calculations discussed later. For many years now a number of RN frigates have been fitted with automatic mechanical or electrical strain gauges (Ref. 11) recording the maximum compressive and tensile deck stresses in each four hour period. Ship's position, speed and sea conditions are taken from the log. Maximum bending stresses have exceeded the $L/20$ standard values by a factor of almost two in some cases. This is not significant while the standard calculation is treated as purely comparative. It does show the difficulties of trying to extrapolate from past experience to designs using new constructional materials (e.g. GRP) or with a distinctly different operating pattern.

THE NATURE OF FAILURE

Stress has never caused any material to fail. Stress is simply a convenient measure of the material behaviour which may 'fail' in many different ways. 'Failure' of a structure might mean permanent strain, cracking, unacceptable deflection, instability, a short life or even a resonant vibration. Some of these criteria are conveniently measured in terms of stress. In defining an acceptable level of stress for a ship, what 'failure' do we have in mind?

Structural failure of the ship girder may be due to one or a combination of (*a*) Cracking, (*b*) Fatigue failure, (*c*) Instability.

It is a fact that acceptable stress levels are at present determined entirely by experience of previous ships in which there have been a large number of cracks in service. The fact that these might have been due to poor local design is at present largely disregarded, suggesting that some poor local design or workmanship somewhere in the important parts of the hull girder is inevitable.

However good the design of local structure and details might be, there is one important influence on the determination of acceptable levels of stress. The material built into the ship will have been rolled and, finally, welded. These processes necessarily distort the material so that high stresses are already built in to the structure before the cargo or sea impose any loads at all. Very little is known about these built-in stresses. Some may yield out, i.e. local, perhaps molecular, straining may take place which relieves the area at the expense of other areas. The built-in stress clearly affects fatigue life to an unknown degree. It may also cause premature buckling. Thus, with many unknowns still remaining, changes to the present practice of a stress level determined by a proliferation of cracks in previous ships, must be slow and cautious. Let us now examine this progress.

REALISTIC ASSESSMENT OF LONGITUDINAL STRENGTH

Study of the simple standard longitudinal strength calculations so far described has been necessary for several good reasons. First, it has conveyed an initial look at the problem and the many assumptions which have had to be made to derive a solution which was within the capabilities of the tools available to naval architects for almost a hundred years. Moreover, the standard has been adequate on the whole for the production of safe ship designs, provided that it was coupled with conventions and experience with similar structure which was known to have been safe. As a comparative calculation it has had a long record of success. Second, it remains a successful method for those without ready access to modern tools or, for those who do enjoy such access, it remains an extremely useful starting point for a process which like any design activity is iterative. That is to say, the structural arrangement is guessed, analysed, tested against standards of adequacy, refined, reanalysed and so on until it is found to be adequate. Analysis by the standard calculation is a useful start to a process which might be prolonged. Third, much of the argument which has been considered up to now remains valid for the new concepts which must now be presented.

Of course, the standard calculation can be performed more readily now with the help of computers. The computer, however, has permitted application of mathematics and concepts of behaviour which have not been possible to apply before. It has permitted an entirely new set of standards to replace the static wave balance and to eradicate many of the dubious assumptions on which it was based. Indeed, it is no longer necessary even to assume that the ship is statically balanced. The basis of the new methods is one of realism; of a moving ship in a seaway which is continuously changing.

During a day at sea, a ship will suffer as many as 10,000 reversals of strain and the waves causing them will be of all shapes and sizes. At any moment, the ship will be subject not to a single $L/20$ wave but a composite of many different waves. Their distribution by size can be represented by a histogram of the numbers occurring within each range of wave lengths. In other words, the waves can be represented statistically. Now the statistical distribution of waves is unlikely to remain constant for more than an hour or two by which time wind, weather and sea state will modify the statistics. Over the life of a ship such sea state changes will have evened out in some way and there will be a lifetime statistical description of the sea which will be rather different from the short-term expectation.

This provides the clue to the new approaches to longitudinal strength. What is now sought as a measure is the likelihood that particular bending moments that the sea can impose upon the ship will be exceeded. This is called the probability of exceedence and it will be different if it is assessed over one hour, four hours, one day or 25 years. What the new standard does is to ensure that there is a comfortably small probability of exceedence of that bending moment which would cause the ship to fail during its lifetime. With some 30 million or so strain reversals during a lifetime, the probability of exceedence of the ship's strength needs to be very small indeed. If the frequency distribution of applied bending moment is that given in Fig. 6.30 and the ship's strength is S, then the probability of failure is

$$p\,(\text{failure}) = \int_S^\infty f(\text{BM})\,\text{d BM}$$

At this stage we have some difficulty in proceeding. While it is absolutely necessary for the student to be aware of these revised methods, much of the mathematics is well beyond the scope of this book, although the concept is not. As a first step, the student is advised to break off from study of this chapter and proceed to consider Chapter 7 on the behaviour of elements of the structure and Chapter 9 for an introduction to the statistics of waves. Chapter 12 in Vol. 2 will provide greater detail in due course. We can then presume upon such understanding and ask only that the reader take on trust a general description of some of the other mathematics with which he will probably not yet be familiar. Thus armed we may consider in turn in terms somewhat more realistic than the single overwhelming trochoidal wave:

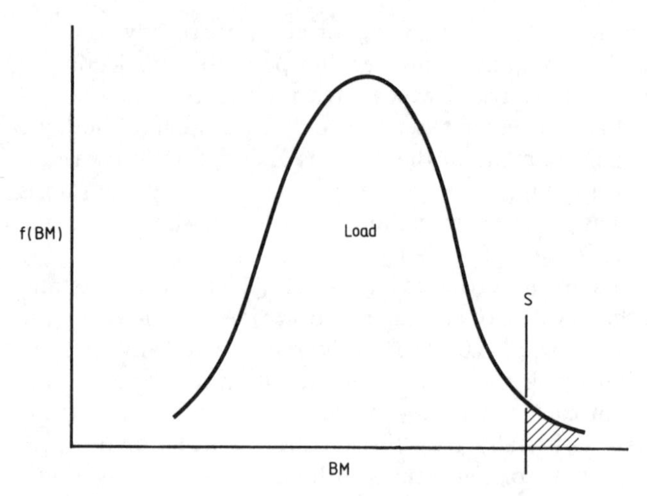

Fig. 6.30

(*a*) loading imposed upon the ship;
(*b*) response of the longitudinal structure;
(*c*) assessment of structural safety.

REALISTIC ASSESSMENT OF LOADING LONGITUDINALLY

It is necessary first to return to the effect of a single wave, not a trochoid which is too inconvenient mathematically but to a sinewave of a stated height and length or frequency. We need to examine the effect of this wave, not upon a rock-like ship presumed to be static, but upon a ship which will move, in particular to heave and to pitch. Movement $q(t)$ of any system with one degree of freedom subject to an excitation $Q(t)$ is governed by the differential equation,

$$\ddot{q}(t) + 2k\,\omega_0\dot{q}(t) + \omega_0 q = Q(t)$$

where k is the damping factor and ω_0 is the natural frequency. If now the input to the system $Q(t)$ is sinusoidal $= x\cos\omega t$, the solution to the equation, or the output, is

$$q(t) = H\,Q(t - \varepsilon)$$

where ε is a phase angle and H is called the response amplitude operator (RAO)

$$\text{RAO} = \frac{1}{\{(\omega_0^2 - \omega^2)^2 + 4k^2\omega_0^2\}^{1/2}}$$

$$= \frac{\text{amplitude of output}}{\text{amplitude of input}}$$

Both H and ε are functions of the damping factor k and the tuning factor ω/ω_0.

This form of solution is a fairly general one and applies when the input is expressed in terms of wave height and the output is the bending moment amidships. The multiplier H is of course more complex but depends upon damping, wave frequency and ship shape, heave position and pitch angle. It is called the bending moment response amplitude operator and may be calculated for a range of wave frequencies (Fig. 6.31).

The calculation of the RAOs—and indeed the heave and pitch of the ship subject to a particular wave—is these days performed by standard computer programs using some applied mathematics which is known generally as strip theory.

One aspect of theoretical hydrodynamics is the study of the shape of the streamlines of a perfect fluid flowing in some constrained manner. The mathematics may hypothesize a source or a sink of energy in the fluid represented by a potential, like the broadcast of radio waves from a transmitter. A mathematical function may represent the size of the potential and streamlines, velocities of flow and pressures may be calculated, showing the effects of such sources and sinks upon the remaining fluid. This is part of potential flow theory.

An array of sources and sinks can be devised to represent the flow past a prismatic body even when there is an air/water interface. Flow past the body, the forces on the body, the consequent movement of the body and the elevation or depression of the interface can all be calculated. A ship can be represented by a series of short prismatic bodies or strips, all joined together and the total forces and bending moments upon such a hull calculated, the different elevations representing the self-generated waves. Waves imposed upon the ship may also be represented in the same way. The potential functions needed for this representation are pulsating ones and interference occurs between waves created and imposed. These are affected by the beam of the ship relative to the wave length of the self-generated waves and this has resulted in several different approaches to strip theory using slightly different but important assumptions.

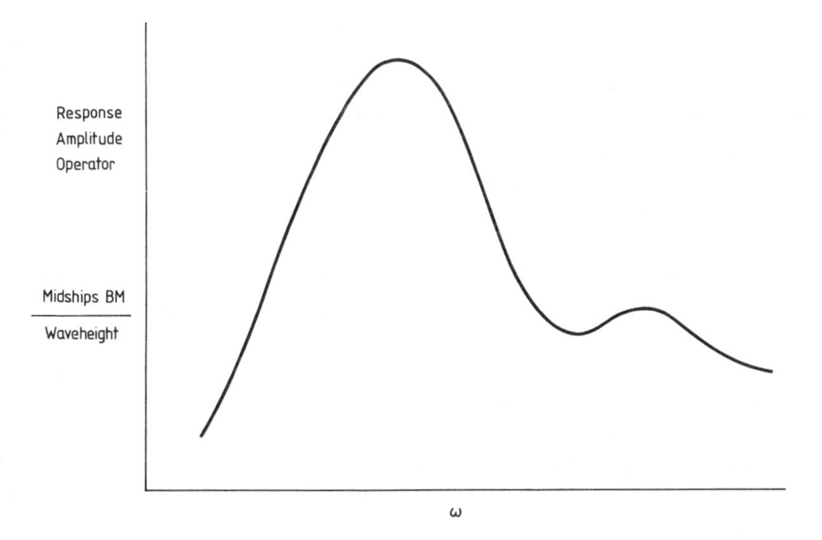

Fig. 6.31

There are other important assumptions concerned with the boundaries. The general term strip theory embraces all such studies, including slender body theory and systematic perturbation analysis.

The mathematics is based upon several assumptions, the most important of which has been that the relationship of bending moment and wave height is linear, i.e. bending moment proportional to waveheight. Another effect of the assumptions is that the mathematics makes the ship wallsided. Despite these obviously incorrect assumptions, the results from strip theory are remarkably accurate and with steady refinements to the theory will improve further.

It is now necessary to combine the effect of all of those waves which constitute the sea as oceanographers have statistically described. For a short time ahead they have found that each maximum hog or sag bending moment can be well represented by a Rayleigh distribution

$$p\,(\mathrm{BM_{max}}) = \frac{\mathrm{BM_{max}}}{m_0} \exp\left(-\frac{\mathrm{BM^2_{max}}}{2m_0}\right)$$

where m_0 is the total energy in the ship response which is the mean square of the response

$$m_0 = \int_0^{\infty} \int_{-\pi}^{\pi} \mathrm{RAO}^2\, S(\omega, \theta)\, \mathrm{d}\theta\, \mathrm{d}\omega$$

S being the total energy of the waves in the direction θ. The probability that $\mathrm{BM_{max}}$ will exceed a value B in any one cycle is given by

$$p\,(\mathrm{BM_{max}} > B) = \exp\left(-\frac{B^2}{2m_0}\right)$$

or the probability of exceedence in n cycles

$$p\,(\mathrm{BM_{max}} > B, n) = 1 - \{1 - \exp\,(-\,B^2/2m_0)\}^n$$

This illustrates the general approach to the problem. It has been necessary to define $S\,(\omega, \theta)$, the spectrum of wave energy which the ship may meet in the succeeding hour or two. Over the entire life of the ship, it may be expected to meet every possible combination of wave height and frequency coming from every direction. Such long-term statistics are described by a two-parameter spectrum agreed by the ISSC which varies slightly for different regions of the world. The procedure for determining the probability of exceedence is a little more complicated than described above but the presentation of the results is similar.

Figure 6.32 shows such a result. At any given probability of exceedence the short time ahead is predicted by a Rayleigh or a Gaussian distribution based on the mean value of the sea characteristics pertaining.

Other approaches to the problem are possible. Some authorities, for example, use as a standard the likelihood that a particular bending moment will be exceeded during any period of one hour (or four hours) during the life of the

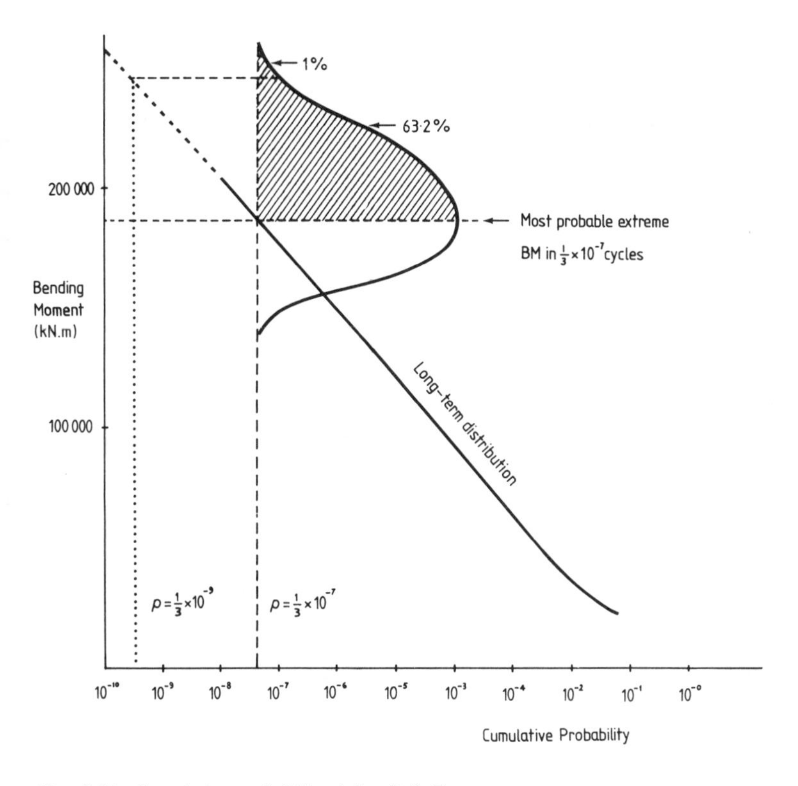

Fig. 6.32 Cumulative probability (after Ref. 7)

ship. This has the advantage of being conveniently compared with measurement of statistical strain gauges installed in ships which record the maximum experienced every hour (or four hours) (Ref. 11).

Figure 6.33 is such a plot on a log logscale which shows linear results that can be easily extrapolated.

We may now capitalize upon the study we made of the simple trochoidal wave. The effective wave height H_e is defined as that trochoidal wave of ship length which by the static standard wave calculation (without the Smith correction) gives the same wave bending moment as the worst that the ship would experience during its lifetime. That which appears to fit frigates very well is

$$H_e = 2\cdot2L^{0\cdot3} \text{ metres}$$

The US Navy uses $H_e = 3\cdot4 \sin \dfrac{\pi L_w}{1,100}$ feet

In metric units, the $1\cdot1\sqrt{L}$ wave is $H_e = 0\cdot607\sqrt{L}$ metres

Figure 6.34 (from Ref. 7) shows various effective wave height formulae compared with frigates at probability of once in a lifetime of 3×10^7 reversals and with merchant ships at 10^8.

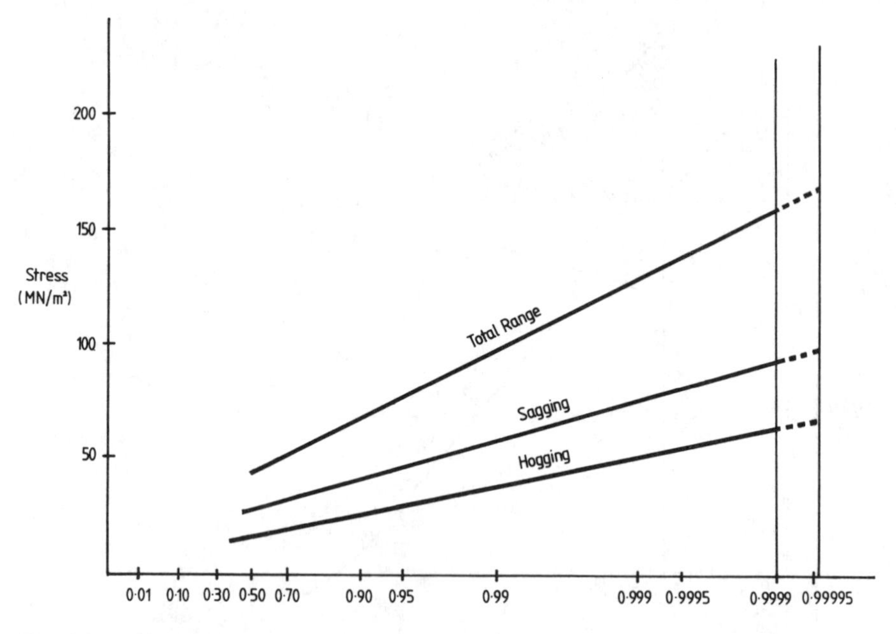

Fig. 6.33 Probability of non-exceedence in a 4-hour period

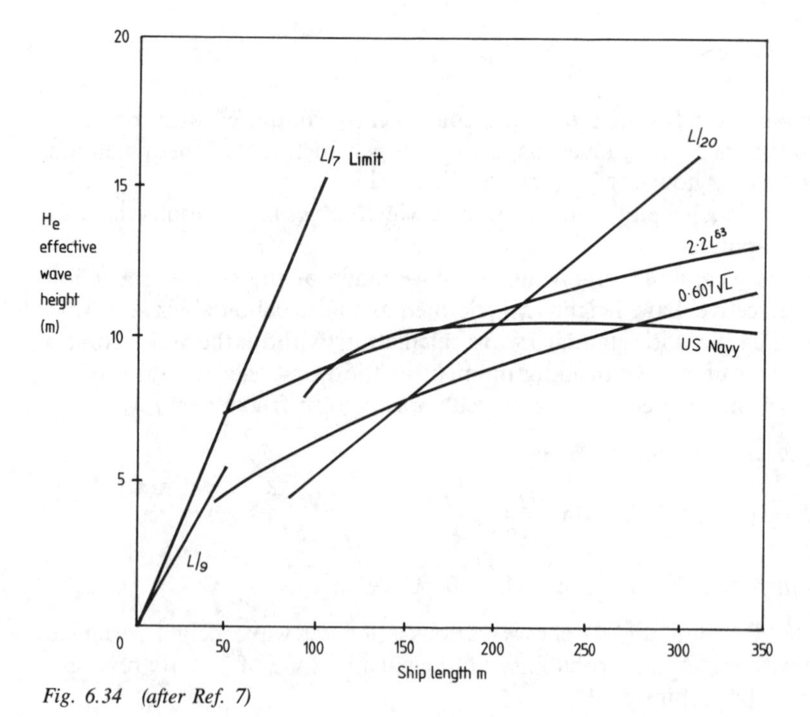

Fig. 6.34 (after Ref. 7)

Comparisons can show that the $L/20$ wave gives 50–70 per cent of the remote expectation for frigates while $0{\cdot}607\sqrt{L}$ runs through the 10^8 spots, overestimating the load for very large ships. As a first estimate of required longitudinal strength therefore a designer may safely use the static standard strength calculation associated with a wave height of $L/9$ up to 50 m, $2{\cdot}2L^{0{\cdot}3}$ around 100 m and $0{\cdot}607\sqrt{L}$ m up to about 300 m.

Let us leave the problems of determining the load for the time being and take a look at the ability of a ship to withstand longitudinal bending.

REALISTIC STRUCTURAL RESPONSE

The totally elastic response of the ship treated as if it were a simple beam has already been considered. Small variations to simple beam theory to account for shear lag were also touched upon. There are some other reasons why a hollow box girder like a ship may not behave like a simple beam. Strain locked in during manufacture, local yielding and local buckling will cause redistribution of the load-bearing capability of each part and its contribution to the overall cross section. If extreme loads are to be just contained, it is necessary to know the ultimate load-bearing ability of the cross section.

Let us first go to one extreme, and it is again necessary to draw upon some results of Chapter 7. It is there explained (p. 259) that the ultimate resistance to bending of a beam after the load has been increased to the point when the beam becomes totally plastic is

$$M_p = \sigma_Y S$$

where S* is the sum of the first moments of area of cross section each side of the neutral axis. This surely is the ultimate strength of a perfect beam which is turned into a plastic hinge. Common sense tells us that such a state of affairs is most unlikely in the large hollow box girder which is the ship. While the tension side might conceivably become totally plastic, the compression side is likely to buckle long before that. Caldwell (Ref. 5) accounted for this in his concept of the ultimate longitudinal resistance to load which formed the basis on which ultimate strength is now assessed. The essence of the method is to build up the total resistance to bending from a summation of the contribution of each element. It is not difficult to imagine that as compressive load in a deck is increased there comes a time when the panels will buckle, shirk their load and throw an increased burden upon their adjacent longitudinal stiffeners. These in turn may buckle as the load is further increased, shirk their contribution and throw an extra burden on 'hard' areas like corners of decks and junctions with longitudinal bulkheads.

This is how C.S. Smith has proposed that realistic assessment of cross sectional resistance should be made, a method which is now coming into general use. Every element which constitutes the effective cross section is first examined and a stress-strain curve is plotted. When the buckling behaviour of the element is embraced, the curves are called load shortening curves and are

*This is a different S from that in the previous section.

usually plotted in non-dimensional form as a family of curves with a range of initial assumed imperfections (Fig. 6.35).

The cross section of the ship girder is then assumed to bend with plane sections remaining plane to take up a radius of curvature R, the cross section rotating through an angle θ (Fig. 6.36). At any distance h above the neutral axis an element n of area A_n will be strained by an amount ε_n. Its load-bearing capacity

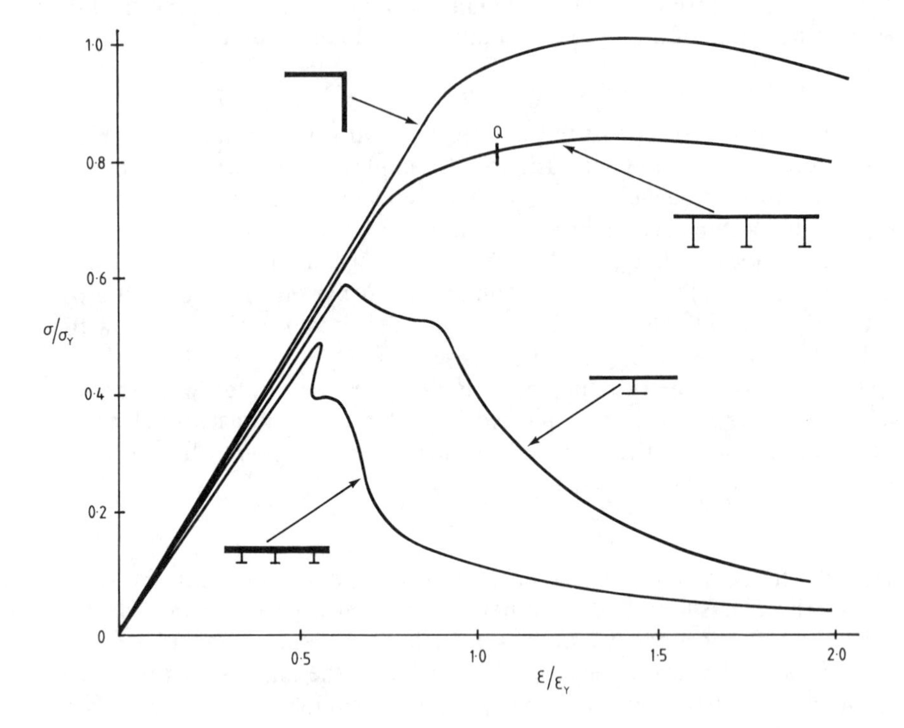

Fig. 6.35

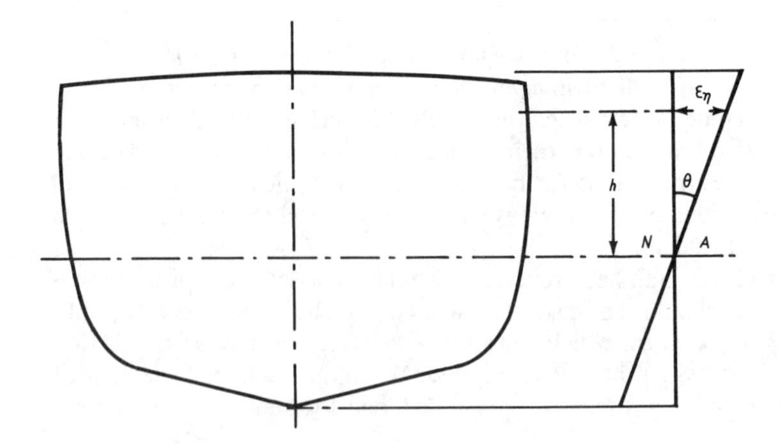

Fig. 6.36

can be picked off the relevant load shortening curve, say, at Q (Fig. 6.35). All such elements, so calculated, will lead to a bending resistance of the cross section

$$M = \sum_n \sigma_n A_n h_n$$

Carried out for a range of values of θ or R, a load shortening curve for the whole cross section can be calculated to provide a good indication of the collapse bending moment (Fig. 6.37). With the heavy structure of the bottom in compression when the ship hogs, the ultimate BM approaches the plastic moment M_p but a ratio of 0·6 in the sagging condition is not unusual.

In the calculation of M, the neutral axis does not stay still except when the behaviour is wholly elastic so that the additional condition for equilibrium of cross section must be imposed.

$$\sum_n \sigma_n A_n = 0$$

For this reason, it is convenient to consider changes to θ in increments which are successively summed and the process is called incremental analysis. Load shortening curves for the elements may embrace any of the various forms of buckling and indeed plastic behaviour whether it be assisted by locked-in manufacturing stresses or not.

While there are some obvious approximations to this method and some simplifications, it is undoubtedly the most realistic assessment of the collapse strength of the ship girder yet. The near horizontal part of Fig. 6.37 clearly represents the failure load, or collapse strength or ultimate strength, whatever it may be called.

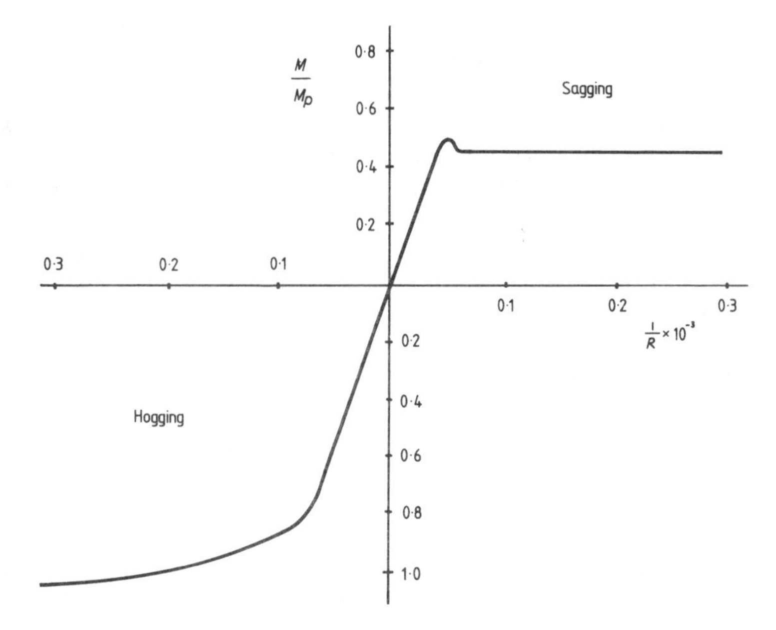

Fig. 6.37

ASSESSMENT OF STRUCTURAL SAFETY

At its simplest, failure will occur when the applied load exceeds the collapse load. It is first necessary to decide what probability of failure is acceptable, morally, socially or economically. History has shown that a comfortable level is a probability of 0·01 or 1 per cent that a single wave encounter will load the ship beyond its strength. With 30 million such encounters during a lifetime—every 10 seconds or so for 25 years—this represents a probability of $0{\cdot}33 \times 10^{-9}$. Some argue that this is altogether too remote and the ship could be weaker. In fact, noting the logarithmic scale, there is not a great deal of difference in the applied load at these very remote occurrences (Fig. 6.32). Because the statistical method is still essentially comparative, all that is necessary is to decide upon one standard and use it to compare new designs with known successful practice or, rarely, failure. This is what all major authorities now do. It is worth taking a further glimpse into the future (Ref. 7).

From the previous section, we have been able to declare a single value of the strength *S* which the loading must exceed for failure to occur. *S* may also be assumed to vary statistically because it may itself possess a variability. The strength, for example, will be affected by the thickness of plating rolled to within specified tolerances because that is what happens in practice. It will be dependent upon slightly different properties of the steel around the ship, upon ship production variations, upon defects, upon welding procedures and initial distortion which differs slightly. All such small variations cannot be precisely quantified at the design stage but the range over which they are likely can be. Instead of a single figure *S* for the strength, a probability density function can be constructed. Figure 6.38 more properly shows the realistic shapes of loading and strength. For bending moment *M*, the shaded area under the load curve to the right of *M* represents the probability that the load will exceed *M*. The shaded area under the strength curve to the left of *M* represents the probability that the ship will not be strong enough to withstand *M*.

We are thus concerned with the shaded overlap at the bottom. If we are studying extreme loads whose probability of occurrence is remote we have to define with some precision the shapes of the very tiny tails of the distributions.

This direct approach to structural reliability is still in its infancy. Instead, most authorities match the two distributions to known mathematical shapes

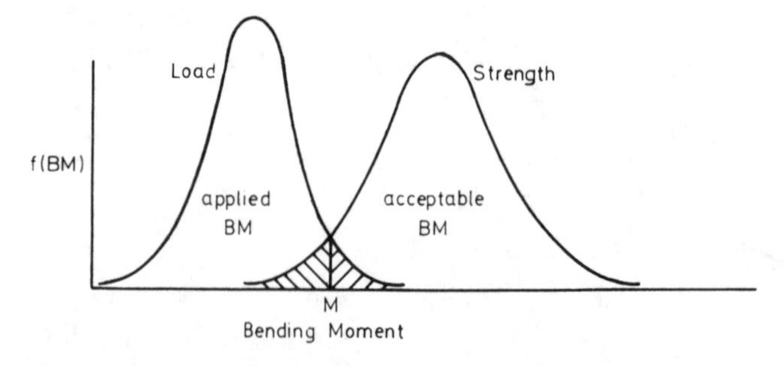

Fig. 6.38

and allow the mathematics to take care of the tails (Ref. 7). The most usual shapes to be assumed are Rayleigh or Gaussian which are defined by two properties only, the mean and the root mean square (or variance). This leads to a relationship between load P and strength S

$$S = \nu_P \nu_S P$$

where ν_P is the partial safety factor deduced from the mean and variance of the applied loading distribution and ν_S is the partial safety factor assessed for the mean and variance of the strength variability. A further partial safety factor is often added, ν_c, which assesses subjectively (i.e. as a matter of judgement) the gravity of failure. This partial safety factor approach is not yet in widespread use, except in building codes for civil engineering.

HYDROELASTIC ANALYSIS

The ship in a seaway is an elastic body which enjoys bodily movements in all six directions and distorts also about and along all three axes. Its distortions affect the load applied by the sea and both structural and hydrodynamic damping affect the problem. The sea itself is a random process.

Until recently a reasonably complete mathematical description and solution of this formidable problem has never been achieved. Bishop and Price have now tackled it with increasing success and a long succession of learned papers and books (Refs 10, 15, etc.) have emerged. They have deduced the dynamic behaviour of the ship in terms of modes of distortion which also embrace such solid body movements as pitch and heave, thus bringing together seakeeping and structural theories. Superposition of each element of behaviour in accordance with the excitation characteristics of the sea enables the total behaviour to be predicted, including even slamming and twisting of the hull.

The mathematics is wildly beyond the scope of this book and much of it has yet to be transformed into readily usable tools. Computer programs are available but require a great deal of data on mass and stiffness distribution which is not available until very late in the design.

This powerful analytical approach has been used to examine the overall stress distribution in large ships (Ref. 19), showing that important problems emerge at sections of the ship other than amidships. Combination of shear force and bending moment cause principal stresses much higher than had been suspected previously. Areas of particular concern are those about 20 per cent of the ship's length from the stern or from the bow, where slamming may further exacerbate matters. A lack of vigilance in the detail design or the production of the structure in these areas could, it has been suggested, have been responsible for some bulk carrier and VLCC fractures.

SLAMMING (see also Chapter 12)

One hydroelastic phenomenon which has been known for many years as slamming has only recently succumbed to theoretical treatment (Ref. 10). When flat areas of plating, usually forward, are brought into violent contact with the water at a very acute angle, there is a loud bang and the ship shudders. The momentum of the ship receives a check and energy is imparted to the ship girder to make it vibrate. Strain records show that vibration occurs in the first mode of flexural vibration imposing a higher frequency variation upon the strain fluctuations due to wave motion. Amplitudes of strain are readily augmented by at least 30 per cent and sometimes much more, so that the phenomenon is an important one.

The designer can do a certain amount to avoid excessive slamming simply by looking at the lines 30–40 per cent of the length from the bow and also right aft to imagine where acute impact might occur. The seaman can also minimize slamming by changes of speed and direction relative to the wave fronts. In severe seas the ship must slow down.

Extreme values of bending moment acceptable by the methods described in this chapter already embrace the augmentation due to slamming. This is because the relationships established between full scale measurements (Ref. 11) and the theory adopted make such allowance.

Material considerations

A nail can be broken easily by notching it at the desired fracture point and bending it. The notch introduces a stress concentration which, if severe enough, will lead to a bending stress greater than the ultimate and the nail breaks on first bending. If several bends are needed failure is by fatigue, albeit, low cycle fatigue.

A stress concentration is a localized area in a structure at which the stress is significantly higher than in the surrounding material. It can conveniently be conceived as a disturbance or a discontinuity in the smooth flow of the lines of stress such as a stick placed in a fast flowing stream would cause in the water flow. There are two types of discontinuity causing stress concentrations in ships:

(*a*) discontinuities built into the ship unintentionally by the methods of construction, e.g. rolling, welding, casting, etc.;
(*b*) discontinuities deliberately introduced into the structural design for reasons of architecture, use, access, e.g. hatch openings, superstructures, door openings, etc.

Stress concentrations cannot be totally avoided either by good design or high standards of workmanship. Their effects, however, can be minimized by attention to both and it is important to recognize the effects of stress concentrations on the ship girder. Many ships and men were lost because these effects were not recognized in the early Liberty ships of the 1940s.

In general, stress concentrations may cause yield, brittle fracture or buckling. There is a certain amount of theory which can guide the designer, but a general understanding of how they arise is more important in their recognition and treatment because it is at the detail design stage that many can be avoided or minimized.

GEOMETRICAL DISCONTINUITIES

The classical mathematical theory of elasticity has produced certain results for holes and notches in laminae. The stress concentration factors at A and B of Fig. 6.39 of an elliptical hole in an infinite plate under uniform tension in the direction of the *b*-axis are given by

$$j_A = \frac{\text{stress}_A}{\sigma} = -1$$

$$j_B = \frac{\text{stress}_B}{\sigma} = 1 + \frac{2a}{b}$$

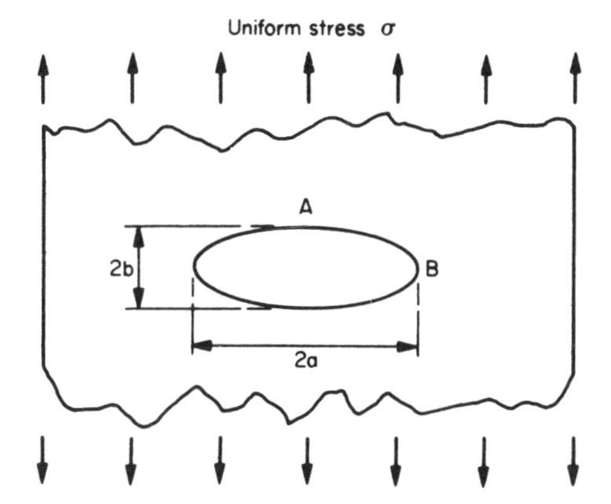

Fig. 6.39

For the particular case of a circular hole, $a = b$ and $j_B = 3$, i.e. the stress at the sides of a circular hole is three times the general tensile stress level in the plate, while at top and bottom there is a compressive stress equal to the general stress level. If a crack is thought of, ideally, as a long thin ellipse, the equation above gives some idea of the level of stress concentration at the ends; a crack twenty times as long as its width, for example, lying across the direction of loading would cause a stress, at the ends, forty-one times the general stress level and yielding or propagation of the crack is likely for very modest values of σ.

A square hole with radiused corners might be represented for this examination by two ellipses at right angles to each other and at 45 degrees to the direction of load. For the dimensions given in Fig. 6.40 the maximum stress concentration factor at the corners is given approximately by

$$j = \frac{1}{2}\sqrt{\frac{b}{r}}\left\{1 + \frac{\sqrt{(2b+2r)}}{\sqrt{b} - \sqrt{r}}\right\}$$

Figure 6.41 shows the effect of the variation of corner radius r on length of side b for a square hole with a side parallel to the direction of stress and for a square hole at 45 degrees. The figure shows

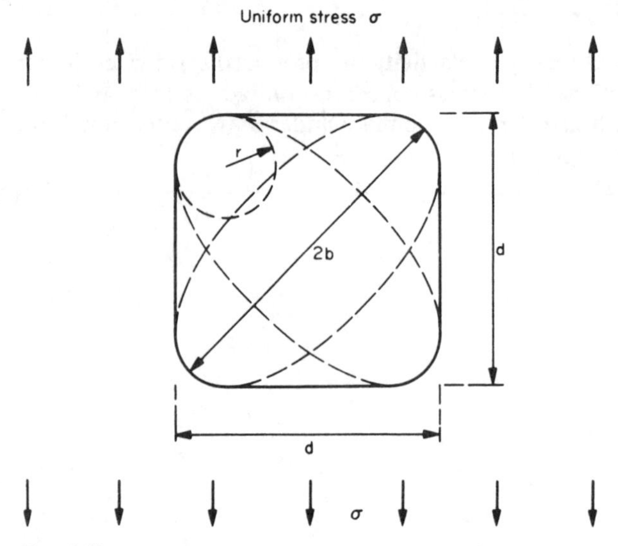

Fig. 6.40

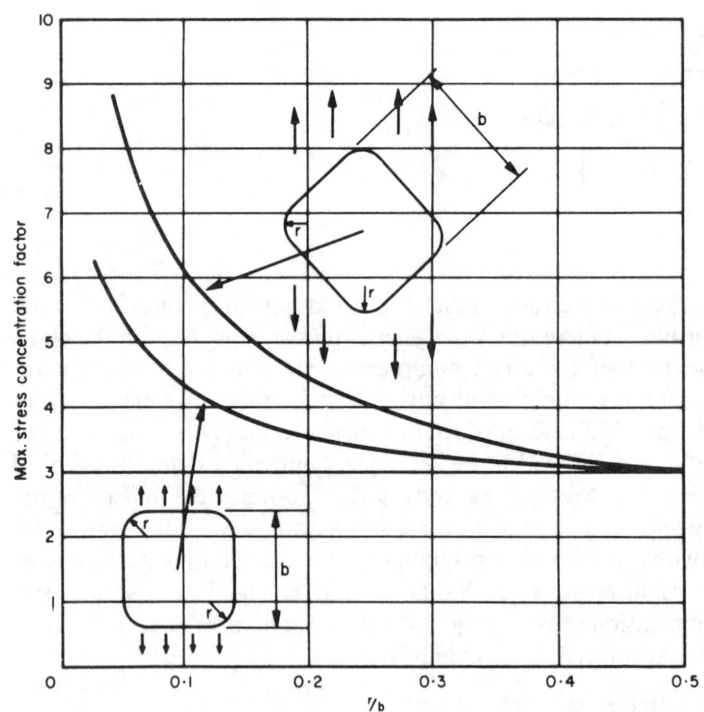

Fig. 6.41

(*a*) that there may be a penalty of up to 25 per cent in stress in failing to align a square hole with rounded corners with the direction of stress;

(*b*) that there is not much advantage in giving a corner radius greater than about one-sixth of the side;

(*c*) that the penalty of corner radii of less than about one-twentieth of the side, is severe. Rim reinforcement to the hole can alleviate the situation.

These results are suitable for large hatches. With the dimensions as given in Fig. 6.42, the maximum stress concentration factor can be found with good accuracy from the expression,

$$j_{\max} = \frac{2-0{\cdot}4^{B/b}}{2-0{\cdot}4^{l/B}}\left\{1+\frac{0{\cdot}926}{1{\cdot}348-0{\cdot}826^{20r/B}}\left[0{\cdot}577-\left(\frac{b}{B}-0{\cdot}24\right)^2\right]\right\}$$

It is of importance that the maximum stress occurs always about 5–10 degrees around the corner and the zero stress 50–70 degrees round. Butts in plating should be made at this latter point. Figure 6.43 shows the results for a hole with

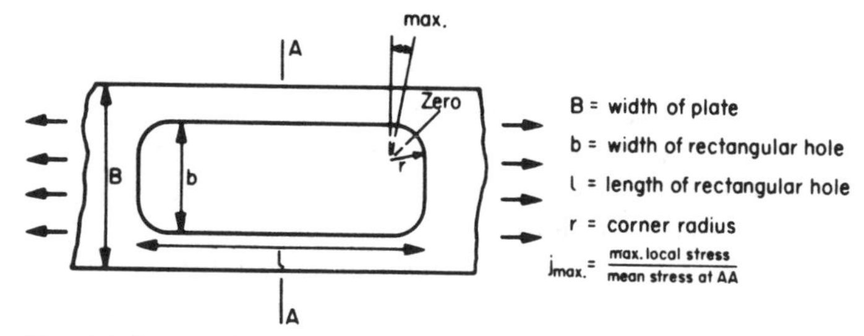

Fig. 6.42 Rectangular hole in finite plate

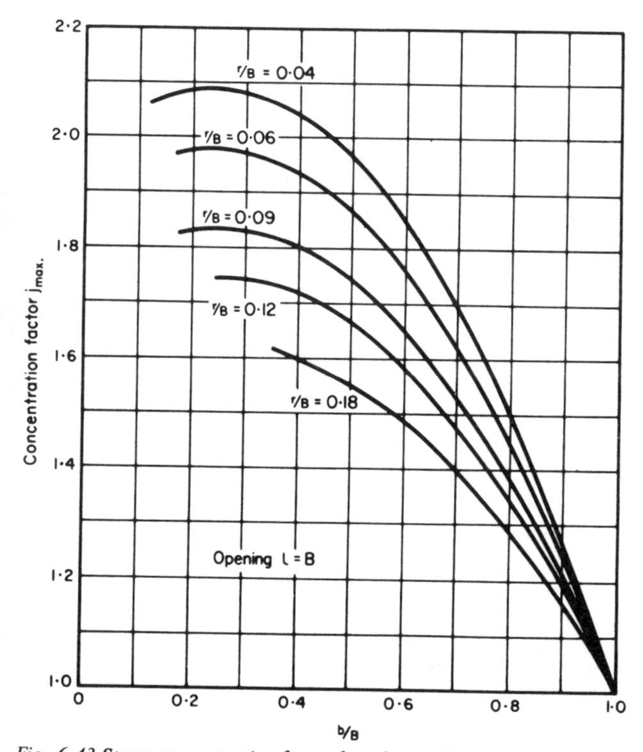

Fig. 6.43 Stress concentration factor for a hole with l = B

$l = B$. Note that the concentration factor is referred not to the stress in the clear plate but to the stress at the reduced section.

BUILT-IN STRESS CONCENTRATIONS

The violent treatment afforded a plate of mild steel during its manufacture, prevents the formation of a totally unstrained plate. Uneven rolling or contraction, especially if cooling is rapid, may cause areas of strain, even before the plate is selected for working into a ship. These are often called 'built in' or 'locked up' stresses (more accurately, strains). Furthermore, during the processes of moulding and welding, more strains are built in by uneven cooling. However careful the welders, there will be some, perhaps minute, holes, cracks and slag inclusions, lack of penetration and undercutting in a weld deposit. Inspection of important parts of the structure will minimize the number of visual defects occurring, and radiography can show up those below the surface, but lack of homogeneity cannot be observed by normal inspection procedures, even though the 'built-in' stress may exceed yield. What then happens to them when the ship is subjected to strain?

In ductile materials, most of the concentrations 'yield out', i.e. the concentration reaches the yield point, shirks further load and causes a re-distribution of stress in the surrounding material. If the concentration is a crack, it may propagate to an area of reduced stress level and stop. If it becomes visible, a temporary repair is often made at sea by drilling a circular hole at its end, reducing the concentration factor. This is a common first aid treatment. There is considerably more anxiety if the material is not so ductile as mild steel, since it does not have so much capacity for 'yielding out'; furthermore, a high yield steel is often employed in places where a general high stress level is expected so that cracks are less able to propagate to areas of reduced stress level and stop. A further anxiety in all materials is the possibility of fatigue, since concentrations at which there is, locally, a high stress level will be able to withstand few reversals.

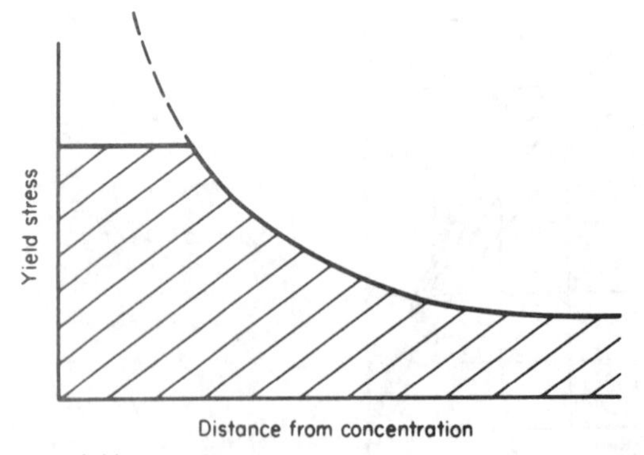

Fig. 6.44 Yielding out

CRACK EXTENSION, BRITTLE FRACTURE

Cracks then, cannot be prevented but can be minimized. It is important that they are observed and rectified before they cause catastrophic failure. They can extend under the action of fatigue or due to brittle fracture. Even in heavy storms fatigue cracks are only likely to increase in length at a rate measured in mm/s. A brittle fracture, however, can propogate at around 500 m/s. Thus brittle fracture is of much greater concern. The loss of Liberty ships has already been mentioned. More recent examples have been the MV *Kurdistan* which broke in two in 1979 (Ref. 16) and the MV *Tyne Bridge* which experiened a 4 m long crack in 1982 (Ref. 17). Some RN frigates damaged in collision in the 'Cod War' in the 1970s exhibited brittle fracture showing that thin plates are not necessarily exempt from this type of failure as had been generally thought up to that time.

The critical factors in determining whether brittle fracture will occur are stress level, length of crack and material toughness, this last being dependent upon temperature and strain rate. The stress level includes the effects of stress concentrations and residual stresses due to the fabrication processes. The latter are difficult to establish but, as an illustration, Ref. 18 showed that in frigates a compressive stress of about 50 MPa is introduced in hull plating by welding the longitudinals, balanced by local regions in the vicinity of the weld where tensile stresses are at yield point.

At low temperatures fracture of structural steels and welds is by cleavage. Once the threshold toughness for crack initiation is exceeded, the energy required for crack extension is so low that it can be provided by the release of stored elastic energy in the system. Unless fracture initiation is avoided structural failure is catastrophic. At higher temperatures fracture initiation is by growth and coalescence of voids. Subsequent crack extension is sustained only by increased load or displacement (Ref. 12). The temperature marking the transition in fracture mode is termed the transition temperature. This temperature is a function of loading rate, structural thickness, notch acuity and material microstructure.

Ideally one would like a simple test that would show whether a steel would behave in a 'notch ductile' manner at a given temperature and stress level. This does not exist because the behaviour of the steel depends upon the geometry and method of loading. For instance, cleavage fracture is favoured by high triaxial stresses and these are promoted by increasing plate thickness. The choice then, is between a simple test like the Charpy test (used extensively in quality control) or a more expensive test which attempts to create more representative conditions (e.g. the Wells Wide Plate or Robertson Crack Arrest tests). More recently the development of linear elastic fracture mechanics based on stress intensity factor, K, has been followed by usable elastic–plastic methodologies based on crack tip opening displacement, CTOD or δ, and the J contour integral, has in principle made it possible to combine the virtues of both types of test in one procedure.

For a through thickness crack of length $2a$ subject to an area of uniform stress, σ, remote from stress concentration the elastic stress intensity factor is given by

$$K = \sigma(\pi a)^{1/2}$$

The value at which fracture occurs is K_c and Ref. 13 proposes that a K_c = 125 MPa $(m)^{1/2}$ would provide a high assurance that brittle fracture initiation could be avoided. The reference also reports on an extensive programme of dynamic CTOD tests. Results are interpreted by using a fracture parameter, J_c, which can be viewed as extending K_c into the elastic–plastic regime, with results presented in terms of K_{Jc} which has the same units as K_c. Approximate equivalents are

$$K_{Jc} = [J_c E]^{1/2} = [2\delta_c \sigma_Y E]^{1/2}$$

It would be unwise to assume that cracks will never be initiated in a steel structure. For example, a running crack may emerge from a weld or heat affected zone unless the crack initiation toughness of the weld procedures meets that of the parent plate. It is prudent, therefore, to use steels which have the ability to arrest cracks. It is recommended that a crack arrest toughness of the material of between 150 and 200 MPa $(m)^{1/2}$ provides a level of crack arrest performance to cover most situations of interest in ship structures.

Ref. 12 recommended that:

(*a*) To provide a high level of assurance that brittle fracture will not initiate, a steel with a Charpy crystallinity less than 70 per cent at 0°C be chosen.
(*b*) To provide a high level of crack arrest capability together with virtually guaranteed fracture initiation avoidance, a steel with Charpy crystallinity less than 50 per cent at 0°C be chosen.
(*c*) For crack arrest strakes a steel with 100 per cent fibrous Charpy fracture appearance at 0°C be chosen.

If the ship is to operate in ice, or must be capable of withstanding shock or collision without excessive damage, steels with higher toughness would be appropriate.

FATIGUE

Provided ships are inspected regularly for cracking, the relatively slow rate of fatigue crack growth means that fatigue is not a cause for major concern in relation to ship safety. If, however, cracks go undetected their rate of growth will increase as they become larger and they may reach a size that triggers brittle fracture. Also water entering, or oil leaking, through cracks can cause problems and repair can be costly. Fatigue, then, is of concern particularly as most cracks occurring in ship structures are likely to be fatigue related. It is also important to remember that fatigue behaviour is not significantly affected by the yield strength of the steel. The introduction of higher strength steels and acceptance of higher nominal stress levels (besides the greater difficulty of welding these steels) means that fatigue may become more prevalent. Thus it is important that fatigue is taken into account in design as far as is possible.

Design for fatigue is not easy—some would say impossible. However, there are certain steps a designer can, and should, take. Experience, and considerable testing, show that incorrect design of detail is the main cause of cracking. Ref.

22 summarized the situation by saying that design for fatigue is a matter of detail design and especially a matter of design of welded connections. Methods used rely very heavily on experimental data. The most common to date has been one using the concept of a nominal stress. Typically for steel the fatigue characteristics are given by a log/log plot of stress range against number of cycles to failure. This *S–N* curve as it is termed takes the form of a straight line with life increasing with decreasing stress range until a value below which the metal does not fatigue. As a complication there is some evidence that in a corrosive atmosphere there is no lower limit. However, in laboratories, tests of welded joints lead to a series of *S–N* lines of common slope. The various joints are classified by number, the number being the stress range (N/mm^2) at 20 million cycles based on a mean test value less two standard deviations which corresponds to a survival probability of 97·7 per cent. As an example, a cruciform joint, K butt weld with fillet welded ends is in Class 71 (Fig. 6.45).

These data relate to constant amplitude loading and they are not too sensitive to mean stress level. However, a ship at sea experiences a varying load depending upon the conditions of sea and loading under which it operates. This is usually thought of in terms of a spectrum of loading and a transfer factor must be used to relate the stress range under spectrum loading to the data for constant amplitude. Reference 22 suggests that, based on testing at Hamburg, a transfer factor of

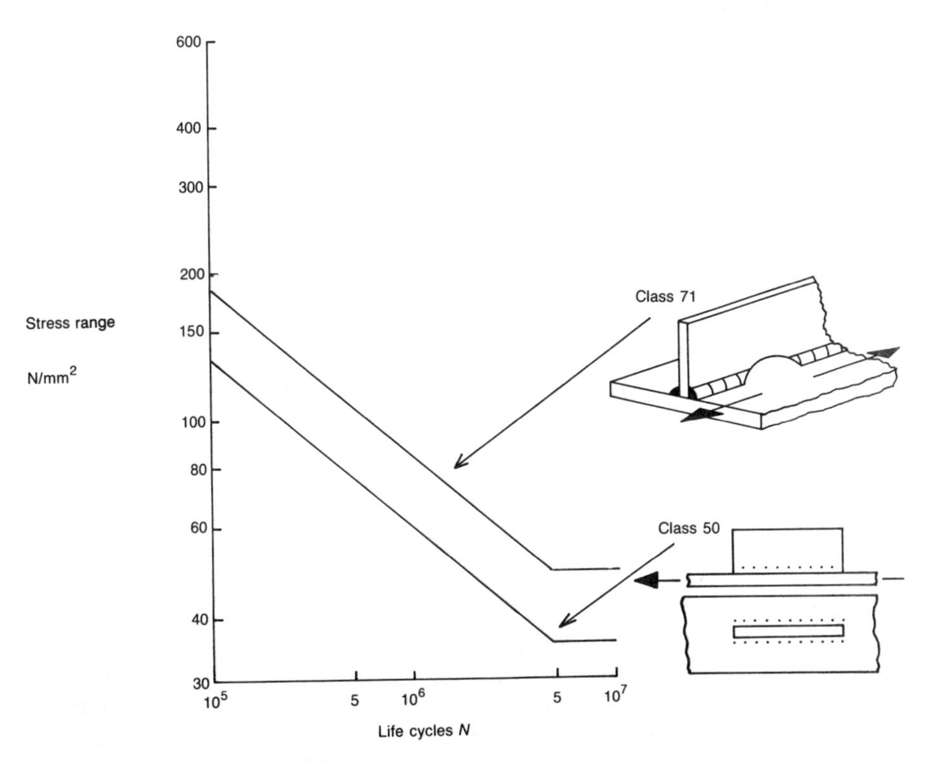

Fig. 6.45 S–N curves (Ref. 22)

4 is appropriate for the range of notch cases existing in ship structures, assuming 20 million cycles as typical of the average merchant ship life. It also recommends a safety factor of 4/3. For a Class 71 detail this gives a permissible stress range of $71 \times 4 \times 3/4 = 213$. This must be checked against the figures derived from the longitudinal strength calculation.

As is discussed in the next chapter, many structures are now analysed by finite element methods. These are capable of analysing local detail on which fatigue strength depends but interpretation of the results is made difficult by the influence of the mesh size used. The smaller the mesh, and the closer one approaches a discontinuity, the higher the stress calculated. The usual 'engineering' solution is to use a relatively coarse mesh and compare the results with the figures accepted in the nominal stress approach described above. The best idea of acceptable mesh size is obtained by testing details which have been analysed by finite element methods and comparing the data for varying mesh size.

Other methods are available but there is not space in a volume such as this to describe them.

DISCONTINUITIES IN STRUCTURAL DESIGN

In the second category of stress concentrations, are those deliberately introduced by the designer. Theory may assist where practical cases approximate to the assumptions made, as, for example, in the case of a side light or port hole. Here, in a large plate, stress concentration factors of about three may be expected. At hatches too, the effects of corner radii may be judged from Figs. 6.41 and 6.43. Finite element techniques outlined in Chapter 7 have extended the degree to which theory may be relied upon to predict the effect of large openings and other major stress concentrations.

Often, the concentrations will be judged unacceptably high and reinforcement must be fitted to reduce the values. One of the most effective ways to do this is by fitting a rim to the hole or curved edge. A thicker insert plate may also help, but the fitting of a doubling plate is unlikely to be effective unless means of creating a good connection are devised.

Designers and practitioners at all levels must be constantly on the watch for the discontinuity, the rapid change of structural pattern and other forms of stress raiser. It is so often bad local design which starts the major failure. Above all, the superposition of one concentration on another must be avoided—butt the reinforcement rim where the concentration is lowest, avoid stud welds and fittings at structural discontinuities, give adequate room between holes, grind the profile smooth at the concentration!

SUPERSTRUCTURES AND DECKHOUSES

These can constitute severe discontinuities in the ship girder. They can contribute to the longitudinal strength but are unlikely to be fully effective. Their effectiveness can be improved by making them long, minimizing changes in plan and profile, extending them the full width of the hull and paying careful attention to their connections to the hull.

The only contact between the upper deck and the superstructure is along the bottom of the superstructure sides through which the strains and forces must be transmitted. Because the upper deck stretches, so also must the lower edge of the superstructure sides thus causing shear forces which tend to distort the superstructure into a shape opposite to that of the hull (Fig. 6.46). The two

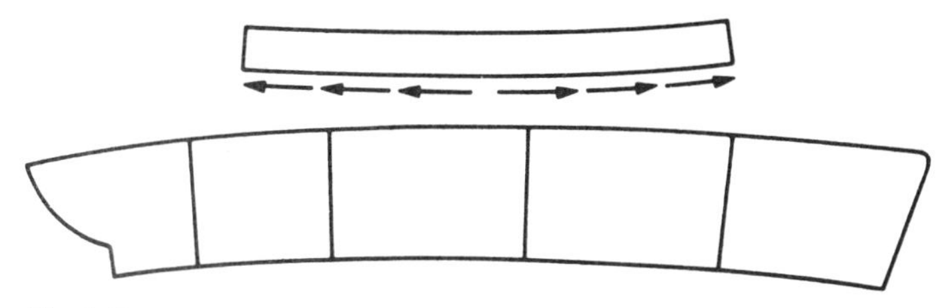

Fig. 6.46

are held together, however, and there must be normal forces on the superstructure having the opposite effect. The degree of these normal forces must depend upon the flexibility of the deck beams and main transverse bulkheads if the superstructure is set in from the ship's sides. The effect of the shear forces will depend on the manner in which the shear is diffused into the superstructure, and the shear lag effects are likely to be more appreciable there than in the main hull girder.

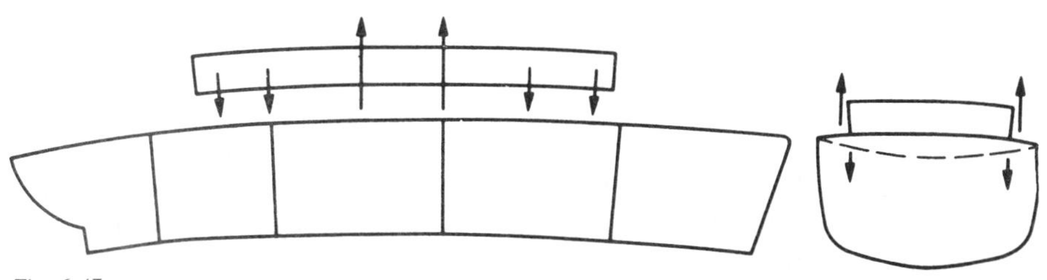

Fig. 6.47

This difficult problem has been satisfactorily solved in Ref. 21. The effects of shear lag are ignored and the results are suitable for the middle portions of long superstructures. Reference 21 embraces the effects of shear diffusion but ignores the concentrated forces from main transverse bulkheads, and it is suitable for short superstructures or for those which extend out to the ship's side. An efficiency of superstructure is defined as

$$\eta = \frac{\sigma_0 - \sigma}{\sigma_0 - \sigma_1}$$

where σ_0 is the upper deck stress which would occur if there were no superstructure present, σ is the upper deck stress calculated and σ_1 is the upper deck stress with a fully effective superstructure. Curves are supplied from which the factors leading to the efficiency may be calculated. As might be expected, the efficiency depends much on the ratio of superstructure length to its transverse dimensions.

The square ends of the superstructure constitute major discontinuities, and may be expected to cause large stress concentrations. They must not be superimposed on other stress concentrations and should be avoided amidships or, if unavoidable, carefully reinforced. For this reason, expansion joints are to be used with caution; while they relieve the superstructure of some of its stress by reducing its efficiency, they introduce stress concentrations which may more than restore the stress level locally. To assist the normal forces, superstructure ends should coincide with main transverse bulkheads.

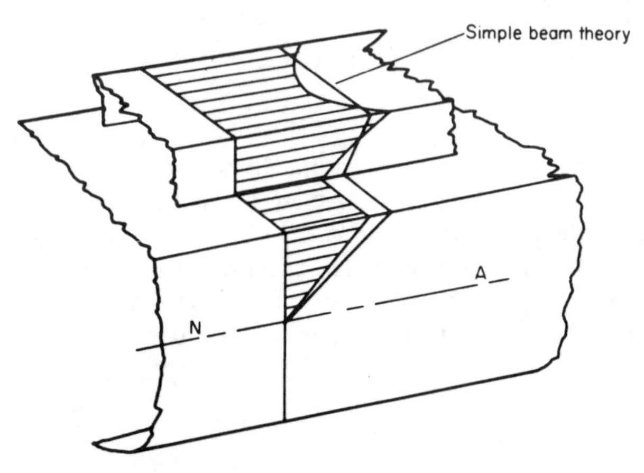

Fig. 6.48 **Direct stress in a superstructure**

Unwanted hull–superstructure interactions can be avoided by using low modulus material in the superstructure such as GRP (Ref. 8) which offers tensile and compressive strengths comparable to the yield strength of mild steel with an *E*-value less than a tenth that of steel. In this case the superstructure will not make any significant contribution to longitudinal strength.

Conclusions

The student may be forgiven if he began this chapter safely assured by the static calculation (sometimes called Reed's method) and ended it aware of some considerable uncertainty. As the proceedings of the International Ship Structures Congress bear witness, this is precisely the current situation. Naval architects are rapidly adopting the statistical approach to ship girder loading and are preparing to marry it to structural reliability when sufficient information about strength variability is to hand. They await the results of the long process of data collection at sea to refine their standards of acceptability.

Problems

1. A landing craft of length 200 ft may be assumed to have a rectangular cross-section of beam of 28 ft. When empty, draughts are 1·5 ft forward and 3·5 ft aft

and its weight is made up of general structure weight assumed evenly distributed throughout the length, and machinery and superstructure weight spread evenly over the last 40 ft. A load of 100 tonf can be carried evenly distributed over the first 160 ft, but to keep the forward draught at a maximum of 2 ft, ballast water must be added evenly to the after 40 ft.

Draw the load, shearing force and bending moment diagrams for the loaded craft when in still water of specific volume of 35 ft^3/tonf. Determine the position and value of the maximum bending moment acting on the craft.

2. State the maximum bending moment and shearing force in terms of the weight and length of a vessel having the weight uniformly distributed and the curve of buoyancy parabolic and quote deck and keel moduli. Where do the maximum shearing force and bending moment occur?

3. The barge shown in the figure floats at a uniform draught of 1 m in sea water when empty.

A heavy weight, uniformly distributed over the middle 5 m of the barge, increases the draught to 2 m. It may be assumed that the buoyancy curves for the barge (loaded and unloaded) and the weight distribution of the unloaded barge are constant over the parallel length of the barge, decreasing linearly to zero at the two ends.

Draw curves of buoyancy, weight, loading, shear force and bending moment for the barge loaded and at rest in salt water.

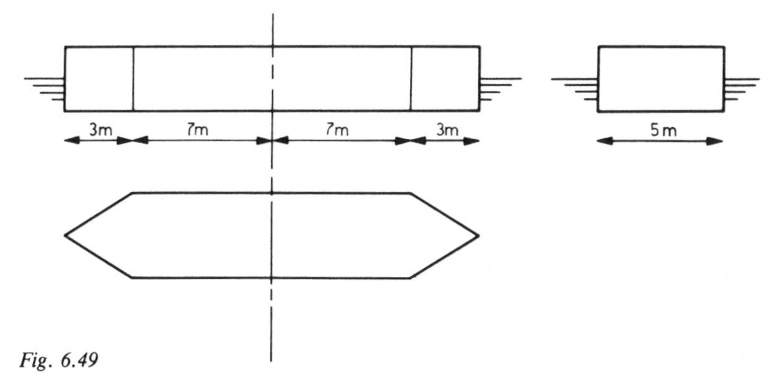

Fig. 6.49

4. A rectangular barge is 80 ft long and has a beam of 14 ft. The weight distribution of the partially loaded barge is shown in the figure below. The barge is floating at rest in still water. Draw curves of loading, shearing force and bending moment, stating the maximum shearing force and bending moment and the positions where they act.

5. A barge, 70 m long, has the cross-section shown. The empty weight is 420 tonnef evenly distributed over the whole length, whereas the load is spread evenly over the middle 42 m of length. Draw the curves of load, shearing force and bending moment for the barge, when it is loaded and balanced on a sea wave whose profile gives draughts, symmetrical about amidships, as follows:

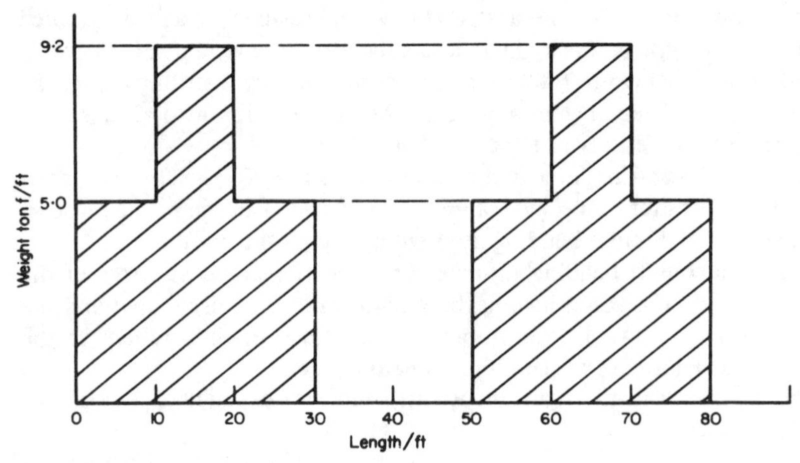

Fig. 6.50

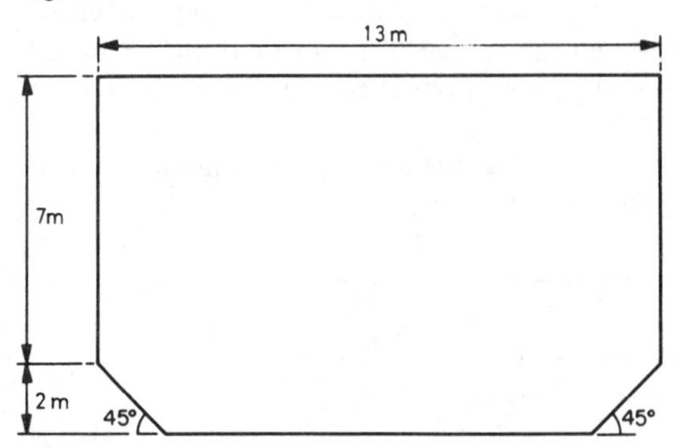

Fig. 6.51

Section	FE	2	3	4	5	Amidships
Draught (m)	5·00	4·55	3·61	2·60	2·10	2·00

6. A steel barge of constant rectangular section, length 72 ft, floats at a draught of 5 ft when loaded. The weight curve of the loaded barge may be regarded as linear, from zero at the two ends to a maximum at the mid-length. The structural section is shown below. If the stress in the deck in still fresh water is not to exceed 1·5 tonf/in^2, estimate the thickness of the plating if this is assumed to be constant throughout.

7. In a calculation of the longitudinal strength for the sagging condition, the following mean ordinates in tonnef/m were found for sectional lengths of a ship each 12 m long, starting from forward:

Section	1	2	3	4	5	6	7	8	9	10	11
Weight	8·3	12·6	24·2	48·2	66·2	70·0	65·1	40·7	23·3	13·0	6·0
Buoyancy	24·8	40·6	39·2	33·6	28·2	30·0	39·6	48·7	47·4	36·0	9·5

Draw the shearing force and bending moment diagrams and state the positions and values of the maxima.

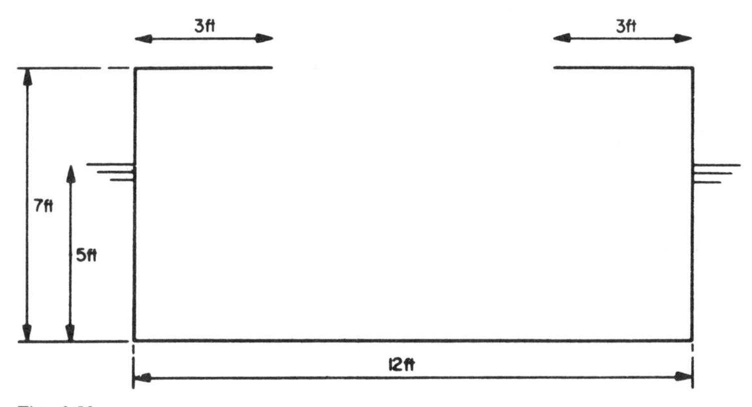

Fig. 6.52

8. A dumb lighter is completely wall-sided, 100 m long, 20 m maximum beam and it floats at a draught of 6 m in still water. The waterline is parabolic and symmetrical about amidships with zero ordinates at the ends. The weight is evenly distributed along its length.

If the vessel is 15 m deep, has a sectional second mt. = 30 m^4 and has a neutral axis 6 m from the keel, find the maximum hogging stress when balanced on an $L/20$ wave.

The coordinates of a trochoid are:

$\frac{x}{L/20}$	0	1	2	3	4	5	6	7	8	9	10
$\frac{z}{L/20}$	0	0·018	0·073	0·161	0·280	0·423	0·579	0·734	0·872	0·964	1·0

9. The following data apply to the half-section modulus of a ship girder 45 ft deep with an assumed neutral axis 20 ft above the keel:

	Area (in^2)	1st moment about ANA (in^2 ft)	2nd moment about ANA (in^2 ft^2)
Above ANA	930	16,020	331,740
Below ANA	1072	14,820	243,460

Find the stresses in the keel and upper deck if the bending moment is 200,000 tonf ft.

10. The data given below applies to a new frigate design. Draw the curves of loading, shearing force and bending moment for the frigate and calculate the maximum stresses in the keel and deck given that:

$$\begin{aligned} \text{2nd moment of area of section about NA} &= 92{,}500\ \text{in}^2\ \text{ft}^2 \\ \text{total depth} &= 34{\cdot}5\ \text{ft} \\ \text{NA above keel} &= 13\ \text{ft} \\ \text{length} &= 360\ \text{ft} \end{aligned}$$

Ordinate	1	2	3	4	5	6	7	8	9	10	11
Buoyancy (tonf/ft)	0·90	8·05	10·16	9·26	6·80	5·80	6·15	7·46	8·91	8·89	6·32
Weight (tonf/ft)	3·8	5·6	5·8	11·9	10·2	9·9	10·4	9·9	5·3	3·6	

11. The particulars of a structural strength section 13 m deep, taken about mid-depth, are as follows:

	Area (m^2)	1st moment (m^3)	2nd moment (m^4)
Above mid-depth	0·33	1·82	12·75
Below mid-depth	0·41	2·16	14·35

On calculation, the maximum stress in the upper deck, for the hogging condition, is found to be 122 N/mm². Determine how many longitudinal girders must be welded to the upper deck to reduce the stress to 100 N/mm² under the same loading conditions, given that

(*a*) area of cross-section of each girder = 0·002 m²
(*b*) c.g. of each girder can be considered to be at the height of the upper deck,
(*c*) second moment of each girder about its own c.g. is negligible.

12. Calculate the deck and keel moduli of the structural section given below. Calculate the maximum stresses due to a sagging bending moment of 37,380 tonf ft.

The centre of area of each girder and longitudinal may be assumed to lie in the plating to which it is attached. Second moments of girders about their own c.gs are negligible.

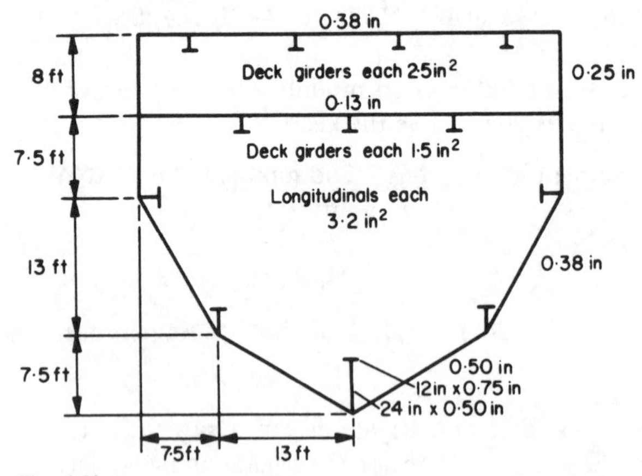

Fig. 6.53

13. The forecastle of a destroyer is to be extended in aluminium alloy, over the midships area by the structure shown below. Details of the existing steel structural midship section are as follows:

Depth, keel to upper deck	20 ft
Neutral axis below upper deck	12·6 ft
Total area of section	750 in^2
Second moment of area about NA	81,000 $in^2\ ft^2$
Calculated upper deck hogging stress	6·85 $tonf/in^2$

Calculate the stress in the deck of the alloy extension and the new keel stress for the same bending moment.

The value of E for alloy should be taken as $\frac{1}{3}$ that for steel

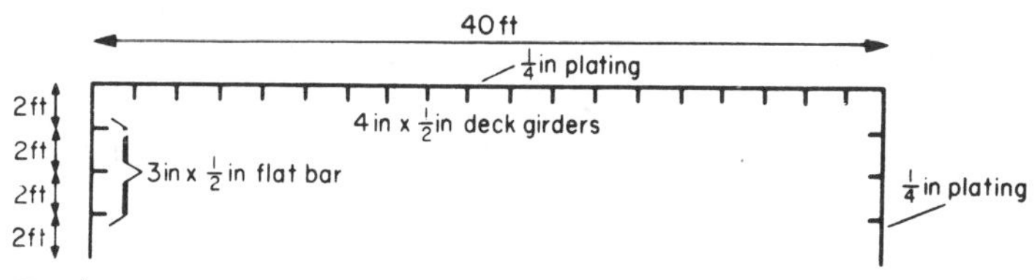

Fig. 6.54

14. An aluminium superstructure made up of a deck 10 m wide supported on two sides 3 m deep, is to be included for strength purposes with the steel structural section whose particulars are as follows:

total area of section	0·545 m^2
second moment of area	10·29 m^4
depth of section	14 m
height of NA above keel	7·20 m

Since the design stresses are already large, it is proposed to replace the steel upper deck by one of aluminium of the same cross-sectional area of 0·09 m^2.

Making reasonable first assumptions and assuming that stiffeners can be taken with the plating, determine the area/metre run of the aluminium superstructure deck and sides (assumed constant for deck and sides), to maintain the same keel stress as before.

Take E for aluminium to be $\frac{1}{3}$ that for steel.

15. A rectangular vessel is 15 m broad and 10 m deep and has deck plating 5 mm thick and sides and bottom 9 mm thick. At a certain section, it is subjected to a hogging bending moment of 65 MN m and a shearing force of 3·15 MN. Calculate, in accordance with classical beam theory,

(*a*) Maximum tensile and compressive stresses,

(*b*) Maximum shearing stress.

16. If, in the previous example, the deck and bottom were 9 mm thick and the sides 5 mm thick, what would then be the stresses?

17. A barge has a rectangular waterline 61 m long with a beam of 5·34 m. When it is floating in calm sea water, the position of the maximum longitudinal bending moment is 6·1 m abaft amidships.

Deduce, from first principles, the influence lines, showing the effect on the maximum still water BM of the addition of a unit concentrated weight at any position along the length of the barge. What is the effect of an addition (*a*) at the extreme after end, (*b*) at the point of maximum BM?

Check that the effect of a load uniformly distributed along the length is zero.

18. The particulars of a structural midship section, 40 ft deep, taken about mid-depth, are as follows:

	Area (in^2)	Moment (in^2 ft)	2nd moment (in^2 ft^2)
above mid-depth	550	9000	175,000
below mid-depth	650	11,400	221,000

Displacement increases would cause deck stresses of 8 tonf/in^2 to arise in the standard sagging condition.

It is proposed to stiffen up the section by the addition of ten girders to the upper deck and five similar longitudinals to the flat bottom. If these stiffeners are to be cut down versions of T bars having a cross-sectional area of 4·5 in^2, determine the actual cross-sectional area required to reduce deck design stress to 7·5 tonf/in^2.

19. A barge is 60 m long and completely symmetrical about amidships. It is constructed of steel.

The following table gives the mean net loading between ordinates and the second moments of area of sections at regular intervals between the forward perpendicular and amidships:

	Amidships										FP
Ordinate	6		5		4		3		2		1
Load (tonnef/m)		−19		−8		7		15		5	
Second moment of area (m^4)	1·77		1·77		1·77		1·60		1·28		0·42

Estimate the breakage due to this loading, which is symmetrical about amidships. E = 209 GN/m^2.

20. The following table gives the loading diagram which was obtained after balancing a ship of length 700 ft in the standard longitudinal strength calculation:

	AE		10		9		8		7		6		5		4		3		2		FP
Load (tonf/ft)		9		17		18		−18		−21		−27		−18		4		24		12	
$I/(10^5\ in^2\ ft^2)$	6		7·5		9		13·5		15		15		13·5		9		7·5		6		4·5

The neutral axis is 24 ft below the strength deck. Find the stress in this deck amidships.

The table also gives the MI of section at the ordinates. Estimate the relative angular movement between a missile launcher at 3 ordinate and a guidance radar amidships ($E = 3 \times 10^7$ lbf/in^2).

21. A ship of length 105 m is 9 m deep. It is moored in the effluent from a power station on a cold night, which results in the keel increasing in temperature by 15°C and the upper deck falling 15°C.

Estimate the breakage resulting. The coefficient of linear expansion of steel is 0·000012 in Celsius units.

References

1. Smith, W. E. Hogging and sagging strains in a seaway as influenced by wave structure, *TINA*, 1883.
2. Beaumont, J. G. and Robinson, D. W. Classification aspects of ship flexibility. *Phil. Trans. Royal Soc. London. A* (1991), 334.

3. Nitta, A., Arai, H. and Magaino, A. *Basis of IACS unified longitudinal strength standard*, Marine Structures 5 (1992).
4. *Rules and Regulations for the classification of ships*, Lloyd's Register of Shipping, 7 volumes, annually.
5. Caldwell, J. B. Ultimate longitudinal strength, *TRINA*, 1965.
6. Mansour, A. and Faulkner, D. On applying the statistical approach to extreme sea loads and ship hull strength, *TRINA*, 1965.
7. Faulkner, D. and Sadden, J. A. Towards a unified approach to ship structural safety, *NA*, Jan. 1979.
8. Smith, C. S. and Chalmers, D. W. Design of ship superstructures in fibre-reinforced plastic, *NA*, May 1987.
9. Toman, R. D. An engineering review of steel and alloy superstructures on surface warships, *TRINA*, 1988.
10. Bishop, R. E. D. and Price, W. G. *Hydroelasticity of ships*, Cambridge U.P.
11. Clarke, J. D. *Wave loading in warships*, Advances in marines structure, Elsevier, London, 1986.
12. Sumpter, J. D. G. *Design against fracture in welded structures*. Ibid.
13. Sumpter, J. D. G., Bird, J., Clarke, J. D. and Caudrey, A. J. Fracture toughness of ship steels, *TRINA*, 1990.
14. Smith, C. S., Anderson, N., Chapman, J. C., Davidson, P. C. and Dowling, P. J. Strength of stiffened plating under combined compression and lateral pressure (and earlier papers), *TRINA*, 1992.
15. Bishop, R. E. D., Price, W. G. and Temeral, P. On the distributions of symmetric shearing force and bending moment in hulls, *TRINA*, 1989.
16. Corlett, E. C. B., Colman, J. C. and Hendy, N. R. KURDISTAN—The anatomy of a marine disaster, *NA*, Jan. 1988.
17. Department of Transport. *A report into the circumstances attending the loss of MV Derbyshire*; Appendix 7, Examination of fractured deckplate of MV *Tyne Bridge*, March 1986.
18. Somerville, W. L., Swan, J. W. and Clarke, J. D. Measurements of residual stresses and distortions in stiffened panels. *Journal of Strain Analysis*, Vol. 12, No. 2, 1977.
19. Bishop, R. E. D., Price, W. G. and Temeral, P. A theory on the loss of the MV *Derbyshire*, *TRINA*, 1991.
20. Aksu, S., Price, W. G. and Temeral, P. Load and stress distribution in bulk carriers and tankers in various loading conditions. International conference on tankers and bulk carriers, *TRINA*, 1992.
21. Caldwell, J. C. The effect of superstructures on the longitudinal strength of ships, *TINA*, 1957.
22. Petershagen, H. *Fatigue problems in ship structures*, Advances in marine structures, Elsevier, 1986.

7 Structural design and analysis

It is not possible within the confines of one chapter—or even a book of this size—to present the naval architect with all that he should know about the design and analysis of the ship's structure. The object of this chapter is to provide an understanding of the structural behaviour of the ship and a recognition of the unit problems. Applied mechanics and mathematics will have provided the tools; now they must be applied to specific problems. The scope and the limitations of different theories must be known if they are to be used with success and the student needs to be aware of the various works of reference. Recognition of the problem and knowledge of the existence of a theory suitable for its solution are exceedingly valuable to the practising engineer. He rarely has the time to indulge an advanced and elegant theory when a simple approach provides an answer giving an accuracy compatible with the loading or the need. On the other hand, if he recognizes that a simple approach is inadequate, he must discard it. It is therefore understanding and recognition which this chapter seeks to provide.

Optimum design is often assumed to mean the minimum weight structure capable of performing the required service. While weight is always significant, cost, ease of fabrication and ease of maintenance are also important. Cost can increase rapidly if non-standard sections or special quality materials are used; fabrication is more difficult with some materials and, again, machining is expensive. This chapter discusses methods for assessing the minimum requirements to provide against failure. The actual structure decided upon must reflect all aspects of the problem.

The whole ship girder provides the background and the boundaries for local structural design. The needs of the hull girder for areas at deck, keel and sides must be met. Its breakdown into plating and stiffener must be determined; there is also much structure required which is not associated with longitudinal strength; finally, there are many particular fittings which require individual design. An essential preliminary to the analysis of any structure or fitting, however, is the assessment of the loading and criterion of failure.

LOADING AND FAILURE

Because it is partly the sea which causes the loading, some of the difficulties arising in Chapter 6 in defining the loading apply here also. The sea imposes on areas of the ship impact loads which have not yet been extensively measured, although the compilation of a statistical distribution of such loading continues. While more realistic information is steadily coming to hand, a loading which is likely to provide a suitable basis for comparison must be decided upon and used to compare the behaviour of previous successful and unsuccessful elements. For example, in designing a panel of plating in the outer bottom, hydrostatic pressure due to draught might provide a suitable basis of comparison, and examination of previous ships might indicate that when the ratio of permanent set to thickness exceeded a certain percentage of the breadth to thickness ratio, extensive cracking occurred.

Some loads to which parts of the structure are subject are known with some accuracy. Test water pressure applied during building (to tanks for example), often provides maximum loads to which the structure is subject during its life; bulkheads cannot be subjected to a head greater than that to the uppermost watertight deck, unless surging of liquid or shift of cargo is allowed for; forces applied by machinery are generally known with some accuracy; acceleration loads due to ship motion may be known statistically. More precise analytical methods are warranted in these cases.

Having decided on the loading, the next step is to decide on the ultimate behaviour, which, for brevity, will be called failure. From the point of view of structural analysis, there are four possible ways of failing, by (*a*) direct fracture, (*b*) fatigue fracture, (*c*) instability, and (*d*) unacceptable deformation.

(*a*) *Direct fracture* may be caused by a part of the structure reaching the ultimate tensile, compressive, shear or crushing strengths. If metallurgical or geometrical factors inhibit an otherwise ductile material, it may fail in a brittle manner before the normally expected ultimate strength. It should be noted, that yield does not by itself cause fracture and cannot therefore be classed as failure in this context;

(*b*) *Fatigue fracture.* The elastic fatigue lives of test specimens of materials are fairly well documented. The elastic fatigue lives of complex structures are not divinable except by test, although published works of tests on similar structural items may give a lead. The history of reversals in practical structures at sea also requires definition. Corrosion fatigue is a special case of accelerated failure under fatigue when the material is in a corrosive element such as sea water. The 'bent nail' fatigue is another special case, in which yield is exceeded with each reversal and the material withstands very few reversals;

(*c*) *Instability.* In a strut, buckling causes an excessive lateral deflection; in a plating panel, it may cause load shirking by the panel or wrinkling; in a cylinder under radial pressure, instability may cause the circumference to corrugate; in a plating stiffener it may cause torsional tripping. Most of these types of instability failure are characterized by a relatively rapid increase in deflection for a small increase in load and would generally be regarded as failure if related to the whole structure; where only part of the structure shirks its load, as sometimes happens, for example, with panels of plating in the hull girder section, overall 'failure' does not necessarily occur.

(*d*) *Deformation.* A particular deflection may cause a physical foul with machinery or may merely cause alarm to passengers, even though there is no danger. Alignment of machinery may be upset by excessive deflection.

Such deformations may be in the elastic or the elasto-plastic range. The stiffness of a structure may cause an excessive amplitude of vibration at a well used frequency. Any of these could also constitute failure.

For each unit of structure in a ship, first the loading must be decided and then the various ways in which it would be judged to have failed must be listed and examined. What are these units of structure?

STRUCTURAL UNITS OF A SHIP

There are four basic types of structure with which the ship designer must deal: (*a*) plating-stiffener combinations, (*b*) panels of plating, (*c*) frameworks, (*d*) fittings.

(*a*) *Plating-stiffener combinations.* The simplest form of this is a single simple beam attached to a plate. Many parallel beams supporting plating constitute a grillage with unidirectional stiffening. Beams intersecting at right angles constitute an orthogonally stiffened grillage. These various units may be initially flat or curved, loaded in any plane and possess a variety of shapes and boundaries.

(*b*) *Panels of plating*. These are normally rectangular and supported at the four edges, subject to normal or in-plane loads. Initially, they may be nominally flat or dished;

(*c*) *Frameworks.* These may be portals of one or more storeys. Frameworks may be constituted by the transverse rings of side frames and deck beams or the longitudinal ring of deck girder, bulkhead stiffeners and longitudinal. They may be circular as in a submarine. Loading may be distributed or concentrated in their planes or normal to their planes.

(*d*) *Fittings.* There is a great variety of fittings in ships the adequacy of whose strength must be checked. Particular ones include control surfaces such as rudders and hydroplanes, shaft brackets and spectacle plates, masts, derricks, davits and machinery seatings.

Let us examine the methods of analysis available for each of these four basic structural units.

Stiffened plating

SIMPLE BEAMS

Very many of the local strength problems in a ship can be solved adequately by the application of simple beam theory to a single stiffener-plating combination. This is permissible if the boundaries of the unit so isolated are truly represented by forces and moments that adjacent units apply to it. Frequently, when there is a series of similarly loaded units, the influence of adjacent units on the edges will be zero; similarity longitudinally might also indicate that the end slopes are zero. These edge constraints have a large influence and in many cases will not be so easily determined. It is important that the deflection of the supporting structure is negligible compared with the deflection of the isolated

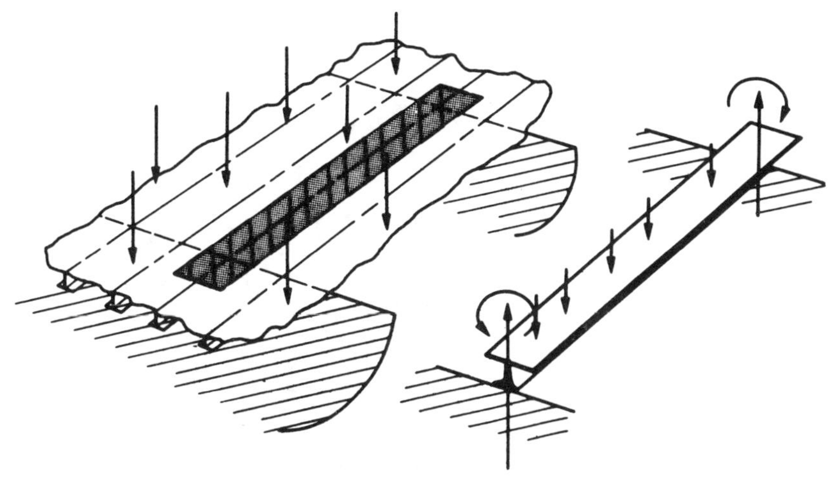

Fig. 7.1

beam, if the unit is to be correctly isolated; this is likely to be true if the supports are bulkheads but not if they are orthogonal beams. A summary of results for common problems in simple beams is given in Fig. 7.2. (See pp. 238 and 239.)

According to the Bernoulli–Euler hypothesis from which the simple theory of bending is deduced, sections plane before bending remain plane afterwards and

$$\sigma = \frac{M}{I}y$$

For many joists and girders this is very closely accurate. Wide flanged beams and box girders, however, do not obey this law precisely because of the manner in which shear is diffused from the webs into the flanges and across the flanges. The direct stresses resulting from this diffusion do not quite follow this law but vary from these values because sections do not remain plane. Distribution of stress across stiffened plating under bending load is as shown by the wavy lines in Fig. 7.3, and this effect is known as the shear lag effect. While the wavy

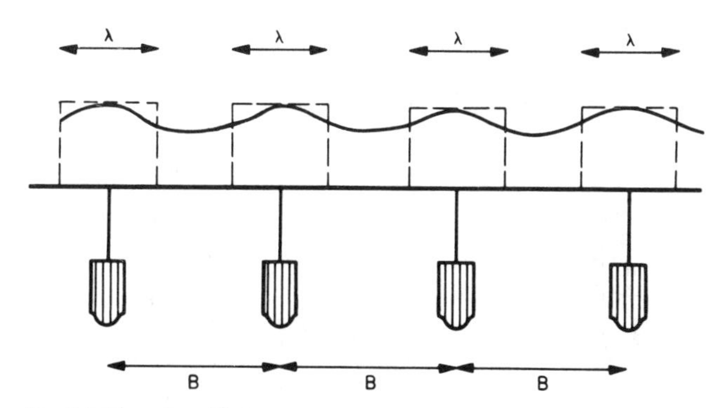

Fig. 7.3 Shear lag effects

Problem	Bending moments	Deflections
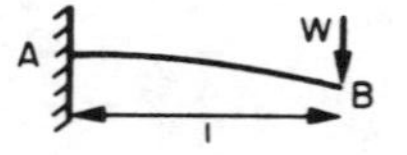 A, W, B, l	$BM_A = Wl$	$\delta_B = \dfrac{Wl^3}{3EI}$
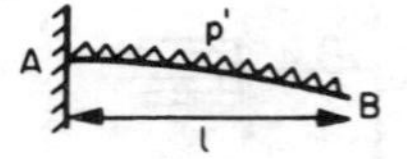 A, p′, B, l	$BM_A = \dfrac{p'l^2}{2}$	$\delta_B = \dfrac{p'l^4}{8EI}$
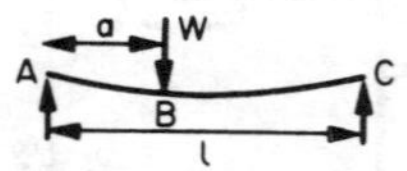 a, W, A, B, C, l	$BM_B = Wa\left(\dfrac{l-a}{l}\right)$	$\delta = \dfrac{Wx}{6EI}\left(\dfrac{l-a}{l}\right)(2al-a^2-x^2)+\dfrac{W}{6EI}[x-a]^3$* if $a = \dfrac{l}{2}$, $\delta_B = \dfrac{Wl^3}{48EI}$ * Disregard last term if $x < a$.
$u = \frac{l}{2}\sqrt{\frac{P}{EI}}$ 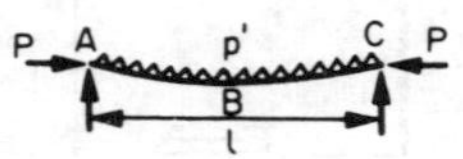 P, A, p′, C, P, B, l	$BM_B = \dfrac{p'l^2}{8} \cdot \dfrac{2(1-\cos u)}{u^2 \cos u}$	$\delta = \dfrac{p'l^2}{4Pu^2}\left\{\dfrac{\cos u(1-2x/l)}{\cos u}-1\right\}-\dfrac{p'x}{2P}(l-x)$ if $P = 0$, $\delta_B = \dfrac{5p'l^4}{384EI}$

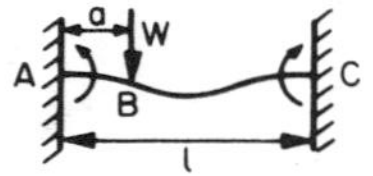	$\mathrm{BM_C} = Wa\left(\frac{l-a}{l}\right)^2$ $\mathrm{BM_A} = W(l-a)\left(\frac{a}{l}\right)^2$ if $a = \frac{l}{2}$. $\mathrm{BM_A} = \mathrm{BM_C} = \frac{Wl}{8}$	$\delta_{\max} = \frac{2Wa^3}{3EI}\left(\frac{l-a}{l+2a}\right)^2$ if $a = \frac{l}{2}$, $\delta_\mathrm{B} = \frac{Wl^3}{192EI}$
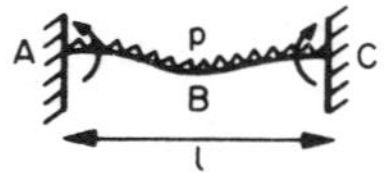	$\mathrm{BM_A} = \mathrm{BM_C} = \frac{p'l^2}{12}$	$\delta = \frac{p'x^2(l-x)^2}{24EI}$ $\delta_\mathrm{B} = \frac{p'l^4}{384EI}$
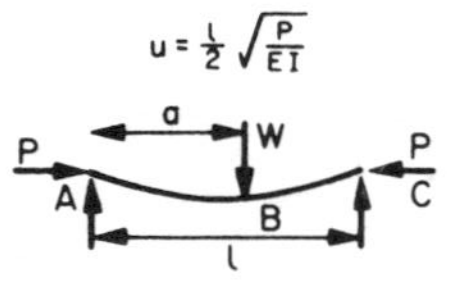 $u = \frac{l}{2}\sqrt{\frac{P}{EI}}$	For AB $\mathrm{BM} = \frac{Wl}{2u} \cdot \frac{\sin\frac{2ux}{l}\sin 2u\left(1-\frac{a}{l}\right)}{\sin 2u}$ max at $x = \frac{\pi l}{4u}$ or a For BC $\mathrm{BM} = \frac{Wl}{2u} \cdot \frac{\sin\frac{2au}{l}\sin 2u\left(1-\frac{x}{l}\right)}{\sin 2u}$ max at $x = l\left(1-\frac{\pi}{4u}\right)$ or a	For AB $\delta = \frac{Wl}{2uP}\left\{\frac{\sin 2u\left(1-\frac{a}{l}\right)\sin\frac{2ux}{l}}{\sin 2u} - \frac{2xu}{l}\left(1-\frac{a}{l}\right)\right\}$ For BC $\delta = \frac{Wl}{2uP}\left\{\frac{\sin\frac{2au}{l}\sin 2u\left(1-\frac{x}{l}\right)}{\sin 2u} - \frac{2au}{l}\left(1-\frac{x}{l}\right)\right\}$

distribution of stress cannot be found without some advanced mathematics, the maximum stress can still be found by simple beam theory if, instead of assuming that all of the plating is partially effective, it is assumed that part of the plating is wholly efective. This effective breadth of flange, λ (Fig. 7.3), is used to calculate the effective second moment of area of cross-section. It is dependent on the type of loading and the geometry of the structure. Because it is quite close to the neutral axis, the effective breadth of plating is not very influential and a figure of thirty thicknesses of plating is commonly used and sufficiently accurate; otherwise $\lambda = B/2$ is used.

There remains, in the investigation of the single stiffener-plating combination, the problem of behaviour under end load. Classical Euler theories assume perfect struts and axial loads which never occur in practice. Many designers use these or the Rankine–Gordon formula which embraces the overriding case of yield, together with a factor of safety often as high as twenty or thirty. This is not a satisfactory approach, since it disguises the actual behaviour of the member. Very often, in ship structures there will be a lateral load, which transforms the problem from one of elastic instability into one of bending with end load. Typical solutions appear in Fig. 7.2. The lateral load might be sea pressure, wind, concentrated weights, personnel load, cargo or flooding pressure. However, there will, occasionally, arise problems where there is no lateral load in the worst design case. How should the designer proceed then?

Practical structures are always, unintentionally, manufactured with an initial bow, due to welding distortion, their own weight, rough handling or processes of manufacture. It can be shown that the deflection of a strut with an initial simple bow y_0 is given by:

$$y = \frac{P_E}{P_E - P} y_0 \quad (y \text{ is the total deflection, including } y_o).$$

$P_E/(P_E - P)$ is called the *exaggeration factor*. P_E is the classical Euler collapse load, $P_E = \pi^2 EI/l^2$, l being the effective length (Fig. 7.4).

The designer must therefore decide first what initial bow is likely; while some measurements have been taken of these in practical ship structures, the designer will frequently have to make a common-sense estimate. Having decided the value, and calculated the Euler load, the designer can find the maximum deflection from the equation above.

The maximum bending moment for a member in which end rotation is not constrained is, of course,

$$\text{Max BM} = Py_{max}$$

In general, so far as end loading is concerned, the assumption of simple support is safe.

GRILLAGES

Consider the effect of a concentrated weight W on two simply-supported beams at right angles to each other as shown in Fig. 7.5. This is the simplest

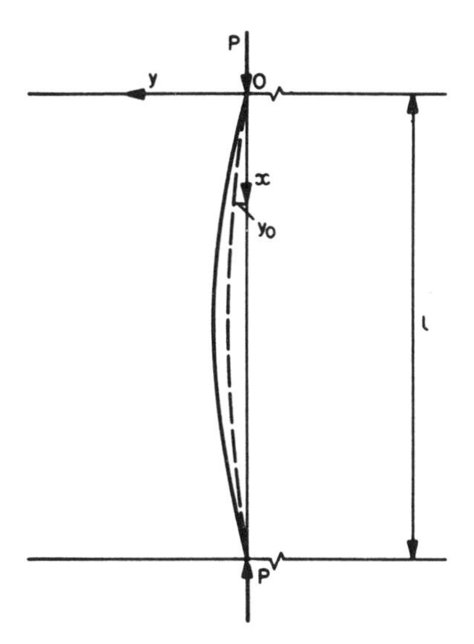

Fig. 7.4

form of grillage. Assume that the beams, defined by suffixes 1 and 2, intersect each other in the middle and that each is simply supported. What is not immediately obvious is how much the flexure of beam 1 contributes to supporting W and how much is contributed by the flexure of beam 2. Let the division of W at the middles be R_1 and R_2, then

$$R_1 + R_2 = W$$

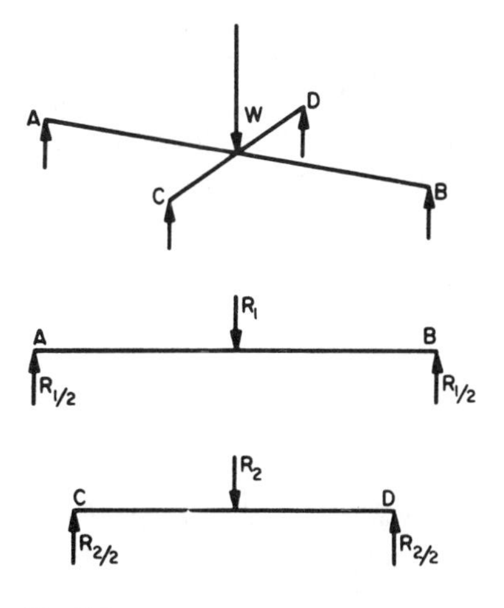

Fig. 7.5

Examining each beam separately, the central deflections are given by

$$\delta_1 = \frac{R_1 l_1^3}{48EI_1}$$

and

$$\delta_2 = \frac{R_2 l_2^3}{48EI_2}$$

But, if the beams do not part company, $\delta_1 = \delta_2$

$$\therefore \quad \frac{R_1 l_1^3}{I_1} = \frac{R_2 l_2^3}{I_2}$$

Since $R_1 + R_2 = W$,

$$\delta = \frac{W l_1^3}{48EI_1} \times \frac{1}{\left(1 + \frac{I_2 l_1^3}{I_1 l_2^3}\right)}$$

and the maximum bending moments in the two beams are

$$\text{Max BM}_1 = \frac{R_1 l_1}{4} = \frac{W l_1}{4\left(1 + \frac{I_2 l_1^3}{I_1 l_2^3}\right)}$$

$$\text{Max BM}_2 = \frac{R_2 l_2}{4} = \frac{W l_2}{4\left(1 + \frac{I_1 l_2^3}{I_2 l_1^3}\right)}$$

This has been an exceedingly simple problem to solve. It is not difficult to see, however, that hand computation of this sort could very quickly become laborious. Three beams in each direction, unaided by symmetry could give rise to nine unknowns solved by nine simultaneous equations. Moreover, a degree of fixity at the edges introduces twelve unknown moments while moments at the intersections cause twist in the orthogonal beams. Edge restraint, uneven spacing of stiffeners, differing stiffeners, contribution from plating, shear deflection and other factors all further contribute to making the problem very difficult indeed.

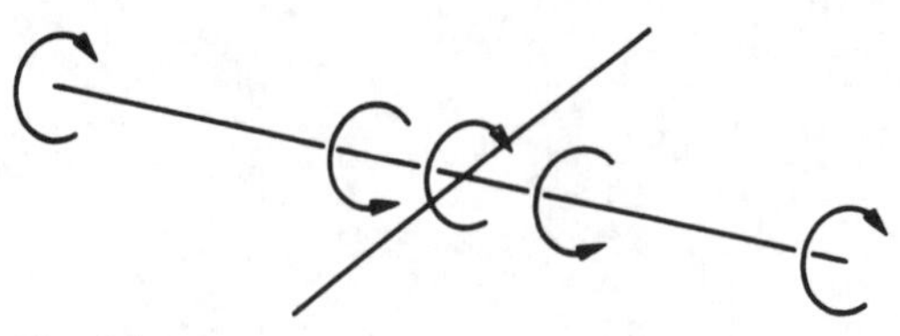

Fig. 7.6

Advanced mathematical theories have been evolved to solve these problems in a variety of ways with various assumptions, and computers have been programmed to turn the theories into data sheets of use to the practising designer.

SWEDGED PLATING

Fabrication costs can be reduced in the construction of surfaces which need not be plane, by omitting stiffeners altogether and creating the necessary flexural rigidity by corrugated or swedged plating. Main transverse bulkheads in a large oil tanker, for example, may have a depth of swedge of 25 cm. Such plating is incorporated into the ship to accept end load in the direction of the swedges as well as lateral loading. It tends to create difficulties of structural discontinuity where the swedge meets conventional stiffeners, for example where the main transverse bulkhead swedges meet longitudinal deck girders.

Buckling of some faces of the plating is possible if the swedges are not properly proportioned and it is this difficulty, together with the shear diffusion in the plating, which makes the application of simple beam theory inadequate. Properly designed, corrugated plating is highly efficient.

COMPREHENSIVE TREATMENT OF STIFFENED PLATING

A remarkable research programme over many years has resulted in the derivation of comprehensive design advice on stiffened plating under combined compression and lateral pressure (Ref. 7). As will be discussed presently, panels or stiffened plating may shirk their duty by buckling so that they do not make their expected contribution to the overall ship's sectional modulus. This shirking, or load shortening, is illustrated in Fig. 7.7 based upon Ref. 7. This shows that the load shortening depends upon:

(*a*) the imperfections of the stiffeners in the form of a bow
(*b*) plating panel slenderness ratio $\beta = (b/t)\sqrt{\sigma_Y/E}$
(*c*) the ratio of stiffener cross section A_s to the overall cross section A ($A_s/A = 0{\cdot}2$ average imperfections of stiffener)

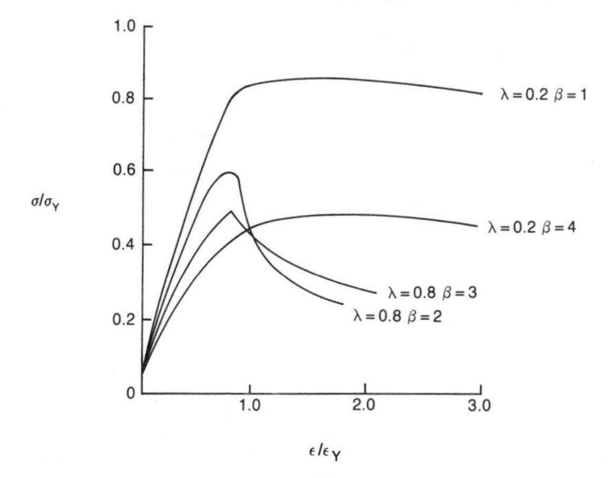

Fig. 7.7 Load shortening curves for panels with tee bar stiffeners

(*d*) the stiffener slenderness ratio

$$\lambda = \frac{l}{\pi k}\sqrt{\frac{\sigma_Y}{E}}$$

where k is the radius of gyration of the plating/stiffener combination and l is the stiffener length.

Note from Fig. 7.7 the sudden nature of the change, occurring especially for high values of λ. A_s/A is also found to be very significant as might be expected. Readers are referred to the reference for a more complete appraisal and also for load shortening curves for plating panels alone like that shown in Fig. 7.37.

Panels of plating

Knowledge of the behaviour of panels of plating under lateral pressure (sometimes called plate elements) has advanced rapidly since 1956. It is important for the designer to understand the actual behaviour of plates under this loading, so that he can select the theory or results most suitable for his application. To do this, consider how a panel behaves as the pressure is increased.

BEHAVIOUR OF PANELS UNDER LATERAL LOADING

Consider, at first, the behaviour of a rectangular panel with its four edges clamped, and unable to move towards each other. As soon as pressure is applied to one side, elements in the plate develop a flexural resistance much like the elements of a simple beam but in orthogonal directions. Theories relating the pressure to the elastic flexural resistance of the plate alone are called *small deflection theories.* As the pressure is increased and deflection of the same order as the plate thickness results, the resistance of the plate to the pressure stiffens because of the influence of membrane tension. This influence

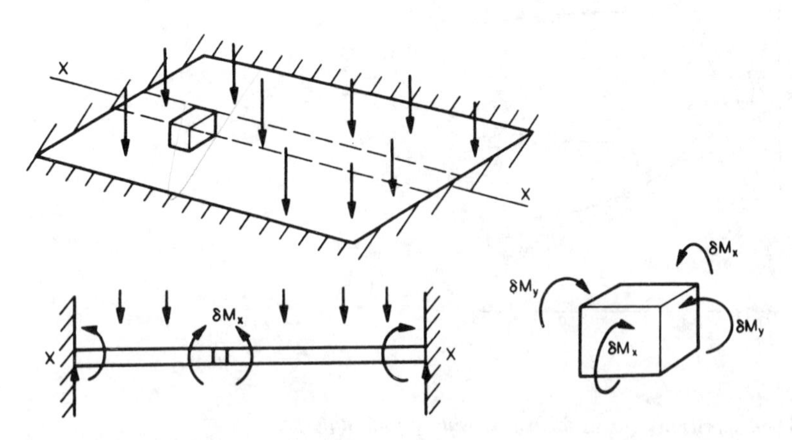

Fig. 7.8 Flexural resistance of a plate

is dependent upon the deflection, since the resistance is due to the resolute of the tension against the direction of the pressure. Clearly, it has but a small influence when the deflection is small. Figure 7.9 shows a typical section through the plate; the orthogonal section would be similar. Elastic theories which take into account both flexural rigidity and membrane tension are called *large deflection theories.*

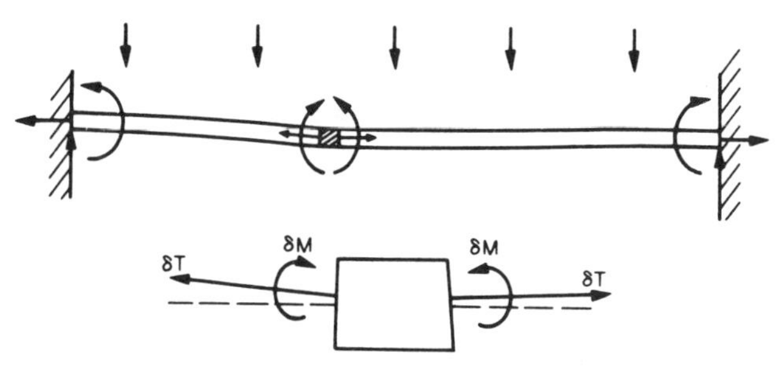

Fig. 7.9 Membrane tension effects in a plate

With a further increase in pressure, yield sets in, deflection increases more rapidly and the effect of membrane tension becomes predominant. The plate is partly elastic and partly plastic and theories relating to the behaviour following the onset of yield are called *elasto-plastic theories.* Yield is first reached at the middles of the two longer sides on the pressure side of the plate; soon afterwards, yield is reached on the other side of the plate and this area of plasticity spreads towards the corners. Plasticity spreads with increase in pressure in the stages illustrated in Fig. 7.10. The plastic areas, once they have developed through the thickness, are called hinges since they offer constant resistance to rotation. Finally, once the hinges have joined to form a figure of eight, distortion is rapid and the plate distends like a football bladder until the ultimate tensile strength is reached.

The precise pattern of behaviour depends on the dimensions of the panel but this description is typical. There can be no doubt that after the first onset of yield, a great deal of strength remains. Unless he has good reasons not to do so, the designer would be foolish not to take advantage of this strength to effect an economical design. Once again, this brings us to an examination of 'failure'. As far as a panel under lateral pressure is concerned, failure is likely to be either fatigue fracture or unacceptable deflection. Deflection considered unacceptable for reasons of appearance, to avoid the 'starved horse' look, might nowadays be thought an uneconomical criterion. In considering fatigue fracture, it must be remembered that any yield will cause some permanent set; removal of the load and reapplication of any lesser load will not increase the permanent set and the plate will behave elastically. A plate designed to yield under a load met very rarely, will behave elastically for all of its life save for the one loading which causes the maximum permanent set. Indeed, initial permanent set caused by welding distortion will permit the plate to behave elastically, thereafter, if this is taken as the maximum acceptable permanent set.

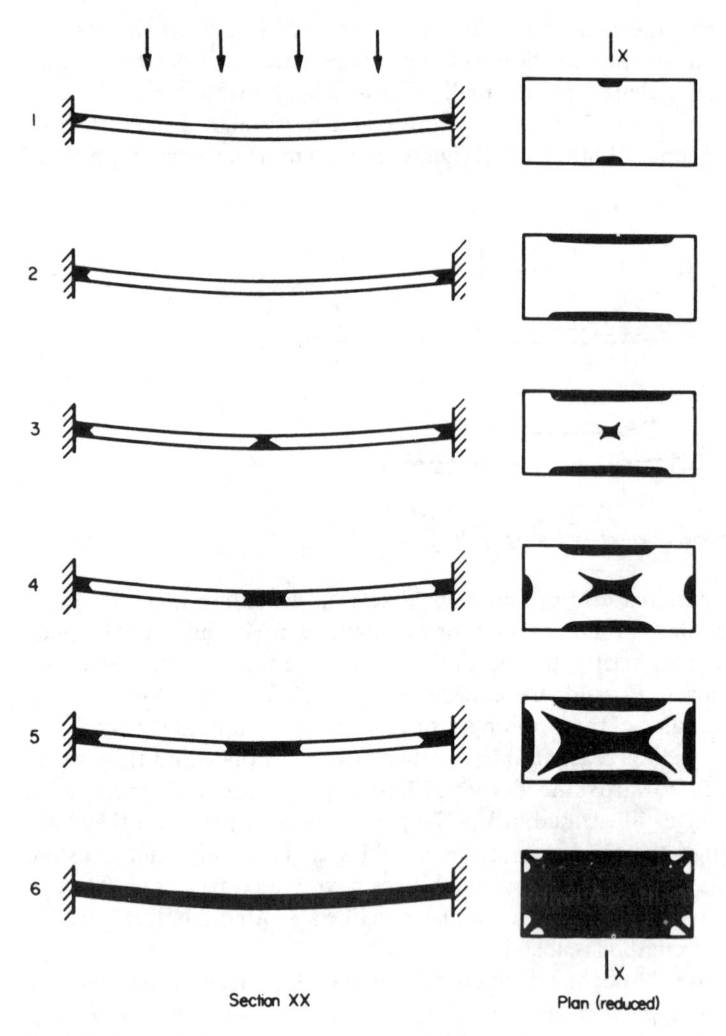

Fig. 7.10 Onset of plasticity in a plate

In ship's structures, small deflection elastic theories would be used for plates with high fluctuating loading such as those opposite the propeller blades in the outer bottom and where no permanent set can be tolerated in the flat keel, or around sonar domes, for example, in the outer bottom. Large deflection elastic theory is applicable where deflection exceeds about a half the thickness, as is likely in thinner panels. Elastoplastic theories are appropriate for large areas of the shell, for decks, bulkheads and tanks.

AVAILABLE RESULTS FOR FLAT PLATES UNDER LATERAL PRESSURE

The three ranges covered by the different theories are illustrated in Fig. 7.11. It is clear that unless the correct theory is chosen, large errors can result.

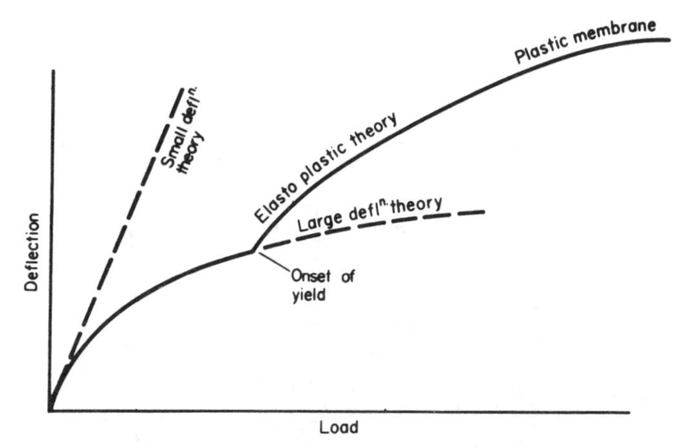

Fig. 7.11 Comparison of plate theories

For plates with totally clamped edges (which are rare):

$$\text{Central deflection} = k_1 \cdot \frac{pb^4}{384D}$$

$$\text{Maximum stress} = k_2 \cdot \frac{p}{2}\left(\frac{b}{t}\right)^2$$

where

$$D = \frac{Et^3}{12(1-\nu^2)}$$

k_1 and k_2 are non-dimensional and the units should be consistent.

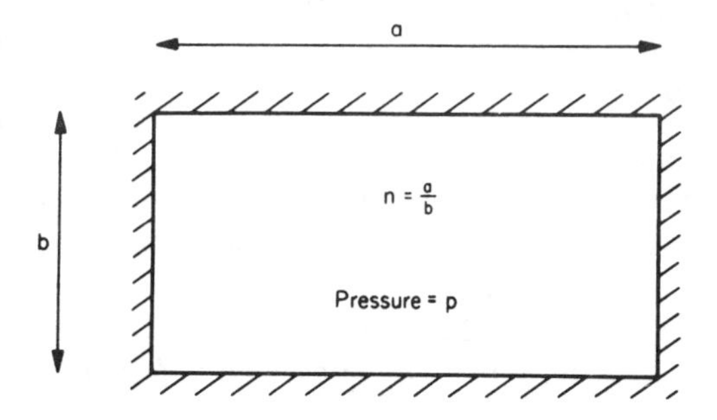

Fig. 7.12

n	1	1·25	1·50	1·75	2	∞
k_1	0·486	0·701	0·845	0·928	0·976	1·0
k_2	0·616	0·796	0·906	0·968	0·994	1·0

Large deflection elastic theory gives results as shown in Fig. 7.13.

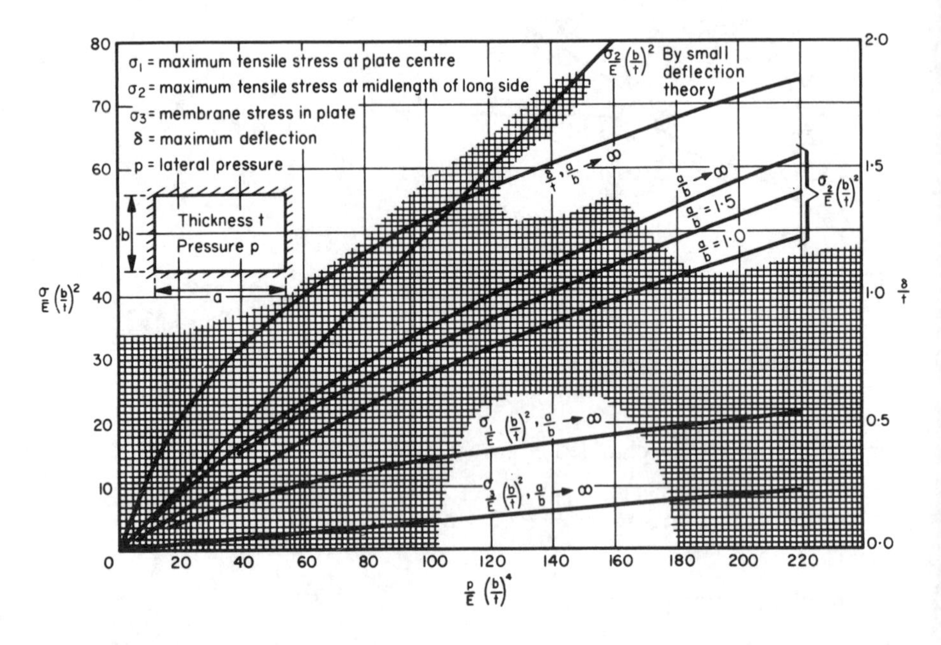

Fig. 7.13 Large deflection results

Elasto-plastic results are based upon the maximum allowable pressure defined, arbitrarily, as the lesser pressure which will cause either

(*a*) a central plastic hinge in a long plate or 1·25 times the pressure which causes yield at the centre of a square plate; or
(*b*) the membrane tension to be two-thirds the yield stress.

The first criterion applies to thick plates and the latter to thin plates. Design curves giving maximum permissible pressure, deflection and permanent set based on these criteria are shown in Fig. 7.14. Results from an important extension to plate theory, in which pressures have been calculated which will not permit any increase in an initial permanent set, i.e. the plate, after an initial permanent set (caused perhaps by welding) behaves elastically, are presented in Fig. 7.15 for long plates.

References 2 and 3 make further important steps in examining the real behaviour of panels forming part of a grillage. These show that pull-in at the edges has an appreciable effect on panel behaviour, and that a panel in a grillage has not the edge constraint necessary to ensure behaviour in the manner of Fig. 7.14. The edge constraint which can give rise to membrane tension arises from the hoop effects in the plane of the boundary. Figure 7.16 gives design curves assuming that the edges of the panel are free to move inwards.

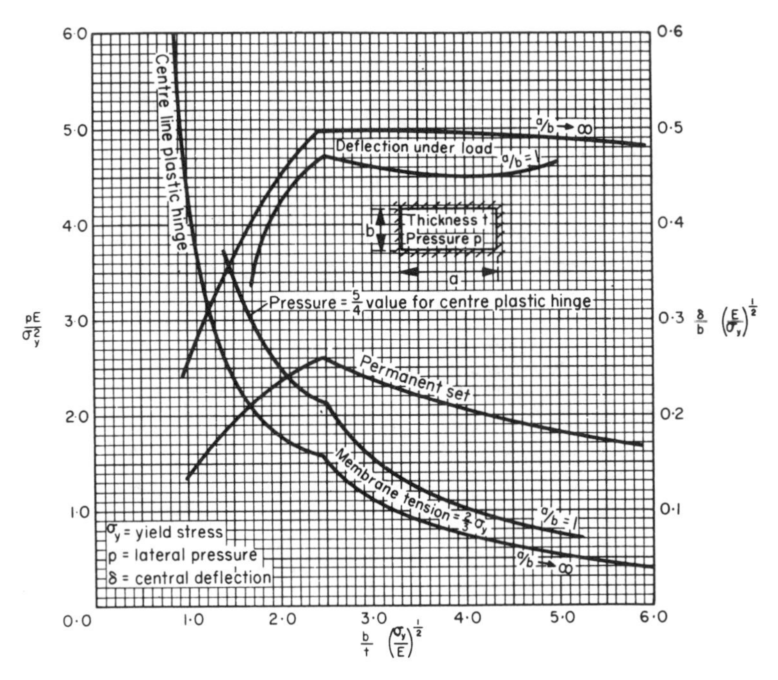

Fig. 7.14 Elasto-plastic results

EXAMPLE. Compare the design pressures suitable for a steel plate 0·56 m × 1·26 m = 6 mm yielding at 278 N/mm² assuming, (*a*) the edges fixed and (*b*) the edges free to move inwards. Calculate also the pressure which will not cause additional permanent set above an initial bow of 2·8 mm due to welding. Had the plate been initially flat, what pressure would have first caused yield?

Solution:

$$\frac{a}{b} = \frac{1{\cdot}26}{0{\cdot}56} = 2{\cdot}25 \quad \frac{b}{t}\left(\frac{\sigma_y}{E}\right)^{\frac{1}{2}} = \frac{0{\cdot}56}{0{\cdot}006}\left(\frac{278}{209{,}000}\right)^{\frac{1}{2}} = 3{\cdot}41 \text{ for which Fig. 7.14}$$

gives

$$\frac{pE}{\sigma_y^2} = 1{\cdot}00 \text{ approx.} \quad \text{and} \quad \frac{\delta}{b}\left(\frac{E}{\sigma_y}\right)^{\frac{1}{2}} = 0{\cdot}224$$

Hence,

$$\text{design pressure} = \frac{1{\cdot}00 \times 278^2}{209{,}000} = 0{\cdot}37 \text{ N/mm}^2$$

Using the same criterion of failure, viz. a permanent set coefficient of 0·224 and interpolating between the two diagrams of Fig. 7.16,

$$\frac{pE}{\sigma_y^2} = 0{\cdot}57 \text{ approx.}$$

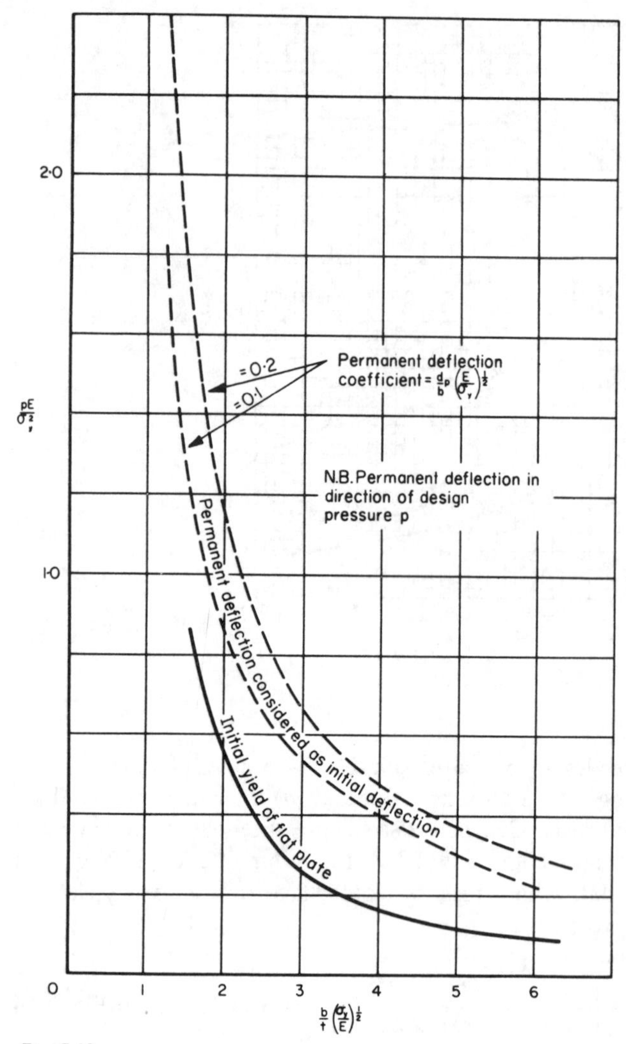

Fig. 7.15

Hence,

$$\text{design pressure} = \frac{0{\cdot}57}{1{\cdot}00} \times 0{\cdot}37 = 0{\cdot}21 \text{ N/mm}^2$$

From Fig. 7.15, with a permanent deflection coefficient of

$$\frac{0{\cdot}0028}{0{\cdot}56}\left(\frac{209{,}000}{278}\right)^{\frac{1}{2}} = 0{\cdot}137, \quad \text{and} \quad \frac{b}{t}\left(\frac{\sigma_y}{E}\right)^{\frac{1}{2}} = 3{\cdot}41,$$

the pressure which will not cause additional permanent set is obtained from $pE/\sigma_y^2 = 0{\cdot}50$, whence

$$p = \frac{0{\cdot}50}{1{\cdot}00} \times 0{\cdot}37 = 0{\cdot}185 \text{ N/mm}^2$$

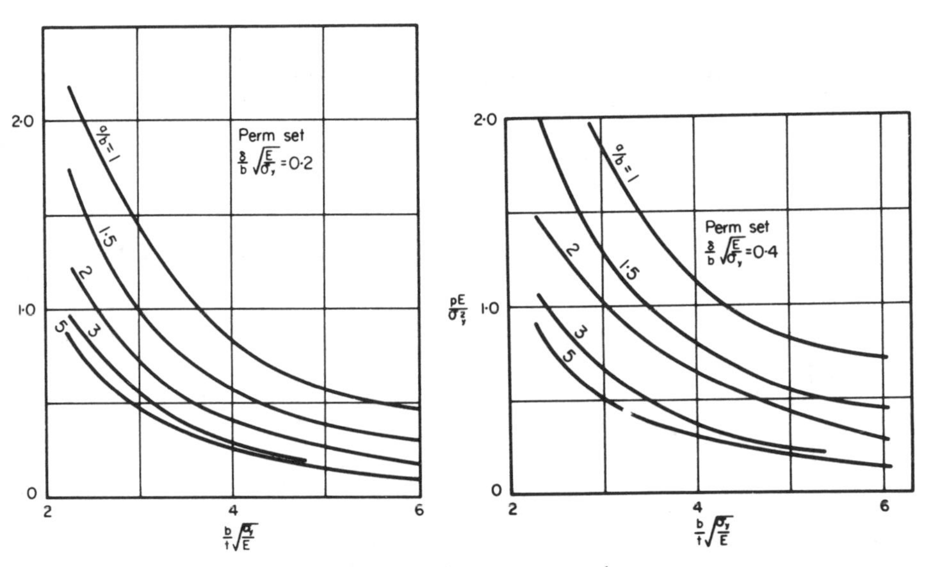

Fig. 7.16 Design curves for panels with edges free to move inwards

To find the pressure at which this plate begins first to yield, try first small deflection theory:

$$\frac{a}{b} = 2{\cdot}25 \quad \text{so that} \quad k_2 = 1{\cdot}0 \text{ approximately}$$

Then

$$278 \text{ N/mm}^2 = \frac{p}{2}\left(\frac{0{\cdot}56}{0{\cdot}006}\right)^2$$

whence

$$p = 0{\cdot}0638 \text{ N/mm}^2$$

Check the deflection at first yield:

$$D = \frac{Et^3}{12(1-\nu^2)} = \frac{209(6)^3\ 10^{-6}}{12(1-0{\cdot}09)} = 0{\cdot}00413 \text{ MN m} \quad \text{and} \quad k_1 = 0{\cdot}98$$

$$\text{deflection} = \frac{0{\cdot}98 \times 0{\cdot}0638(0{\cdot}56)^4}{384 \times 0{\cdot}00413} = 3{\cdot}9 \text{ mm}$$

Small deflection theory thus gives a deflection about 65 per cent of the thickness and is probably adequate. However, let us take this example to illustrate the use of large deflection theory. Because the stress is divided into two parts, direct calculation of the pressure to cause yield is not possible, and two trial pressures must be taken: referring to Fig. 7.13,

$$\frac{\sigma}{E}\left(\frac{b}{t}\right)^2 = \frac{278}{209{,}000}\left(\frac{0{\cdot}56}{0{\cdot}006}\right)^2 = 11{\cdot}6$$

Now, for

$$p = 0{\cdot}06 \text{ N/mm}^2 \qquad \frac{p}{E}\left(\frac{b}{t}\right)^4 = \frac{0{\cdot}06}{209{,}000}\left(\frac{0{\cdot}56}{0{\cdot}006}\right)^4 = 21{\cdot}8$$

whence, from Fig. 7.13,

$$\frac{\sigma_3}{E}\left(\frac{b}{t}\right)^2 = 1{\cdot}2 \quad \text{and} \quad \frac{f_2}{E}\left(\frac{b}{t}\right)^2 = 10{\cdot}1$$

thus, $\sigma_3 = 28{\cdot}8$ and $\sigma_2 = 242{\cdot}3$ so that the total stress is 271 N/mm². Similarly, for $p = 0{\cdot}07$ N/mm²

$$\frac{p}{E}\left(\frac{b}{t}\right)^4 = \frac{0{\cdot}07}{0{\cdot}06} \times 21{\cdot}8 = 25{\cdot}4$$

$$\frac{\sigma_3}{E}\left(\frac{b}{t}\right)^2 = 1{\cdot}3 \quad \text{and} \quad \frac{f_2}{E}\left(\frac{b}{t}\right)^2 = 11{\cdot}2$$

so that $\sigma = 31{\cdot}2 + 268{\cdot}7 = 300$ N/mm²

By interpolation, the pressure first to cause yield of 278 N/mm² is 0·062 N/mm². This example typifies the use of Fig. 7.13 for large deflection elastic theory; in this particular case, it confirms that small deflection theory was, in fact, sufficiently accurate.

BUCKLING OF PANELS

Buckling of panels in the direction of the applied lateral pressure is known as snap through buckling. This is likely where the initial permanent set is $1\frac{1}{2}$–3 times the thickness.

Buckling due to edge loading has been dealt with on a theoretical basis. For a panel simply supported at its edges, the critical buckling stress is given by

$$\sigma_c = \frac{k\pi^2 E}{12(1-\nu^2)}\left(\frac{t}{b}\right)^2$$

where k is given by Fig. 7.17. When a plate does buckle in this way, it may not

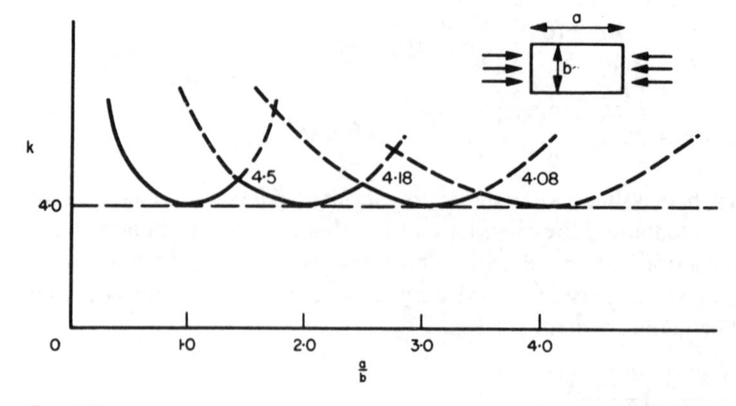

Fig. 7.17

be obvious that it has occurred; in fact, the middle part of the panel shirks its load which is thrown on to the edge stiffeners. It is common practice to examine only the buckling behaviour of these stiffeners associated with a width of plating equal to thirty times its thickness.

Buckling due to shear in the plane of the plate causes wrinkling in the plate at about 45 degrees. Such a failure has been observed in the side plating of small ships at the sections of maximum shear. The critical shear stress is given by

$$\tau_c = kE\left(\frac{t}{b}\right)^2$$

where $k = 4{\cdot}8 + 3{\cdot}6\,(b/a)^2$ for edges simply supported or $k = 8{\cdot}1 + 5{\cdot}1(b/a)^2$ for edges clamped. A more truly representative examination of panel behaviour under biaxial compression and lateral pressure is given in Ref. 8. This important culmination of many years of research enables a designer to determine optimum panel shapes and to take into account initial strains and imperfections in the plating.

Frameworks

Analysis of the three-dimensional curved shape of the hull between main bulkheads is the correct approach to the determination of its strength. Results of a generalized analysis of such a three-dimensional grillage are not yet available to the practising engineer and, when they are, may not be easy to use. The reduction of the problem to two-dimensional strips or frames therefore remains important. In reducing the problem to two-dimensional frameworks, it is necessary to be aware that approximations are being made.

There are, in general, three types of plane framework with which the ship designer is concerned:

(*a*) orthogonal portals
(*b*) ship-shape rings
(*c*) circular rings.

Portals arise in the consideration of deckhouses, superstructures and similar structures and may have one or more storeys. If the loading in the plane of the portal is concentrated, the effect of the structure perpendicular to the plane of the portal is likely to be one of assistance to its strength. If the loading is spread over many portals, one of which is being isolated for analysis, the effects of the structure perpendicular to its plane will be small, from considerations of symmetry unless there is sidesway when the in-plane stiffness of the plating will be appreciable. Thus, the reduction of the problems to two-dimensional frameworks is, in general, pessimistic and safe, although each problem should be examined on its merits.

Ship-shape rings arise by isolating a transverse slice of the ship, comprising bottom structure, side frames and deck beams, together with their associated plating. Treatment of the complete curved shell in this manner is likely to be highly pessimistic because the effect of longitudinal structure in keeping this ring to shape must be considerable. These longitudinals connect the ring to transverse bulkheads with enormous rigidity in their own planes. The calculations are, nevertheless, performed to detect the likely bending moment distribution around the ring so that material may be distributed to meet it. Transverse strength calculations are, therefore, generally comparative in nature except, perhaps, in ships framed predominantly transversely.

Circular rings occur in submarines and other pressure vessels, such as those containing nuclear reactors.

METHODS OF ANALYSIS

There are many good textbooks available, which explain the various analytical tools suitable for frameworks in more detail than is possible here. Standard textbooks on structural analysis treat the more common framed structures met in ship design. A brief summary of four methods of particular use to the naval architect may, however, be worthwhile. These methods are: (*a*) moment distribution, (*b*) slope-deflection, (*c*) energy methods, and (*d*) limit design methods.

(*a*) The *moment distribution* of Hardy Cross is particularly suitable for portal problems where members are straight and perpendicular to each other. Bending moment distribution is obtained very readily by this method but slopes and deflection are not obtained without the somewhat more general approach of Ref. 4.

The moment distribution process is one of iteration in methodical sequence as follows:

(i) all joints of the framework are assumed frozen in space, the loading affecting each beam as if it were totally encastre, with end fixing moments;
(ii) one joint is allowed to rotate, the total moment at the joint being distributed amongst all the members forming the joint according to the formula $(I/l)/(\Sigma I/l)$; the application of such a moment to a beam causes a carry-over of one-half this value (if encastre at the other end) to the far end, which is part of another joint;
(iii) this joint is then frozen and a half of the applied moment is carried over to each remote end (sometimes, the carry-over factor is less than one-half—indeed, when the remote end is pinned, it is zero);
(iv) the process is repeated at successive joints throughout the framework until the total moments at each joint are in balance.

This process prevents sidesway of the framework which occurs unless there is complete symmetry. This is detected by an out-of-balance moment on the overall framework. As a second cycle of operations therefore, sufficient side force is applied to liquidate this out-of-balance without allowing joint rotation, thus causing new fixing moments at the joints. These are then relaxed by repeating the first cycle of operations and so on. A simple example of a moment distribution is given presently, but for a full treatment, the student is referred to standard textbooks on the analysis of frameworks.

(*b*) *Slope deflection analysis* is based upon the fundamental equation

$$M = EI\frac{\mathrm{d}^2 y}{\mathrm{d}x^2}$$

Thence, the area of the M/EI-curve, $= \int \frac{M}{EI}\,\mathrm{d}x$, gives the change of slope,

dy/dx. Integrated between two points in a beam, $\int \frac{M}{EI} dx$ gives the difference in the slopes of the tangents at the two points. Further, $\int \frac{Mx}{EI} dx$ between A and B, i.e. the moment of the M/EI-diagram about a point A gives the distance AD between the tangent to B and the deflected shape as shown in Fig. 7.18. These two properties of the M/EI-diagram are used to determine the distribution of bending moment round a framework.

Consider the application of the second principle to a single beam AB subjected to an external loading and end fixing moments M_{AB} and M_{BA} at which the slopes are θ_{AB} and θ_{BA}, positive in the direction shown in Fig. 7.18.

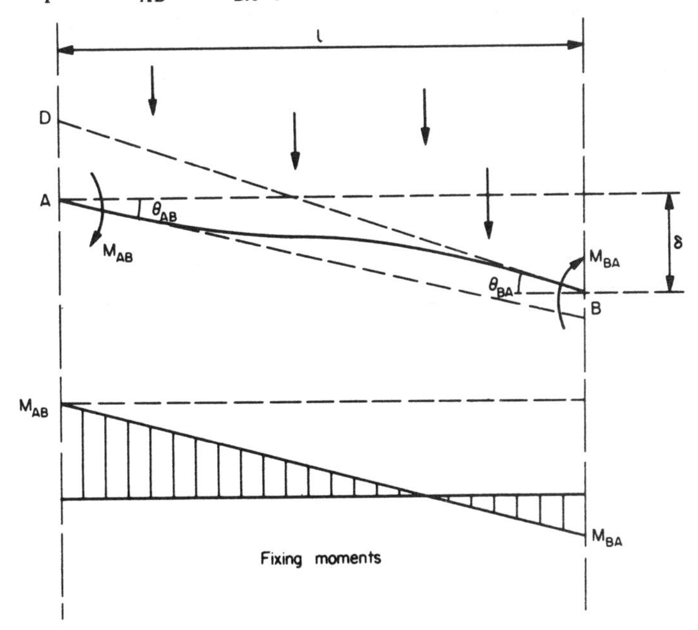

Fig. 7.18

Let the first moment of the free bending moment diagram (i.e. that due to the external loading assuming the ends to be pinned) about the ends A and B be respectively m_A and m_B. Then, taking moments of the M/EI-diagram about A,

$$\frac{M_{AB}}{EI}\frac{l^2}{2} - \frac{(M_{AB}+M_{BA})}{2EI}\frac{2l^2}{3} + \frac{m_A}{EI} = -l\theta_{BA} + \delta$$

i.e.

$$\theta_{BA} = -\frac{l}{6EI}\left(M_{AB} - 2M_{BA} + \frac{6m_A}{l^2}\right) + \frac{\delta}{l}$$

Taking moments about B,

$$\frac{M_{AB}}{EI}\frac{l^2}{2} - \frac{(M_{AB}+M_{BA})}{2EI}\frac{l^2}{3} + \frac{m_B}{EI} = l\theta_{AB} - \delta$$

i.e.

$$\theta_{AB} = \frac{l}{6EI}\left(2M_{AB} - M_{BA} + \frac{6m_B}{l^2}\right) + \frac{\delta}{l}$$

These expressions are fundamental to the slope deflection analysis of frameworks. They may be applied, for example, to the simple portal ABCD of Fig. 7.19, expressions being obtained for the six slopes, two of which will be zero and two pairs of which (if C and B are rigid joints) will be equated. On eliminating all these slopes there will remain five equations from which the five unknowns, M_{AB}, M_{BA}, M_{CD}, M_{DC} and δ can be found. This method has the advantage over moment distribution of supplying distortions, but it becomes arithmetically difficult when there are several bays. It is suitable for computation by computer where repetitive calculations render a program worth writing. Sign conventions are important. Reference 18 gives some applications of slope deflection analysis to transverse strength.

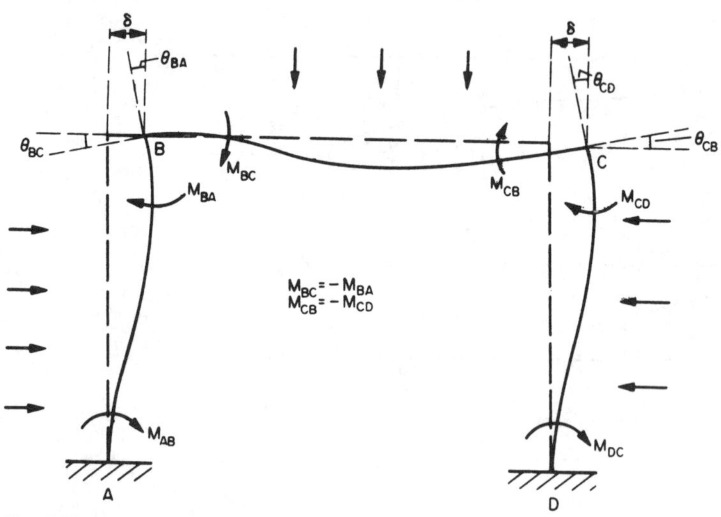

Fig. 7.19

(c) The *energy method* most useful in the context of this chapter is based on a theorem of Castigliano. This states that the partial derivative of the total strain energy U with respect to each applied load is equal to the displacement of the structure at the point of application in the direction of the load:

$$\frac{\partial U}{\partial P} = \delta_P$$

The expressions for strain energies U due to direct load, pure bending, torsion and shear are:

Direct load P, member of cross-sectional area A,

$$U = \int \frac{P^2}{2AE} \mathrm{d}x$$

Bending moment M, curved beam of second moment I,

$$U = \int \frac{M^2}{2EI} \, ds$$

Torque T, member of polar second moment J,

$$U = \int \frac{T^2}{2CJ} \, dx$$

Shearing force S, element of cross-sectional area A, (Fig. 7.20)

$$U = \int \frac{S^2}{2AC} \, dx = \int \frac{\tau^2 A \, dx}{2C}$$

The strain energy in a cantilever of rectangular cross-section $a \times b$, for example, with an end load W is given by

$$U = \int_0^l \frac{M^2}{2EI} \, dx + \int_0^{ab} \int_0^l \frac{\tau^2}{2C} \, dA \, dx$$

Now shear stress τ varies over a cross-section according to the expression $\tau = (SA/Ib)\bar{y}$. For the rectangular cantilever then,

$$\tau = \frac{6W}{a^3 b}\left(\frac{a^2}{4} - y^2\right) \quad \text{and} \quad dA = b \, dy.$$

Then

$$U = \int_0^l \frac{W^2 x^2 \, dx}{2EI} + 2\int_0^{a/2} \int_0^l \frac{18W^2}{a^6 bC}\left(\frac{a^4}{16} - \frac{a^2 y^2}{2} + y^4\right) dy \, dx$$

$$= \frac{W^2 l^3}{6EI} + \frac{3W^2 l}{5abC}$$

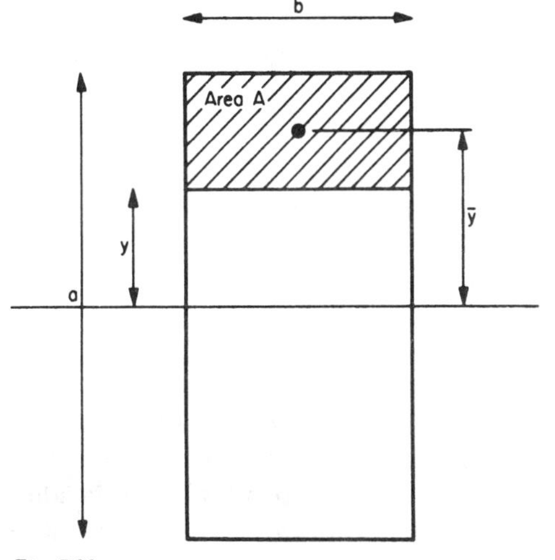

Fig. 7.20

End deflection

$$\delta = \frac{\partial U}{\partial W} = \frac{Wl^3}{3EI} + \frac{6Wl}{5abC}$$

The first expression is the bending deflection as given in Fig. 7.2, and the second expression is the shear deflection.

In applying strain energy theorems to the ring frameworks found in ship structures, it is common to ignore the effects of shear and direct load which are small in comparison with those due to bending. Confining attention to bending effects, the generalized expression becomes

$$\delta = \frac{\partial U}{\partial P} = \int \frac{M}{EI} \frac{\partial M}{\partial P} \mathrm{d}s$$

Applying this, by example, to a simple ship-shape ring with a rigid centre line bulkhead which can be replaced by three unknown forces and moments, these can be found from the three expressions, since all displacements at B are zero:

$$0 = \frac{\partial U}{\partial H} = \int \frac{M}{EI} \frac{\partial M}{\partial H} \mathrm{d}s$$

$$0 = \frac{\partial U}{\partial V} = \int \frac{M}{EI} \frac{\partial M}{\partial V} \mathrm{d}s$$

$$0 = \frac{\partial U}{\partial M} = \int \frac{M}{EI} \frac{\partial M}{\partial M_{\mathrm{B}}} \mathrm{d}s,$$

summed for members BC and CDE.

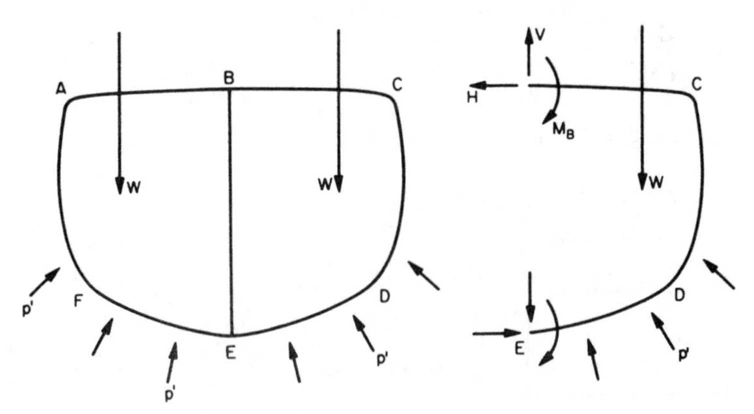

Fig. 7.21

Expressions for the bending moment M in terms of V, H, M_{B} and the applied load can be written down for members BC and CDE, added, and the above expressions determined. Hence V, H and M_{B} can be found and the bending moment diagram drawn. This method has the advantage that it deals readily with frames of varying inertia, the integrations being carried out by Simpson's rule. Like the slope deflection method, the arithmetic can become formidable.

(*d*) The *limit design method* is called also plastic design or collapse design. This method uses knowledge of the behaviour of a ductile material in bending beyond the yield point. A beam bent by end couples M within the elastic limit has a cross-section in which the stress is proportional to the distance from the neutral axis,

$$M = \sigma Z \quad \text{where } Z = \frac{I}{y}, \quad \text{the section modulus.}$$

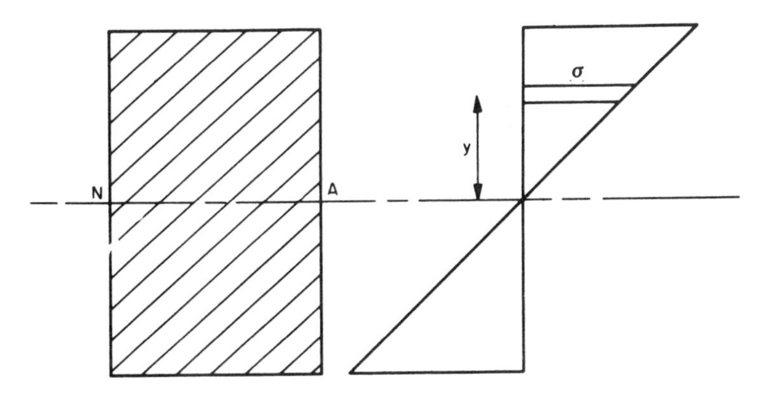

Fig. 7.22

Bent further, yield is reached first in the outer fibres and spreads until the whole cross-section has yielded, when the plastic moment, $M_p = \sigma_y S$ where S is the addition of the first moments of area of each side about the neutral axis and is called the plastic modulus. The ratio S/Z is called the *shape factor* which has a value 1·5 for a rectangle, about 1·2 for a rolled steel section and about 1·3 for a plating-stiffener combination. When a beam has yielded across its section, its moment of resistance is constant and the beam is said to have formed a plastic hinge.

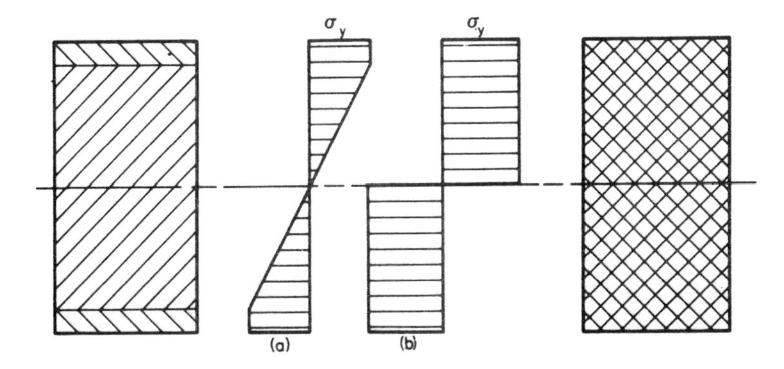

Fig. 7.23

The least load which forms sufficient plastic hinges in a framework to transform it into a mechanism is called the collapse load. A portal, for example, which has formed plastic hinges as shown in Fig. 7.24 has become a mechanism and has 'collapsed'. Let us apply this principle to an encastre beam (Fig. 7.25).

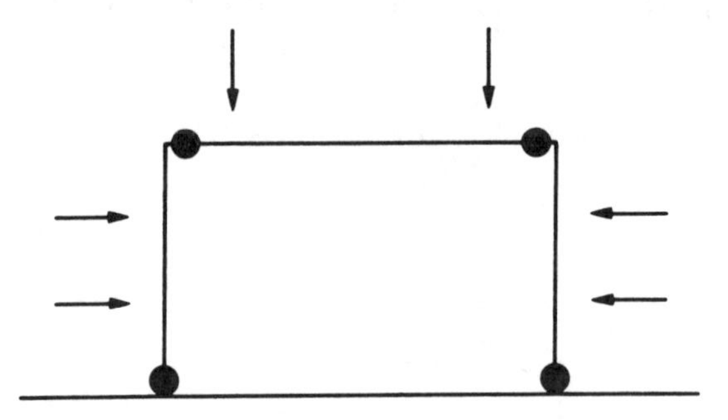

Fig. 7.24

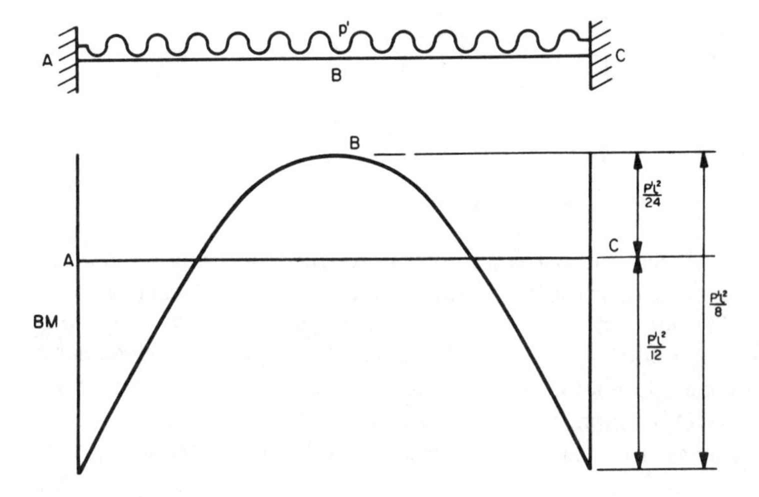

Fig. 7.25

If yield is assumed to represent 'failure' by an elastic method of design, the maximum uniformly distributed load that the beam can withstand is (from Fig. 7.2)

$$p' = \frac{12Z\sigma_y}{l^2}$$

Using the definition of failure for the plastic method of design, however, collapse occurs when hinges occur at A, B and C, and they will all be equal to $\frac{1}{2}p'(l^2/8)$, i.e.

$$p' = \frac{16S\sigma_y}{l^2}$$

If the shape factor for the beam section is taken as 1·2, the ratio of these two maximum carrying loads is 1·6, i.e. 60 per cent more load can be carried after the onset of yield before the beam collapses.

The plastic design method is often more conveniently applied through the principle of virtual work whence the distance through which an applied load moves is equated to the work done in rotating a plastic hinge. In the simple case illustrated in Fig. 7.26, for example,

$$W\frac{l\theta}{2} = M_p\theta + M_p\theta + M_p(2\theta) = 4M_p\theta$$

i.e. collapse load, $W = 8M_p/l$. This method can be used for finding the collapse loads of grillages under concentrated load; in this case, various patterns of plastic hinges may have to be tried in order to find the least load which would cause a mechanism (Fig. 7.27).

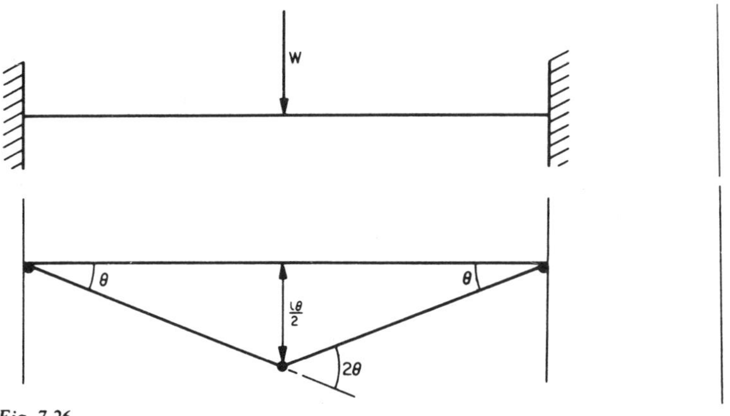

Fig. 7.26

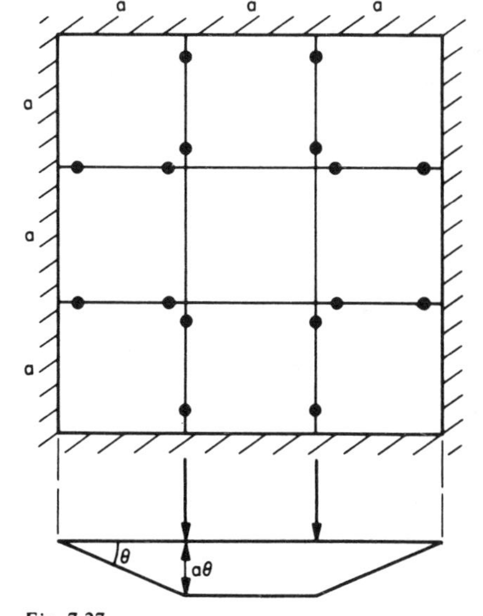

Fig. 7.27

Knowledge of the collapse loads of superstructures under nuclear blast or extreme winds is necessary, and the limit design method is the only way of calculating them. The method is suitable for the determination of behaviour of other parts of a ship structure subject to once-in-a-lifetime extreme loads such as bulkheads. Use of the method with a known factor of safety (or load factor), can ensure normal behaviour in the elastic range and exceptional behaviour in the plastic range—it is indeed, the only method which illustrates the real load factor over working load.

ELASTIC STABILITY OF A FRAME

The type of elastic instability of major concern to the designer of plated framed structures in a ship is that causing tripping, i.e. the torsional collapse of a stiffener sideways when the plating is under lateral load. Tripping is more likely,

(*a*) with unsymmetrical stiffener sections,
(*b*) with increasing curvature,
(*c*) with the free flange of the stiffener in compression, rather than in tension,
(*d*) at positions of maximum bending moment, especially in way of concentrated loads.

Recommended spacing, l, for tripping brackets is summarized in Fig. 7.28 for straight tee stiffeners and for curved tee stiffeners for which R/W is greater than 70.

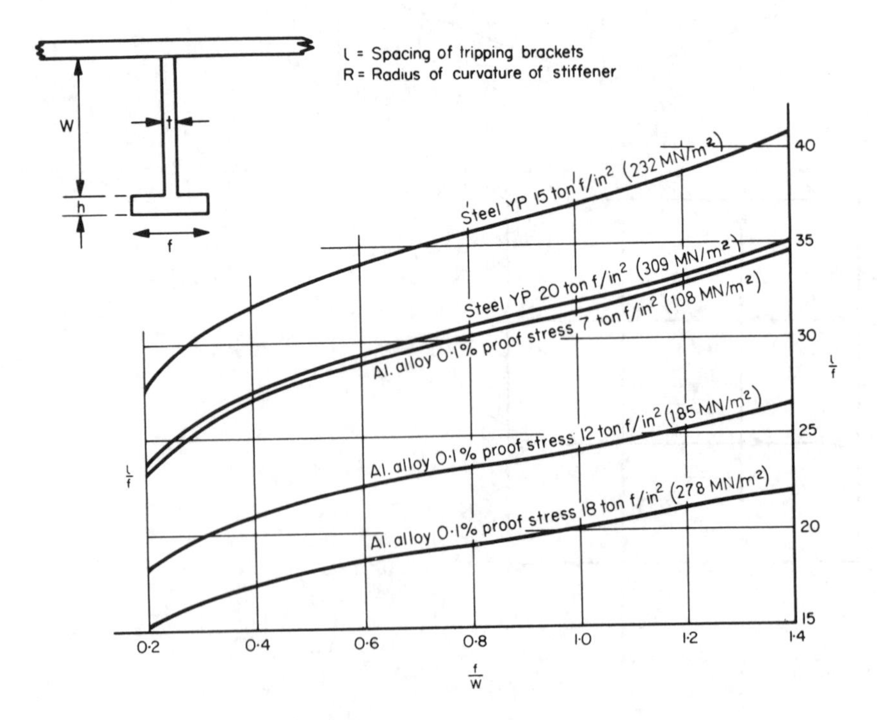

Fig. 7.28 Spacing of tripping brackets

For curved stiffeners for which R/W is less than 70, Fig. 7.28 may be used by putting $l' = Rl/70W$. For straight unsymmetrical stiffeners l should not exceed $8W$. For curved unsymmetrical stiffeners for which R is more than $4f^2h/t^2$ (unsymmetrical stiffeners should not otherwise be used), $l = (RW/5)^{\frac{1}{2}}$ if the flange is in compression and $(2RW/5)^{\frac{1}{2}}$ if in tension.

The elastic stability of a circular ring frame under radial and end compression is the basis of important investigations in the design of submarines. Theoretically a perfectly circular ring under radial compression will collapse in a number of circumferential corrugations or lobes. Under additional end loading, a ring stiffened cylinder may collapse by longitudinal corrugation; this load, too, alters the number of circumferential lobes due to radial pressure. Elastic instability of the whole ring stiffened cylinder between bulkheads is also possible. Finally, built-in distortions in a practical structure have an important effect on the type of collapse and the magnitude of the collapse loads. These problems involve lengthy mathematics and will not be pursued in this book (see Ref. 5). The government of the diving depth of the submarine by consideration of such elastic stability should, nevertheless, be understood.

END CONSTRAINT

The degree of rotational end constraint has more effect on deflections than it has on stresses. End constraints in practical ship structures approach the clamped condition for flexural considerations, provided that the stiffeners are properly continuous at the joint. The degree of rotational end constraint of a member is due to

(*a*) the stiffness of the joint itself. It is relatively simple to produce a joint which can develop the full plastic moment of the strongest of the members entering the joint; no more is necessary;
(*b*) the effects of the other members entering the joint. These can be calculated to give the actual rotational stiffness pertaining to the member.

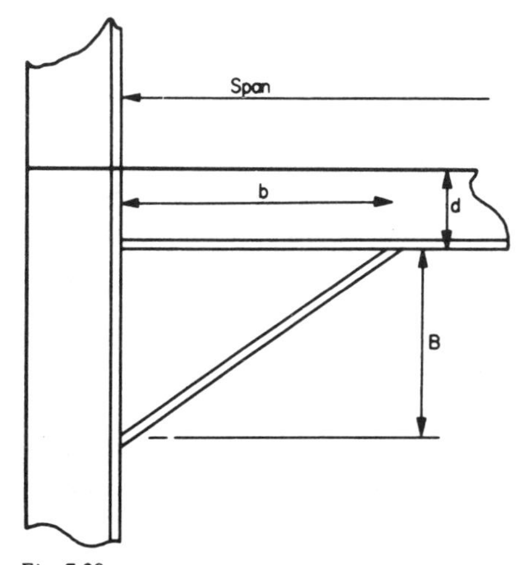

Fig. 7.29

A square joint can provide entirely adequate stiffness. Brackets may be introduced to cheapen fabrication and they also reduce, slightly, the effective span of the member. The reduction in span (Fig. 7.29) is

$$b' = \frac{b}{1+d/B}$$

EXAMPLE 2. The gantry for an overhead gravity davit comprises, essentially, a vertical deck edge stanchion 10 ft high and an overhead member 15 ft long sloping upwards at 30 degrees to the horizontal, inboard from the top of the stanchion. The two members are of the same section rigidly joined together and may be considered encastre where they join the ship.

A weight of 12 tonf acts vertically downwards in the middle of the top member. Compare the analyses of this problem by (*a*) arched rib analysis, (*b*) slope deflection, and (*c*) moment distribution.

Solution (*a*): Arched rib analysis (Castigliano's theorem)

For a point P in AB distance x from A, BM $= M_0 - Hx \sin 30 - Vx \cos 30 +$ $12 \cos 30[x-7{\cdot}5]$, where the term [ ] is included only if positive, i.e.,

$$M = M_0 - \frac{Hx}{2} - V\frac{13x}{15} + 12\times\frac{13}{15}\left[x-\frac{15}{2}\right]$$

$$\frac{\partial M}{\partial M_0} = 1, \qquad \frac{\partial M}{\partial H} = -\frac{x}{2}, \qquad \frac{\partial M}{\partial V} = -\frac{13x}{15}$$

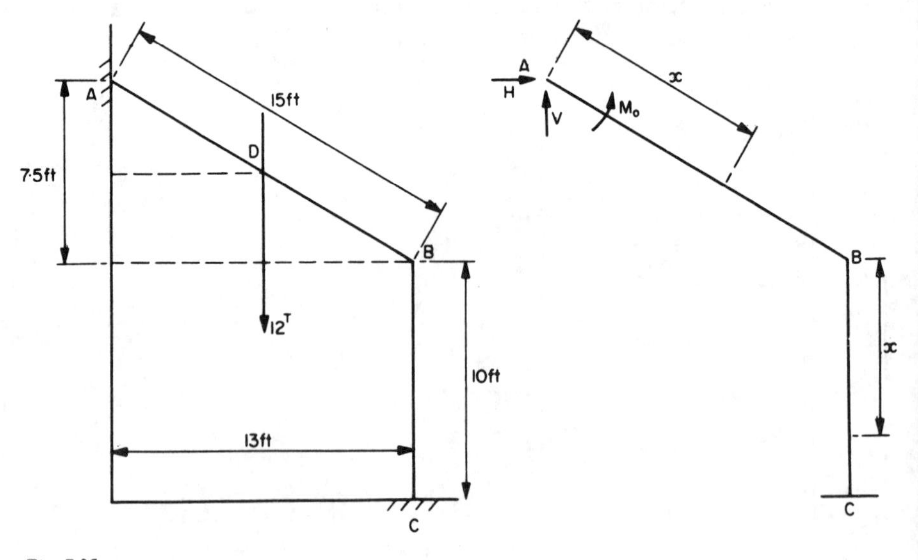

Fig. 7.30

Hence, taking care to integrate the Macaulay's brackets between the correct limits,

$$\int M\frac{\partial M}{\partial M_0}\,\mathrm{d}x = 15M_0-56{\cdot}3H-97{\cdot}5V+292{\cdot}5$$

$$\int M\frac{\partial M}{\partial H}\,\mathrm{d}x = -56{\cdot}3M_0+281{\cdot}3H+487{\cdot}5V-1828$$

$$\int M\frac{\partial M}{\partial V}\,\mathrm{d}x = -97{\cdot}6M_0+487{\cdot}5H+845{\cdot}0\mathrm{V}-3169$$

Similarly, for BC,

$$M = M_0-H\frac{15}{2}-13V+78-Hx$$

$$\frac{\partial M}{\partial M_0} = 1, \qquad \frac{\partial M}{\partial H} = -\left(\frac{15}{2}+x\right), \quad \frac{\partial M}{\partial V} = -13$$

Hence, again, by integrating

$$\int M\frac{\partial M}{\partial M_0}\,\mathrm{d}x = 10M_0-125H-130V+780$$

$$\int M\frac{\partial M}{\partial H}\,\mathrm{d}x = -125M_0+1646H+1625V-9750$$

$$\int M\frac{\partial M}{\partial V}\,\mathrm{d}x = -130M_0+1625H+1690V-10{,}140$$

Adding the similar expressions for AB and BC and equating to zero gives three simultaneous equations, whence

$$V = 5{\cdot}92, \qquad H = 1{\cdot}71, \qquad M_0 = 23{\cdot}4$$

Putting these values back into the expressions for bending moment gives

$$M_{\mathrm{D}} = 21{\cdot}5, \qquad M_{\mathrm{B}} = 11{\cdot}6, \qquad M_{\mathrm{C}} = -5{\cdot}5$$

The BM diagram can thus be plotted.

A good deal of algebra has, of course, been omitted above.

Solution (*b*): Slope deflection (Fig. 7.31)

$$\text{Moment of free BM diagram about A or B} = \frac{1}{2}\times 39\times\frac{15^2}{2}$$

$$= 2195 = m_{\mathrm{A}} = m_{\mathrm{B}}$$

From the slope deflection equations derived on pp. 248–9.

$$\theta_{AB} = 0 = \frac{15}{6EI}\left(2M_{AB} - M_{BA} + \frac{6 \times 2195}{225}\right)$$

$$\theta_{BA} = -\frac{15}{6EI}\left(M_{AB} - 2M_{BA} + \frac{6 \times 2195}{225}\right)$$

$$\theta_{BC} = \frac{10}{6EI}(2M_{BC} - M_{CB})$$

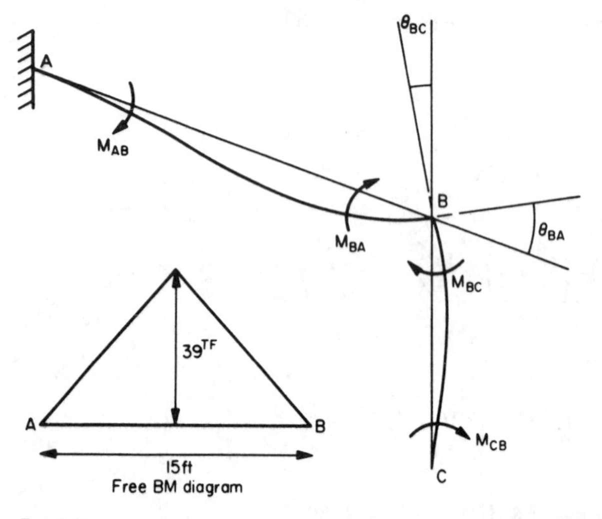

Fig. 7.31

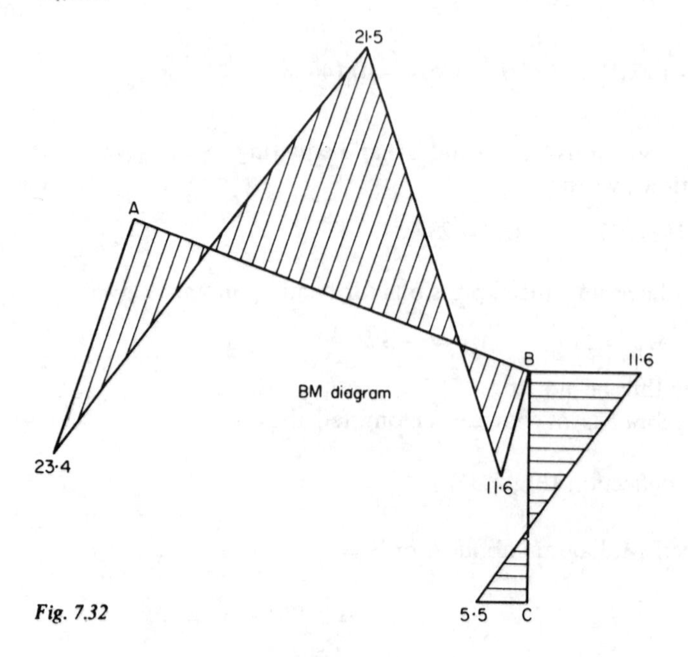

Fig. 7.32

$$\theta_{CB} = 0 = -\frac{10}{6EI}(M_{BC} - 2M_{CB})$$

Now $\theta_{BA} = \theta_{BC}$. Eliminating θ and solving the simultaneous equations gives

$$M_{AB} = -23{\cdot}4$$

$$M_{BA} = 11{\cdot}7 = -M_{BC}$$

$$M_{CB} = -5{\cdot}9$$

Solution (*c*): Moment distribution, in the manner described in the text

	A	B		C
	M_{AB}	M_{BA}	M_{BC}	M_{CB}
Fixing moment $\frac{Wl}{8}$	19·5	−19·5	0	0
Distribution factors $\frac{I/l}{\Sigma I/l}$	—	$\frac{2}{5}$	$\frac{3}{5}$	—
Carry over factors	$\frac{1}{2}$			$\frac{1}{2}$
Fix	19·5	−19·5	0	0
Distribute		7·8	11·7	
Carry over	3·9			5·8
Totals	23·4	−11·7	11·7	5·8

From any of the above solutions the BM diagram can be sketched. (Fig. 7.32).

EXAMPLE 3. An engine room bulkhead 40 ft high is propped on one side by three equally spaced intermediate decks. Stiffeners are uni-directional and, in order to be continuous with longitudinals, must be equally spaced 2 ft 6 in. apart.

Once-in-a-lifetime loading of flooding up to 20 ft above the compartment crown is judged necessary. Top and bottom may be assumed encastre.

Compare the section moduli of stiffeners and plating thicknesses required by fully elastic and elasto-plastic analyses.

Solution: Assume, for simplicity that the hydrostatic loading is evenly distributed between each deck with a value equal to the value at mid-deck.

$$\text{Pressure A to B} = (20+5)\,\text{ft} \times \frac{64\,\text{lbf}}{\text{ft}^3} \times \frac{1\,\text{ft}^2}{144\,\text{in}^2} = 11{\cdot}1\,\text{lbf/in}^2$$

$$\text{Load per foot on each stiffener A to B} = 11{\cdot}1\frac{\text{lbf}}{\text{in}^2} \times 30\,\text{in.} \times \frac{12\,\text{in.}}{1\,\text{ft}}$$

$$\times \frac{1\,\text{tonf}}{2240}\,\text{lbf} = 1{\cdot}79\,\text{tonf/ft}$$

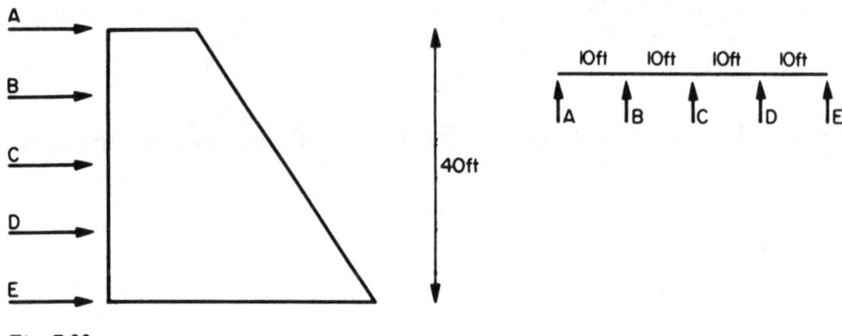

Fig. 7.33

Similarly, pressures and loads on other spans are

	AB	BC	CD	DE
p, press.	11·1	15·5	20·0	24·4 lbf/in²
p', load/ft	1·79	2·50	3·22	3·93 tonf/ft

Plating thickness: Use large deflection elastic theory, the results of which are given in Fig. 7.13. The aspect ratio $a/b = 120/30 = 4{\cdot}0$ so use curves for $a/b = \infty$. Maximum stress is at the plate edge given by the addition of σ_2 and σ_3 and it is necessary to use trial values of thickness. For AB,

$$\frac{p}{E}\left(\frac{b}{t}\right)^4 = \frac{11{\cdot}1}{30\times10^6}\left(\frac{30}{0{\cdot}2}\right)^4 \quad \text{for trial value of } t = 0{\cdot}2 \text{ in.}$$

$$= 187{\cdot}5 \quad \text{for which, with } a/b = \infty,$$

$$\sigma = \frac{30\times10^6}{150^2}\times\frac{(55{\cdot}0+8{\cdot}0)}{2240} = 37{\cdot}5 \text{ tonf/in}^2$$

For a second trial value of $t = 0{\cdot}3$ in., similarly, with $(b/t)^4 p/E = 35{\cdot}8$

$$\sigma = \frac{30\times10^6}{100^2}\times\frac{(15{\cdot}5+2{\cdot}0)}{2240} = 23{\cdot}4 \text{ tonf/in}^2$$

The value of σ_3 the membrane tension, is low and is having only a small effect. Adopting small deflection theory therefore, for a yield stress of 16 tonf/in²,

$$16 \text{ tonf/in}^2 = 1{\cdot}0\times\frac{11{\cdot}1}{2}\frac{\text{lbf}}{\text{in}^2}\times\left(\frac{30}{t}\right)^2\times\frac{1 \text{ tonf}}{2240 \text{ lbf}}$$

$$\therefore \quad t = 30\left(\frac{11{\cdot}1}{2\times16\times2240}\right)^{\frac{1}{2}} = 0{\cdot}37 \text{ in.}$$

on small elastic theory, plating thicknesses required are

AB	BC	CD	DE
0·37	0·44	0·50	0·55 in.

Implicit in these figures is an unknown factor of safety over 'failure'. Let us now adopt the elasto-plastic definition of failure and choose, deliberately, a load factor of 2·0. Referring to Fig. 7.14 for span DE,

$$\frac{pE}{\sigma_y^2} = \frac{2 \times 24{\cdot}4 \times 30 \times 10^6}{(16 \times 2240)^2} = 1{\cdot}14$$

for which

$$\frac{b}{t}\left(\frac{\sigma_y}{E}\right)^{\frac{1}{2}} = 2{\cdot}95$$

so that

$$t = \frac{30}{2{\cdot}95}\left(\frac{16 \times 2240}{30 \times 10^6}\right)^{\frac{1}{2}} = 0{\cdot}35 \text{ in.}$$

Carrying out similar exercises for all bays,

Bay	AB	BC	CD	DE
$\frac{pE}{\sigma_y^2}$	0·52	0·72	0·93	1·14
$\frac{b}{t}\left(\frac{\sigma_y}{E}\right)^{\frac{1}{2}}$	5·00	4·10	3·40	2·95
t	0·21	0·25	0·30	0·35

A corrosion margin, say 40 per cent, now needs to be added to give thicknesses suitable for adoption. By this method, a saving has been achieved by adopting deliberate standards which are disguised by the elastic approach. Savings do not always occur, but the advantages of a rational method remain.

To determine the size of stiffeners, assumed constant for the full depth, elastic analysis is probably best carried out by moment distribution. Since the deck heights are equal at 10 ft each,

$$\text{distribution factors for B, C and D} = \frac{I/l}{\Sigma I/l} = \tfrac{1}{2}$$

The fixing moments, i.e. the residuals at the ends of AB are due to a mean uniform pressure of 11·1 lbf/in². Hence, with ends held encastre (Fig. 7.2)

$$M_{AB} = -M_{BA} = -\frac{pl^2b}{12} = -\frac{11{\cdot}1(120)^2(30)}{12^2\, 2240} = -14{\cdot}9 \text{ tonf ft}$$

Similarly, due to pressures of 15·5, 20·0 and 24·4 lbf/in²

$$M_{BC} = -M_{CB} = -20{\cdot}8 \text{ tonf ft}$$
$$M_{CD} = -M_{DC} = -26{\cdot}8 \text{ tonf ft}$$
$$M_{DE} = -M_{ED} = -32{\cdot}7 \text{ tonf ft}$$

The resultant moments at joints B, C and D are 5·9, 6·0 and 5·9 tonf ft. Since distribution factors are $\frac{1}{2}$, each member takes approximately 3 tonf ft at each joint. Carry-over factors are each $\frac{1}{2}$ so that relaxing the 3 tonf ft moment at end

B of AB induces 1·5 tonf ft moment at A. This process is repeated for end B of BC, end C of BC, etc. The process is then repeated until the residuals at the ends of the members at each joint are substantially the same but of opposite sign, i.e. the joint as a whole is balanced. This process is worked through in the table below.

	A M_{AB}	B M_{BA}	M_{BC}	M_{CB}	C M_{CD}	D M_{DC}	M_{DE}	E M_{ED}
Distribution factors $\frac{I/l}{\Sigma I/l}$		$\frac{1}{2}$	$\frac{1}{2}$	$\frac{1}{2}$	$\frac{1}{2}$	$\frac{1}{2}$	$\frac{1}{2}$	
Fixing moments $\frac{p'l^2}{12}$	−14·9	+14·9	−20·8	20·8	−26·8	26·8	−32·8	32·8
Distribute	0	+3·0	+3·0	+3·0	+3·0	+3·0	+3·0	0
Carry over (all factors $\frac{1}{2}$)	+1·5	0	+1·5	+1·5	+1·5	+1·5	0	1·5
Distribute	0	−0·7	−0·7	−1·5	−1·5	−0·7	−0·7	0
Carry over	−0·4	0	−0·7	−0·4	−0·4	−0·7	0	−0·4
Totals	−13·8	17·2	−17·7	23·4	−24·2	29·9	−30·5	33·9
Distribute	0	+0·3	+0·3	+0·4	+0·4	+0·3	+0·3	0
Carry over	0·1	0	0·2	0·1	0·1	0·2	0	0·1
Totals	−13·7	17·5	−17·2	23·9	−23·7	30·4	−30·2	34·0

Maximum bending moment occurs in DE = 34·0 tonf ft.
For a maximum yield stress of 16·0 tonf/in².

$$Z = \frac{I}{y} = \frac{34 \times 12}{16} = 25{\cdot}5 \text{ in}^3$$

Using a plastic method of analysis, the lowest span, which must be the critical span, will become a mechanism when plastic hinges occur at D, E and near the middle of this span (Fig. 7.33). Taking a load factor of 2·0, approx.

$$M_p = \frac{\frac{1}{2} \times p'l^2}{12} \times 2{\cdot}0 = \frac{3{\cdot}93 \times 100}{24} \times 12 \times 2{\cdot}0 = 393 \text{ tonf in.}$$

$$\therefore S = \frac{393}{16{\cdot}0} = 24{\cdot}6 \text{ in}^3$$

If the shape factor is taken as 1·3, $Z = 19{\cdot}0 \text{ in}^3$. Thus, some savings occur by adopting plastic design methods for both plating and stiffeners. Because the loading is presumed to occur once in a lifetime, there is no reason why the smaller figures should not be adopted, provided that there are no other criteria of design.

While the theoretical considerations should follow, generally, the elasto-plastic and plastic methods illustrated in this example, there are other features which need to be brought out in the design of main transverse bulkheads. It will often be found economical to reduce scantlings of both stiffeners and plating towards the top, where it is warranted by the reduction in loading. Continuity of bulkhead stiffener, bottom longitudinal and deck girder is of extreme importance and will often dictate stiffener spacings. In ships designed to withstand the effects of underwater explosion, it is important to incorporate thicker plating around the edges and to permit no piercing of this area by pipes or other stress raisers; the passage of longitudinal girders must be carefully compensated; fillet connections which may fail by shear must be avoided.

Finite element techniques

The displacement δ of a simple spring subject to a pull p at one end is given by $p = k\delta$ where k is the stiffness. Alternatively, $\delta = fp$ where f is the flexibility and $f = k^{-1}$.

If the forces and displacements are not in the line of the spring or structural member but are related to a set of Cartesian coordinates, the stiffness will differ in the three directions and, in general

$$\mathbf{p}_1 = \mathbf{k}_{11}\delta_1+\mathbf{k}_{12}\delta_2$$
$$\text{and } \mathbf{p}_2 = \mathbf{k}_{21}\delta_1+\mathbf{k}_{22}\delta_2$$

This pair of equations is written in the language of matrix algebra

$$\mathbf{P} = \mathbf{K}\,\mathbf{d}$$

P is the complete set of applied loads and **d** the resulting displacements. **K** is called the stiffness matrix and is formed of such factors $\mathbf{k}_{11}$ which are called member stiffness matrices (or sub-matrices). For example, examine a simple member subject to loads p_X and p_Y and moments m at each end causing displacements δ_X, δ_Y and θ.

For equilibrium,

$$m_1+m_2+p_{Y_2}l = 0 = m_1+m_2-p_{Y_1}l$$

also $p_{X_1}+p_{X_2} = 0$

For elasticity,

$$p_{X_1} = -p_{X_2} = \frac{EA}{l}\,(\delta_{X_1}-\delta_{X_2})$$

From the slope deflection analysis on page 246 *et seq*. can be obtained

$$m_1 = \frac{6EI}{l^2}\delta_{Y_1}+\frac{4EI}{l}\theta_1-\frac{6EI}{l^2}\delta_{Y_2}+\frac{2EI}{l}\theta_2$$

These equations may be arranged

$$\begin{bmatrix} p_{X_1} \\ p_{Y_1} \\ m_1 \end{bmatrix} = \begin{bmatrix} \frac{EA}{l} & 0 & 0 \\ 0 & \frac{12EI}{l^3} & \frac{6EI}{l^2} \\ 0 & \frac{6EI}{l^2} & \frac{4EI}{l} \end{bmatrix} \begin{bmatrix} \delta_{X_1} \\ \delta_{Y_1} \\ \theta_1 \end{bmatrix} + \begin{bmatrix} -\frac{EA}{l} & 0 & 0 \\ 0 & -\frac{12EI}{l^3} & \frac{6EI}{l^2} \\ 0 & -\frac{6EI}{l^2} & \frac{2EI}{l} \end{bmatrix} \begin{bmatrix} \delta_{X_2} \\ \delta_{Y_2} \\ \theta_2 \end{bmatrix}$$

i,e. $$\mathbf{p} = \mathbf{k}_{11}\,\boldsymbol{\delta}_1+\mathbf{k}_{12}\,\boldsymbol{\delta}_2$$

This very simple example is sufficient to show that a unit problem can be expressed in matrix form. It also suggests that we are able to adopt the very powerful mathematics of matrix algebra to solve structural problems which would otherwise become impossibly complex to handle. Furthermore, computers can be quite readily programmed to deal with matrices. A fundamental problem is concerned with the inversion of the matrix to discover the displacements arising from applied loads, viz.

$$\mathbf{d} = \mathbf{K}^{-1}\mathbf{P}$$

Now strains are related to displacements,

$$\boldsymbol{\epsilon} = \mathbf{Bd}$$

For plane strain for example,

$$\boldsymbol{\epsilon} = \begin{bmatrix} \epsilon_x \\ \epsilon_y \\ \gamma_{zy} \end{bmatrix} = \begin{bmatrix} \dfrac{\partial u}{\partial x} \\ \dfrac{\partial v}{\partial y} \\ \dfrac{\partial u}{\partial y} + \dfrac{\partial v}{\partial x} \end{bmatrix} = \mathbf{Bd}$$

and stress is related to strain,

$$\boldsymbol{\sigma} = \mathbf{D}\,\boldsymbol{\epsilon}$$

D is a matrix of elastic constants which, for plane stress in an isotropic material is

$$\mathbf{D} = \frac{E}{1-\nu^2}\begin{bmatrix} 1 & \nu & 0 \\ \nu & 1 & 0 \\ 0 & 0 & \frac{1}{2}(1-\nu) \end{bmatrix}$$

There are other relationships which are valuable such as the transformation matrix which changes reference axes. These together form the set of tools required for the solution of structural problems using finite element techniques. This brief description can do little more than explain the concept and the student must examine standard textbooks. Finite element analysis is approached broadly as follows:

(a) The structure is divided up by imaginary lines meeting at nodes, forming finite elements which are often triangular or rectangular and plane (but may be irregular and three-dimensional.)
(*b*) For each element, a displacement function is derived which relates the displacements at any point within the element to the displacements at the nodes. From the displacements strains are found and from the strains, stresses are derived.
(*c*) Forces at each node are determined equivalent to the forces along the boundaries of the element.
(*d*) Displacements of elements are rendered compatible with their neighbour's (this is not often totally possible).
(*e*) The whole array of applied loads and internal forces are arranged to be in equilibrium.

It is not within the scope of this book to describe how this analysis is carried out. It requires a good knowledge of the shorthand of matrix algebra and draws upon the work described in this chapter concerning various unit problems of beams, panels, grillages and frameworks and also the concepts of relaxation techniques and minimum strain energy.

It places in the hands of the structural analyst, a tool of enormous power and flexibility. There is no longer any need to make the assumption of simple beam theory for the longitudinal strength of the ship; the ship girder may now be built

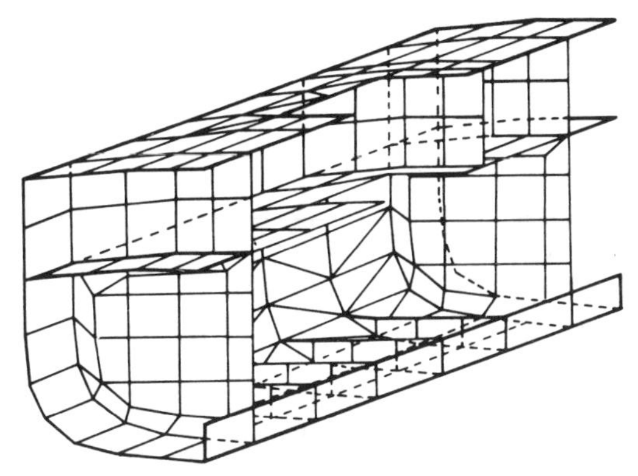

Fig. 7.34

up from finite elements (Fig. 7.34) and the effects of the loads applied by sea and gravity determined. Indeed, this is now the basis for the massive suites of computer programs available for the analysis of total ship structure. The effects of the sea spectra are translated by strip theory into loads of varying probability and the effects of those loadings upon a defined structure are determined by finite element analysis. It is not yet a perfect tool. Moreover, it is a tool of analysis and not of design which is often best initiated by cruder and cheaper methods before embarking upon the expense of these programs.

Realistic assessment of structural elements

The division of the ship into small elements which are amenable to the types of analysis presented earlier in this chapter remains useful as a rough check upon more advanced methods. There is now becoming available a large stock of data and analytical methods which do not have to adopt some of the simplifying assumptions that have been necessary up to now. This is due in large measure to the widespread use of finite element techniques and the computer programs written for them. Experimental work has carefully sifted the important parameters and relevant assumptions from the unimportant, so that the data sheets may present the solutions in realistic forms most useful to the structural designer. Once again the designer has to rely on information derived from computer analysis which he cannot check, so that the wise will need to fall back from time to time upon simple analysis of sample elements to give him confidence in them.

As explained in Chapter 6, the elemental behaviour of the whole ship cross section may be integrated to provide a knowledge of the total strength. Judgement remains necessary in deciding what elements should be isolated for individual analysis. This judgement has been assisted by extensive experimentation into box girders of various configurations under end load (e.g. Ref. 9). Figure 7.35 shows the cross section of a typical specimen.

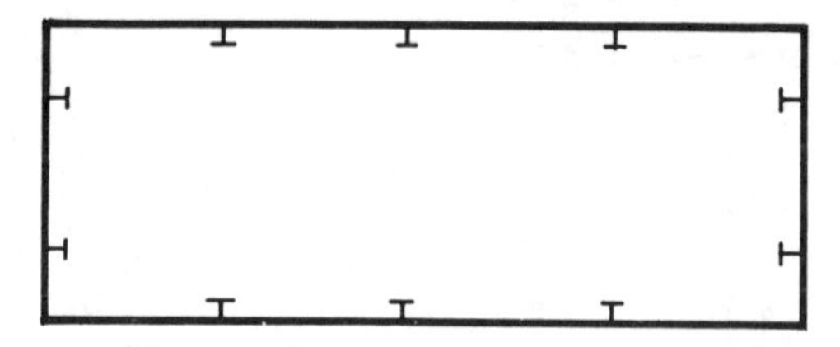

Fig. 7.35

The wisdom of generations of ship designers has steadily evolved a structure which:

(*a*) Has more cross sectional area in the stiffener than in the plating.
(*b*) Has longitudinals more closely spaced than transverses.
(*c*) Favours quite deep longitudinals, preferably of symmetrical cross section.

Such structures tend to provide high collapse loads in compression and an efficient use of material. What happens as the load is steadily increased is first a buckling of the centres of panels midway between stiffeners. Shirking of load by the panel centres throws additional load upon each longitudinal which, as load is further increased, will finally buckle in conjunction with the strip of plate to which it is attached. This throws all of the load upon the 'hard' corners which are usually so stiff in compression that they remain straight even after plasticity has set in. It has been found that these hard corners behave like that in conjunction with about a half the panel of plating in each direction. Thus the elements into which the box girder should be divided are plating panels, longitudinals with a strip of plating and hard corners.

Finite element analysis of these elements is able to take account of two factors which were previously the subject of simplifying assumptions. It can account for built-in manufacturing strains and for initial distortion. Data sheets or standard programs are available from software houses for a very wide range of geometry and manufacturing assumptions to give the stress-strain (or more correctly the load shortening curves) for many elements of structure. These can be integrated into total cross sectional behaviour in the manner described in Chapter 6.

By following the general principles of structural design above—i.e. close deep longitudinals, heavy transverse frames and panels longer than they are wide—two forms of buckling behaviour can usually be avoided:

(*a*) Overall grillage buckling of all plating and stiffeners together; this is likely only when plating enjoys most of the material and stiffeners are puny.
(*b*) Tripping of longitudinals by sideways buckling or twisting; this is likely with stiffeners, like flat bars which have low torsional stiffness.

The remaining single stiffener/plating behaviour between transverse frames may now be examined with varying material geometry and imperfections.

Data sheets on the behaviour of panels of plating shown in Figs 7.36 and 7.37 demonstrate firstly the wisdom of the general principles enunciated above and secondly how sensitive to imperfections they are. With moderate initial distortion and average built-in strains some 50–80 per cent of the theoretical yield of a square panel can be achieved (Fig. 7.36). Long panels as in transversely framed ships (Fig. 7.37) achieve only 10 per cent or so.

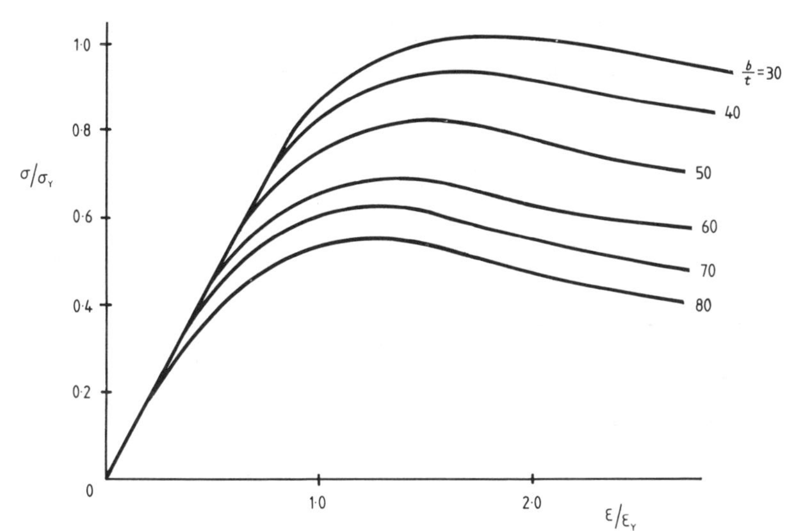

Fig. 7.36 (after Ref. 9)

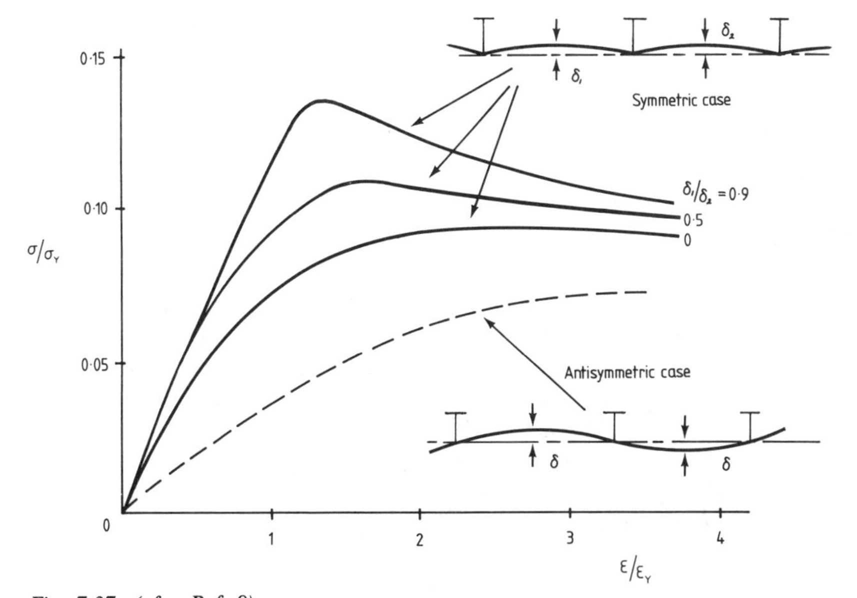

Fig. 7.37 (after Ref. 9)

Data sheets for stiffened plating combinations are generally of the form shown in Fig. 7.38. On the tension side the stress/strain relationship is taken to be of the idealized form for ductile mild steel. In compression, the effects of progressive buckling of flanges is clearly seen. Sometimes the element is able to hold its load-bearing capacity as the strain is increased; in other cases the load-bearing capacity drops off from a peak in a form called catastrophic buckling.

Because the finite element analysis is able to account for the separate panel buckling the width of the associated plating may be taken to the mid panel so that the 30 t assumption is no longer needed. Moreover, many of the elemental data sheets now available are for several longitudinals and plating acting together.

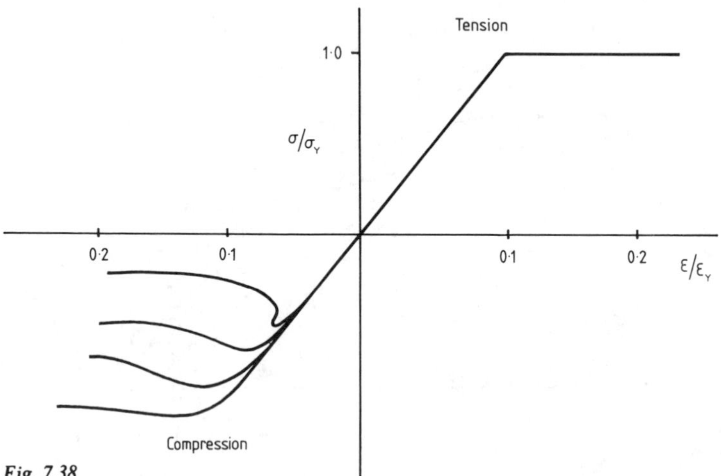

Fig. 7.38

Fittings

CONTROL SURFACES

Control surfaces in a ship are those which, through hydrodynamic lift and drag forces, control the motion of the ship. They include rudders, stabilizers, hydroplanes, hydrofoils and tail fins. The forces and their effects on the ship are discussed at length in other chapters. In this chapter, we are concerned with the strength of the surface itself, the spindle or stock by which it is connected to operating machinery inside the ship and the power required of that machinery.

Those control surfaces which are cantilevered from the ship's hull are the most easily calculable because they are not redundant structures. The lift and drag forces for ahead or astern motions (given in Chapter 13) produce a torque T and a bending moment M at section Z of the stock. If the polar moment of inertia of the stock is J, the maximum shear stress is given by

$$\tau_{max} = \frac{D}{2J}(M^2 + T^2)^{\frac{1}{2}}$$

Permissible maximum stresses are usually about 3·0 tonf/in^2 for mild steel (45

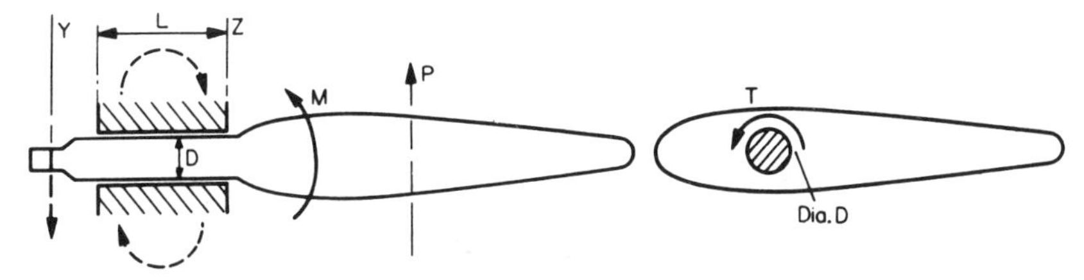

Fig. 7.39 Forces and moment on a control surface

MPa) and 4·5 tonf/in² for best quality forged steel (70 MPa). At section Y of the stock, torque required for the machinery to which the stock is subjected is augmented by the friction of the sleeve, whence

$$T^1 = T + P\mu\frac{D}{2}$$

The value of the coefficient of friction μ is about 0·2. The power required of the machinery is given by T^1 multiplied by the angular velocity of rotation of the surface: for a stabilizer fin this might be as high as 20 deg/s, while for a rudder it is, typically, 2–3 deg/s.

When there are pintle supports to the rudder, the reactions at the pintles and sleeve depend upon the wear of the bearings that is permitted before renewal is demanded. The effects of such wear may be assumed as given in Fig. 7.40.

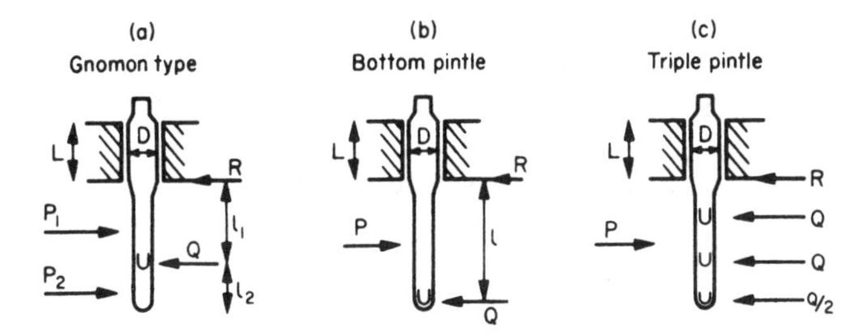

Fig. 7.40 Forces assumed acting on rudders

Permitted bearing pressure at the sleeve which is usually about $\frac{1}{2}$ in. thick is about 1500 lbf/in² (10·3 MPa),

$$\frac{R}{L(D+2t)} \text{ should not exceed } 1500 \text{ lbf/in}^2 \text{ (10·3 MPa)}$$

At the pintles, classification societies take about the same bearing pressure. Ship department takes a somewhat larger figure.

$$\frac{Q}{ld} \text{ should not exceed } 5600 \text{ lbf/in}^2 \text{ (38·6 MPa).}$$

The shearing stress in the steel pintle due to Q should not exceed the figures used for the stock itself. Structure of the rudder is analysed by conventional

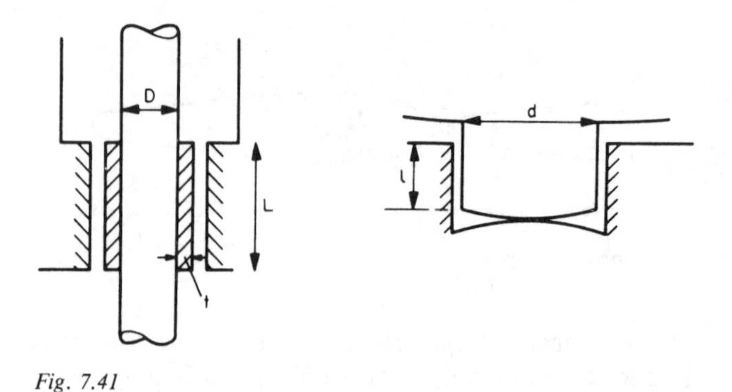

Fig. 7.41

means, the stock continuation being regarded as a beam subject to the loads of Fig. 7.40. Because erosion can be severe the acceptable stresses are low.

SHAFT BRACKETS

A shaft bracket houses the long shaft bearing adjacent to the propeller or partway along an exposed shaft, and comprises essentially a cylinder attached to the hull by two arms of hydrofoil section. The arms may be angled slightly to the axis of the cylinder to align with the hydrodynamic flow, but it is bad practice from the point of view of strength to crank the arms so that they enter the hull normally. Shaft brackets are either fabricated or forged. Spectacle plates are forgings or castings in the shape of a pair of spectacles connecting twin shafts with a centre skeg by single arms; they generally form the extremities of the shaft swell plating.

A hydrofoil shape commonly used in warships is given in Fig. 7.42.

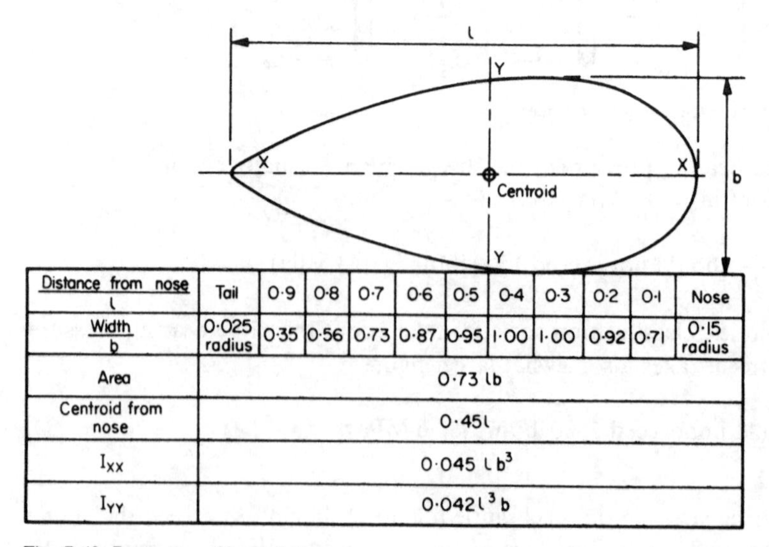

Distance from nose / l	Tail	0·9	0·8	0·7	0·6	0·5	0·4	0·3	0·2	0·1	Nose
Width / b	0·025 radius	0·35	0·56	0·73	0·87	0·95	1·00	1·00	0·92	0·71	0·15 radius
Area	$0{\cdot}73\, lb$										
Centroid from nose	$0{\cdot}45l$										
I_{XX}	$0{\cdot}045\, lb^3$										
I_{YY}	$0{\cdot}042 l^3 b$										

Fig. 7.42 Properties of hydrofoil section

Perfect propellers in open water would not throw any load other than weight on the shaft brackets. In practice, loads may be thrown on to the brackets by,

(*a*) weights of propeller and shafting, accentuated by ship motion;
(*b*) gyroscopic effects under conditions of ship motion, including turning;
(*c*) variation in hydrodynamic forces on the blades;
(*d*) centrifugal forces when a blade is lost;
(*e*) whirling of the shaft.

So far as (*c*) is concerned, since the propeller operates in the wake of the shaft bracket, the opening between the arms should be kept as large as possible to reduce the magnitude of the wake. Also since each blade experiences a change in hydrodynamic load as it passes through the wake of each shaft bracket arm, the angle between the arms must not be equal to the angle between the propeller blades. (*a*) to (*c*) are difficult to estimate with any accuracy. Investigation of each of these effects has led to a comparative type calculation in which the largest of the following three forces is assumed to act at the centre of the propeller,

(*a*) F_1, based on the force needed to shear the steel propeller shaft,

$$F_1 = \frac{\pi}{10}\frac{(d_1^4 - d_2^4)}{d_1^2} \text{ tonf}$$

(*b*) F_2, based on the centrifugal force when a blade is lost,

$$F_2 = \frac{WN^2D}{10^5} \text{ tonf}$$

(*c*) F_3, based on the accentuated weight of the propeller

$$F_3 = 9W \text{ tonf}$$

Where d_1 and d_2 are external and internal shaft diameters in inches, W tonf and D ft are the weight and diameter of the propeller, and N is the full power revolutions per minute of the shaft. The largest of these forces, F, is assumed to cause a rotating force in the plane of the arm axes and a moment $0{\cdot}8Fh$, the reduction of 20 per cent assumed to account for the flexibility of the shaft.

Thorough analysis of many shaft brackets has shown that three of the effects of this force are important (Fig. 7.42)

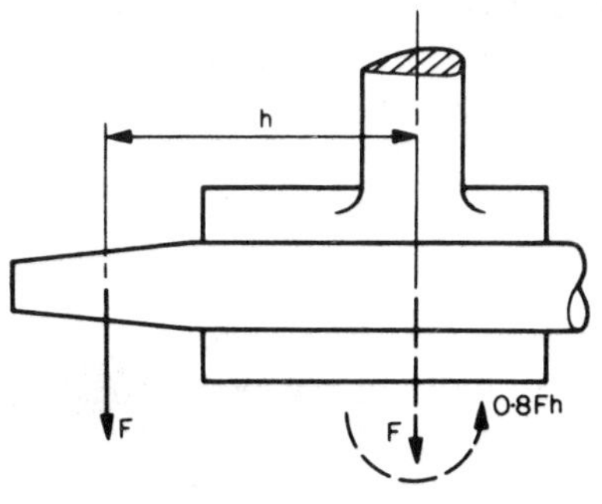

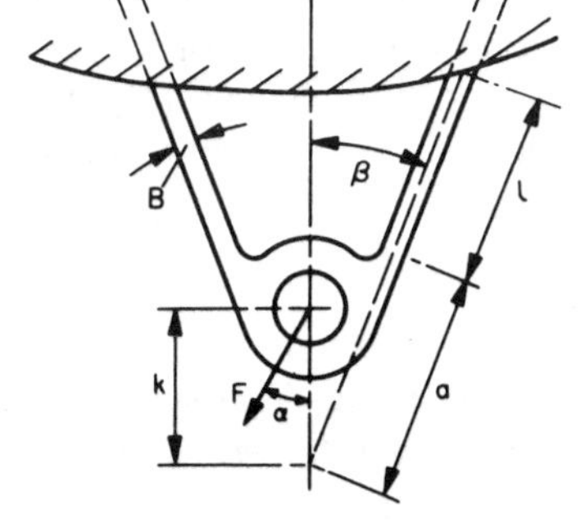

Fig. 7.43

(*a*) longitudinal bending of an arm about YY,
(*b*) transverse bending of an arm about XX,
(*c*) axial tension or compression in an arm.

The values of these can be determined by a somewhat laborious analysis whose important results are summarized in Fig. 7.44. Maximum stress due to transverse bending will occur out of phase with those due to the longitudinal bending and the axial stress. It is not unusual to adopt the pessimistic assumption that they all occur together.

	Longitudinal bending moment M_L	Transverse bending moment M_T	Axial loading stress
Symbols	$M_1 = \pm\frac{1}{2}(0{\cdot}8Fh)\sec\beta$ $M_2 = \pm\frac{u}{2}(0{\cdot}8Fh)\,\mathrm{cosec}\,\beta$ $u = \frac{l(a/2+l/3)}{a^2+al+l^2/3}$		A = arm cross-sectional area
Value	$M_1\cos\alpha + M_2\sin\alpha$	$\frac{1}{2}Fku\sin\alpha$	
Max value occurring at	$(M_1^2+M_2^2)^{\frac{1}{2}}$ $\tan\alpha = \pm u\cot\beta$	$\frac{1}{2}Fku$ $\alpha = \pi/2$	$F/A\,\mathrm{cosec}\,2\beta$

Fig. 7.44

Although maximum stresses due to each plane of bending occur at the extremities of the neutral axes, the combined maxima occur in between these points at various points around the hydrofoil section. If σ_L and σ_T are the individual maxima due to longitudinal and transverse bending, the combined maximum stress for the hydrofoil of Fig. 7.42 is given approximately by

$$\sigma_{max} = 0{\cdot}5\sigma_L + 0{\cdot}9\sigma_T \quad \text{provided that} \quad \sigma_T > \tfrac{2}{8}\sigma_L$$

This should be augmented by that axial stress of the same sign and the total stress should not be permitted to exceed 14 tonf/in^2 if the arms are fabricated in mild steel or 18 tonf/in^2 if forged (216 and 278 MPa respectively).

That the procedures above are pragmatic and comparative is obvious. An excellent design method for P brackets made in composite material is given in Ref. 6. This is applicable in part also to A brackets and should be used in preference to the empirical process given above when less conservative design is important.

Problems

1. A bottom longitudinal is in an area subject to an end compressive stress of 100 N/mm^2 and is at a still water draught of 4 m. With a width of steel plating of 28 cm assumed to be fully effective, estimate the maximum stress under the end load and a lateral load due to twice the draught. Spacing of longitudinals

is 36 cm and 1 m (simply supported), plating thickness is 4 mm and the T bar stiffener has the following properties

flange (cm)	depth (cm)	c.s. area (cm^2)	MI (toe) (cm^4)	MI (N.A.)(cm^4)
4	8	5·0	196·0	30·0

2. A transverse floor in a double bottom is 53 cm deep and 1·6 cm thick. Adjacent floors are 3 m apart and the outer bottom is 1·9 cm thick and the inner bottom is 0·8 cm thick. Each floor is supported by longitudinal bulkheads 4 m apart and is subjected to a uniformly distributed load from the sea.

Calculate the effective modulus of section.

3. A tubular strut of length 3 m external diameter 8 cm cross-sectional area 16 cm^2 and moment of inertia about a diameter 150 cm^4 when designed for an axial working load of 0·1 MN may be assumed to be pin-ended. It has an initial lack of straightness in the form of a half sine wave, the maximum (central) departure from the straight being 8 mm; find (*a*) the ratio of the maximum working stress to the yield stress of 270 MPa and (*b*) the ratio of the working load to the load which would just cause yield (E = 209 GN/m^2).

4. A square grillage is composed of six equally spaced beams of length 10 ft, three in each direction. The beams are fixed at their ends and are each acted on by a uniformly distributed load of 2 tonf/ft throughout the length. The central beam in each direction has a moment of inertia of 36 in^4. The others have the same moment of inertia which need not be that of the central beams; what should this moment of inertia be in order that no beam shall exert a vertical force on another? If, in fact, each of these four beams has a moment of inertia of 18 in^4, what is the central deflection?

5. Four girders, each 12 m long, are arranged two in each of two orthogonal directions to form a square grillage of nine equal squares. The ends of all girders are rigidly fixed and the joints where the girders intersect are capable of transmitting deflection and rotation. All four members are steel tubes with polar MI of 0·02 m^4.

Find the deflection at an intersection when a uniformly distributed load of 2 MN/m is applied along each of two parallel girders and the two other girders are unloaded. (Modulus of rigidity is $\frac{3}{8}$ of Young's modulus.)

6. Calculate by the plastic design method, the 'collapse' load of a grillage in which all members are tubes of mean diameter 8 in., thickness $\frac{1}{8}$ in. and of steel of yield point 16 $tonf/in^2$. The grillage is 32 ft × 21 ft, consisting of three members 21 ft long spaced 8 ft apart in one direction, intersecting at right angles two members 32 ft long spaced 7 ft apart. The ends of all members are fixed. At each joint there is an equal concentrated load perpendicular to the plane of the grillage. Ignore torsion.

7. Three parallel members of a grillage are each 4 m long and equally spaced 2 m apart. They are intersected by a single orthogonal girder 8 m long. All ends are simply supported and each member has a second moment of area of 650 cm^4. Each member is subject to a uniformly distributed load of 15,000 N/m. Calculate the maximum deflection. Neglect torsion. E = 20·9 MN/cm^2.

8. A long flat panel of plating is 40 in. wide and subject to a uniform pressure of

25 lbf/in^2. What thickness should it be

(*a*) so that it nowhere yields,
(*b*) to permit a membrane stress of (28/3) tonf/in^2, E = 13,500 tonf/in^2, σ = 0·3 and yield stress = 14 tonf/in^2.

9. An alloy steel plate 1·50 m × 0·72 m has an initial bow due to welding of 3 mm in the middle. It is 6 mm thick and made of steel with a Y.P. of 248 MPa.

What pressure would this plate withstand without causing further permanent set? If the plate had been considered initially flat, what pressure would small deflection theory have given to have caused yield? What would have been the deflection?

10. The yield stress of an alloy plate is 30 tonf/in^2 and Young's modulus is 8000 tonf/in^2. The plate is 60 in. × 20 in. × 0·20 in. thick and its edges are constrained against rotation. Calculate the approximate values of

(*a*) the pressure first to cause yield;
(*b*) the elasto-plastic design pressure assuming edges restrained from inward movement;
(*c*) the design pressure assuming edges free to move in and a permanent set coefficient of 0·2;
(*d*) the pressure which the plate could withstand without further permanent set if bowed initially by 0·10 in., the edges assumed restrained.

11. A boat is supported from the cross bar of a portal frame at points 6 ft from each end. The vertical load at each point of support is 2 tonf. The portal is 10 ft high and 30 ft long and is made of thin steel tube 8 in. in diameter and $\frac{1}{8}$ in. thick; the steel has a yield point of 18 tonf/in^2.

If both feet are considered encastre, find the load factor over collapse in the plane of the portal. Ignore the effects of axial load.

Find also, by a moment distribution method, the ratio of the maximum working stress to the yield stress.

12. When a temporary stiffener is welded diametrically across a completed ring frame of a submarine, bad fitting and welding distortion cause a total diametral shrinkage of 8 mm.

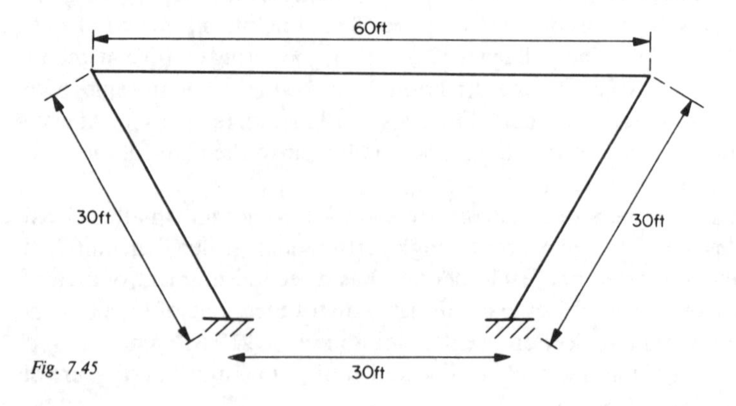

Fig. 7.45

The mean diameter of the steel ring is 6·1 m. The MI of the frame section is 5000 cm^4 and the least section modulus of the ring is 213 cm^3. What is the greatest stress induced in the frame?

13. Due to careless building, the ring frame shown in Fig. 7.45 is left unsupported while being built into a ship. The deck beam is an 8 in. T bar of 10 lbf/ft run with an MI of 34 in^4, while the ribs are 6 in. T bars of 8 lbf/ft run with an MI of 17 in^4. Members are straight and the ribs rigidly attached to the double bottom. The neutral axes of section for the deck beams and ribs are respectively 6 in. and 4·5 in. from the toe.

Assuming that the frame does not buckle out of its plane, find the maximum stress due to its own weight.

14. Light alloy superstructure above the strength deck of a frigate may be treated as a symmetrical two-dimensional single storey portal as shown. The MI of AD, AE and DH is 5,000 cm^4 and of BF and CG is 2330 cm^4 All joints are rigid except F and G where the elasticity of the deck causes carry-over factors of $\frac{1}{4}$ from B to C. Find the bending moment distribution in the structure when a load of 19,200 N/m is carried evenly along AD.

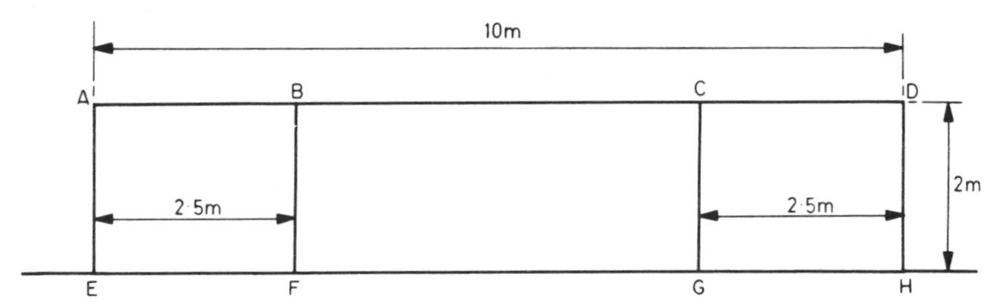

Fig. 7.46

15. The structure of a catapult trough is essentially of the shape shown. G and H are fixed rigidly but the rotary stiffnesses of A and F are such as to allow one-half of the slope which would occur were they to be pinned.

The mid-point of CD is subjected to a vertical force of 120 tonf as shown.

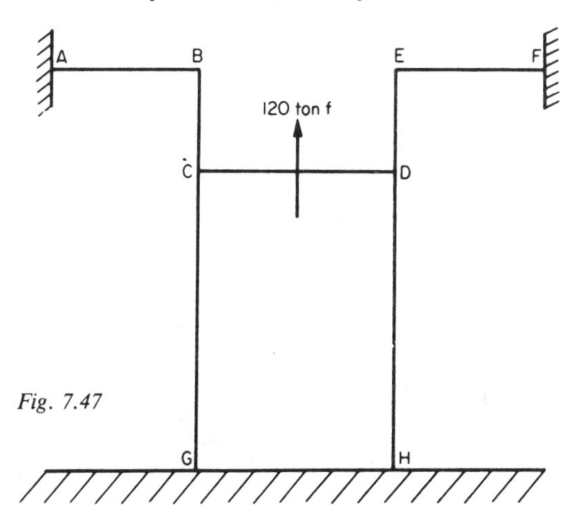

Fig. 7.47

Calculate the distribution of bending moment in the structure both by moment distribution and slope deflection analysis,

Members	AB, EF	BC, ED	CG, DH	CD
Length (ft)	3·0	2·0	6·0	4·0
$I\,(\text{in}^4)$	180	140	60	160

16. Calculate the collapse load of a rectangular portal having 3 m stanchions and a 8 m cross bar all made of tubes 20 cm in diameter, 3 mm thick and having a yield stress of 250 MPa.

The lower feet are pinned and the upper corners rigid. Equal loads are applied horizontally inwards at the middle of each stanchion and vertically downwards at the middle of the crossbar. (Neglect the effect of end compression on the plastic moment of resistance.)

17. The balanced rudder shown (Fig. 7.48) has a maximum turning angle of 35 degrees and is fitted immediately behind a single propeller. What torque and bending moment are applied to the rudder stock at the lower end of the sleeve bearing when the rudder is put fully over at a ship's speed of 26 knots?

In the force equation, take the constant $k = 0{\cdot}100$. Also if the length of an elemental strip of the rudder surface, drawn at right angles to the centre line of the stock is l, then assume the centre of pressure of the strip to be $0{\cdot}32l$ from the leading edge.

18. Limiting the shear stress to 4·5 tonf/in^2, what should be the diameter of the stock, if solid, of the previous question? If the coefficient of friction at the sleeve is 0·20 and the rate of turn at 35 degrees is 3 degrees per second, what power is required of the steering gear?

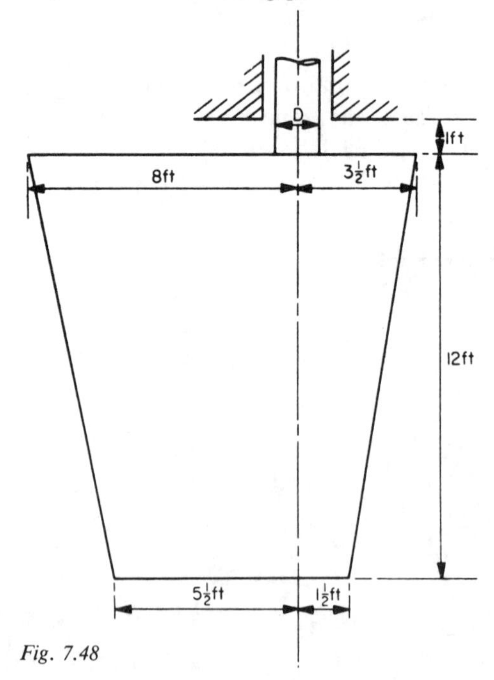

Fig. 7.48

19. For a symmetrical shaft bracket with arms of conventional shape, $\beta = 18°$, $a = 5{\cdot}2$ ft, $l = 5{\cdot}9$ ft, $k = 44{\cdot}0$ in., $h = 61{\cdot}6$ in. The arm section is 27 in. × 4·5 in.
Find the maximum stress in each arm near the barrel for a load of 65 tonf at the propeller.

References

1. Clarkson, J. *The elastic analysis of flat grillages*. Cambridge U.P., 1965.
2. Clarkson, J. Uniform pressure tests on plates with edges free to slide inwards, *TRINA*, 1962.
3. Aalami, B. and Chapman, J. H. Large deflection behaviour of ship plate panels under lateral pressure and in-plane loading, *TRINA*, 1972.
4. Southwell, R. *Relaxation methods in engineering sciences*. Oxford U.P., 1940.
5. Kendrick, S. B. The buckling under external pressure of ring stiffened cylinders, *TRINA*, 1965.
6. Shenoi, R. A., Gordon, A., Hamson, K., Humphries, A., Violette, F. and Dodkins, A. Ships' P Brackets in composite materials, *TRINA*, 133A, 1991.
7. Smith, C. S., Anderson, N., Chapman, J. C., Davidson, P. C. and Dowling, P. J. Strength of stiffened plating under combined compression and lateral pressure, *TRINA* 134A, 1992.
8. Davidson, P. C., Chapman, J. C., Smith, C. S. and Dowling, P. J. The design of plate panels subject to biaxial compression and lateral pressure, *TRINA* 134A, 1992.
9. Dow, R. S., Hugill, R. C., Clarke, J. D. and Smith, C. S. Evaluation of ultimate ship hull strength, *TSNAME Extreme loads symposium*, Oct. 1981.

8 Launching and docking

It is probably the enormity of the problem of moving a ship into and out of the water that most catches the imagination. The majesty of a launch is hardly less awesome than the sight of a ship from the bottom of a dock. Neither operation can be achieved without sound practical knowledge guided by some basic theory. This chapter is concerned with some important theoretical considerations associated with launching and docking.

Launching

A ship is best built under cover for protection from the elements leading to better quality and less disruption of schedules. Because of the capital costs it was not until the 1970s that covered slipways became fairly common. Partly to offset these costs a conveyor belt method of construction was pioneered in Scandinavia. The ship is built section by section and gradually pushed from the building hall into the open. A dry dock can be used and flooded up when the ship is sufficiently far advanced. A dry dock is too valuable a capital asset to be so long engaged and the method is generally confined to very large vessels where normal launching presents special difficulties. Occasionally, slipways are partially or wholly below high water level and the site is kept dry by doors which, on launch day are opened to let the water in and provide some buoyancy to the hull. More often, the hull is constructed wholly above water and slid into the water when ready. Very small hulls may be built on a cradle which is lowered on wheels down a ramp into the water, but ships over about 100 tonf displacement must be slid on greased ways under the action of gravity.

Usually, the ship slides stern first into the water because this part of the ship is more buoyant, but bow first launches are not unknown. The theory presently described is applicable to both. Less often, the ship is slid or tipped sideways into the water and the considerations are then rather different.

In a well ordered, stern first launch, sliding ways are built around the ship, and shortly before the launch the gap separating them from the fixed groundways is filled by a layer of grease to which the weight of the ship is transferred from the building blocks.

There may be one, two or even four ways; generally, there are two although in Holland one way with propping ways at the sides are common. The ways are inclined and, often, cambered. Movement is prevented until the desired moment by triggers which are then knocked away to allow the ship to move under gravity down the inclined ways. After it begins to enter the water, buoyancy builds up at the stern until it reaches a value sufficient to pivot the entire ship about the fore poppet (i.e. the forward end of the launching cradle). The ship continues down the ways until it is floating freely having slid away from, or dropped off the end of, the ways. If launched in a restricted waterway, its progress into the water is impeded by drags which are arranged so as to gradually bring the vessel to a stop before it strikes the far bank.

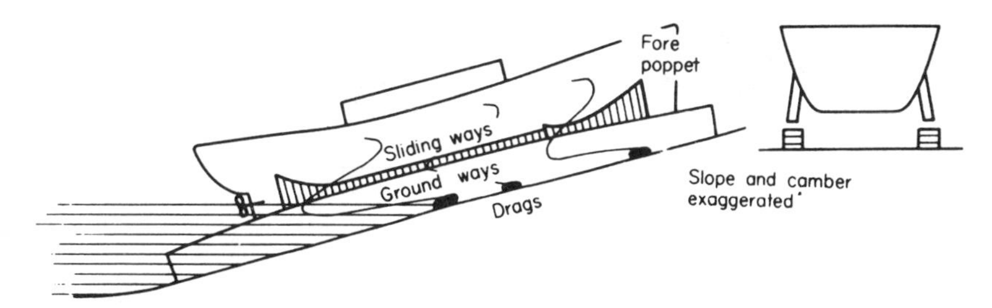

Fig. 8.1 A stern first launch

Consider what might go wrong with this procedure. The grease might be too slippery or not slippery enough. It might be squeezed out by the pressure. Instead of the stern lifting, the ship might tip the wrong way about the end of the groundways and plunge. The forefoot might be damaged by dropping off the end of the ways or it may dig into the slip when the stern has lifted. The ship might be insufficiently strong locally or longitudinally or the ways may collapse. The braking effect of the drags might be too much or too little. The ship might, at some instant, be unstable.

Calculations are carried out before arranging the launch to investigate each one of these anxieties. The calculations predict the behaviour of the ship during launch, to enable a suitable high tide to be selected and to arrange the ship and slip to be in a proper and safe condition.

LAUNCHING CURVES

A set of six curves is prepared to predict the behaviour of the ship during launch. They are curves, plotted against distance of travel down the slip, of

(i) Weight, W
(ii) Buoyancy, Δ
(iii) Moment of weight about fore poppet, Wa
(iv) Moment of buoyancy about fore poppet, Δd
(v) Moment of weight about after end of groundways, Wb
(vi) Moment of buoyancy about after end of groundways, Δc

The geometry of the ship at any position in its travel down the ways is illustrated in Fig. 8.2 for the ways in contact. A distributed load $(W - \Delta)$ along the ways is not shown; after the stern lifts this becomes concentrated all at the fore poppet. Conventionally, the moment (v) is drawn positive when anticlockwise and (vi) is drawn positive when clockwise. A typical set of launching curves is shown in Fig. 8.3.

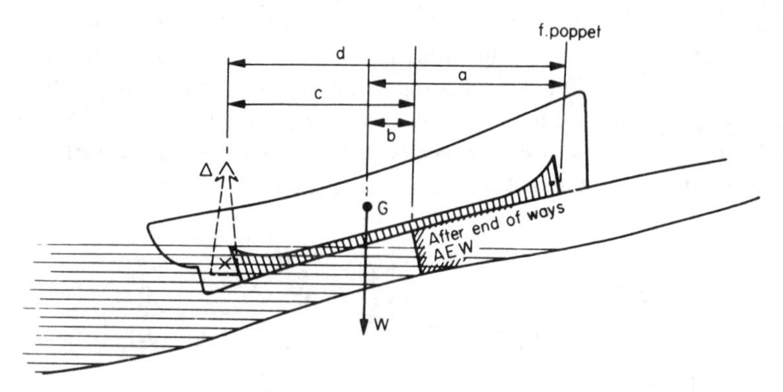

Fig. 8.2 Ship and way geometry

The important features of these curves are as follows:

(*a*) at the point at which the moment of buoyancy about the fore poppet equals the moment of weight about the fore poppet, the stern lifts;
(*b*) the difference between the weight and buoyancy curves at the position of stern lift, is the maximum force on the fore poppet;
(*c*) the curve of moment of buoyancy about the after end of the ways must lie wholly above the curve of moment of weight; the least distance between the two curves of moment about the after end of ways, gives the least moment against tipping about the end of ways;
(*d*) crossing of the weight and buoyancy curves before the after end of ways, indicates that the fore poppet will not drop off the end of the ways.

These curves answer some of the anxieties about the launch directly, but certain other investigations are necessary. From the difference between the

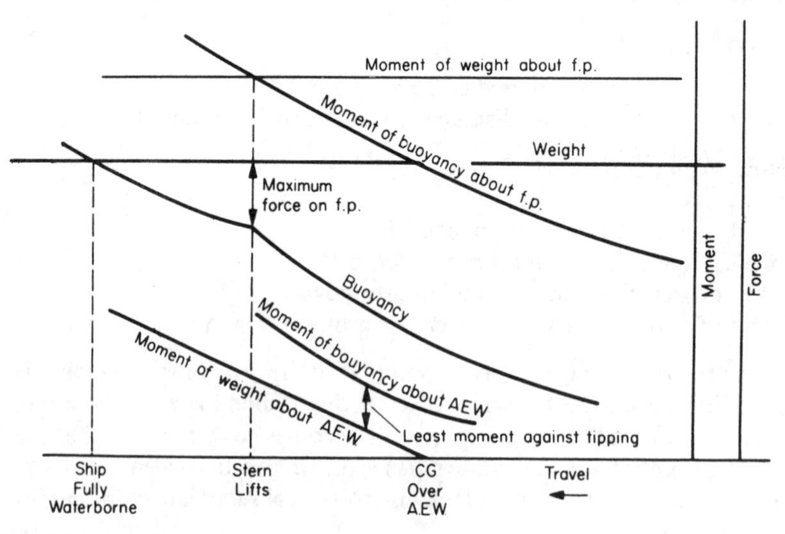

Fig. 8.3 Typical launching curves

weight and buoyancy curves before stern lift, the grease pressures can be determined. This difference at stern lift (and for *Queen Mary* for example, it was 8325 tonf) giving the maximum force on the fore poppet, enables the poppet and internal strengthening of the ship to be devised; it also causes a loss of stability. From all of these features is judged the adequacy of the height of tide and the length of ways. Indeed, curves are usually constructed for more than one height of water to determine the minimum acceptable height. How are the curves constructed?

CONSTRUCTION OF LAUNCHING CURVES

The curve of weight results usually from the weights weighed into the ship plus an estimate of what remains to be built in during the period between calculation and launch. Centre of gravity position is similarly estimated and the two moment of weight curves produced.

Buoyancy and centre of buoyancy at any position of travel is determined by placing the relevant waterline over a profile of the ship with Bonjean curves drawn on and integrating in the usual fashion. For this calculation, dynamic considerations are ignored and the ship is assumed to travel very slowly. While the correct waterline can be determined by sliding a tracing of the ship over a drawing of the slip, it is more accurately found from the geometry of the situation. Let

α = the initial slope of the keel
L = length between perpendiculars
e = distance of the fore poppet abaft the FP
h = initial height of the fore poppet above water
β = declivity of groundways, i.e. slope of chord
f = camber of ways of length K
r = radius of camber
t = distance abaft the fore poppet

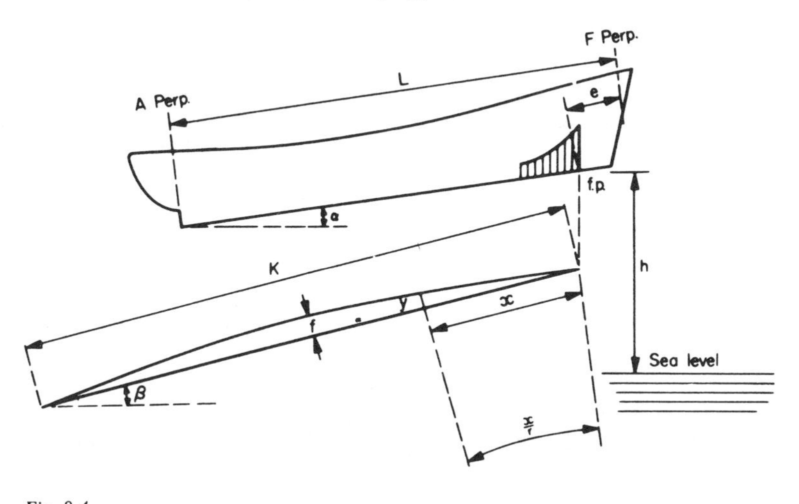

Fig. 8.4

From simple mensuration, the camber of an arc of a circle is given by

$$f = \frac{K^2}{8r} \quad \text{so that} \quad (f-y) = \frac{(K-2x)^2}{8r}$$

$$\therefore \quad y = \frac{K^2}{8r} - \frac{(K-2x)^2}{8r} = x\frac{(K-x)}{2r}$$

After travelling a distance x, the fore poppet is raised y above the chord and the keel has moved through an angle x/r. The height of the fore poppet above the water is now therefore, approximately

$$h-\beta x+y = h-\beta x+\frac{x}{2r}(K-x),$$

while the height above water of a point t abaft the fore poppet is

$$h-\beta x+y-t\left(\alpha+\frac{x}{r}\right) = h-\beta x+\frac{x}{2r}(K-x)-t\left(\alpha+\frac{x}{r}\right)$$

If there is no camber, r is, of course, infinite.

If, in this expression t is put equal to $-e$, it will give the height of the keel at the FP above water or, when it is negative, below water, i.e. the draught at the FP. If t is put equal to $L-e$, the expression gives the negative draught at the AP. Thus, for a given travel down the ways x, a waterline can be drawn accurately on the Bonjean profile from which buoyancy and centre of buoyancy can be determined.

This is satisfactory when the ways are fully in contact, but after stern lift, while the draught at the fore poppet can be determined in this way, trim cannot. Now at all times after stern lift the moments of weight and buoyancy about the fore poppet must be equal. For the position of travel under consideration, the fore poppet draught is found as already described; buoyancy and moment of buoyancy are then calculated for several trims about the fore poppet and a curve plotted as shown in Fig. 8.5. Where the curves of moment about fore poppet cut, there is the correct trim; buoyancy can be read off and the position of centre of buoyancy calculated. Having determined the correct trim, the passage of the forefoot to assess the clearance from the slip can be drawn.

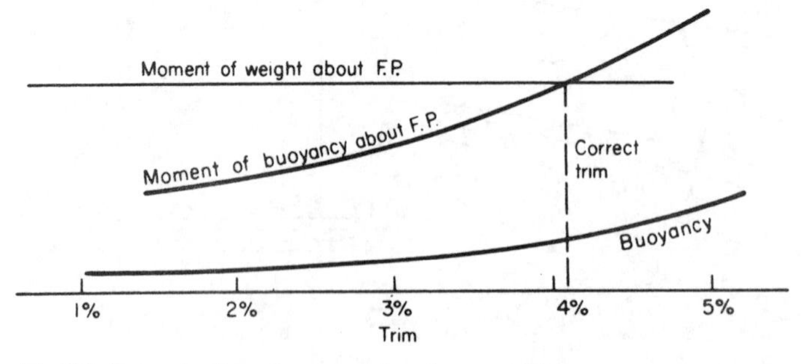

Fig. 8.5 Determination of correct trim after stern lift

If the launching curves indicate that the fore poppet will drop off the ways, adequate depth of water must be allowed under the forefoot for the momentum of the drop to prevent damage to the forefoot. What is adequate will depend on the amount of the excess of weight over buoyancy, the inertia of the ship and the damping effects of the water; it is best determined from an examination of previous launches.

GROUNDWAYS

Typically, the declivity of the groundways is 1 in 20 and the camber a half a metre in a groundway length of 300 m from fore poppet to after end (or 18 in. in 1000 ft). The radius corresponding to this camber is 22,500 m. Originally, camber was probably meant to offset the sinkage of the slip as the ship's weight grew. It has another important effect in rotating the ship to dip the stern deeply into the water; this increases the buoyancy force and causes an earlier stern lift than would be the case without camber. This increases the moment against tipping but also increases the load on the fore poppet.

The total load on the groundways is the difference between weight and buoyancy $W - \Delta$. Dividing by the length in contact gives a mean load per unit length and this, divided by the width of ways gives a mean pressure. The maximum total load on the ways is the initial one, W, and experience has shown that the mean pressure associated with this load should not, for many greases, exceed about 2·5 tonf/ft² (27 tonnef/m²) or the grease tends to get squeezed out. Both the ship, which has an uneven weight distribution, and the ways are elastic and they are separated by grease; what is the true distribution of pressure along the length has never been measured. Ignoring the moment causing the small angular acceleration of the ship due to camber, it is known that, before stern lift (Fig. 8.2),

(*a*) $W - \Delta$ = total load on ways,
(*b*) $Wa - \Delta d$ = moment of way load about fore poppet.

Some appreciation of the load distribution at any instant can be obtained by assuming that the distribution is linear, i.e. that the curve of load per unit length p' against length is a trapezoid. If the length of ways remaining in contact is l at any instant and the loads per unit length at the fore poppet and after end of ways are, respectively, p'_f and p'_a, the conditions (*a*) and (*b*) become

$$(a)\quad W - \Delta = \tfrac{1}{2}l(p'_f + p'_a)$$

$$(b)\quad Wa - \Delta d = p'_f\frac{l^2}{2} + \frac{1}{3}l^2(p'_a - p'_f)$$

Solving these equations for p'_f and p'_a,

$$p'_f = 4\frac{W-\Delta}{l} - 6\frac{Wa - \Delta d}{l^2}$$

$$p'_a = 6\frac{Wa - \Delta d}{l^2} - 2\frac{W-\Delta}{l}$$

This is a satisfactory solution while p'_f and p'_a are positive, and the load per unit length can be represented by a trapezium. When $(Wa-\Delta d)/(W-\Delta)$ is greater than $\frac{2}{3}l$, p'_f becomes negative and when $(Wa-\Delta d)/(W-\Delta)$ is less than $\frac{1}{3}l$, p'_a becomes negative. There cannot be a negative load so for these conditions, the trapezoidal presumption is not permissible. It is assumed, instead, that the distribution is triangular whence, for

$$\frac{Wa-\Delta d}{W-\Delta} > \frac{2}{3}l \quad \text{then} \quad p'_a = \frac{2(W-\Delta)^2}{3l(W-\Delta)-3(Wa-\Delta d)}$$

and for

$$\frac{Wa-\Delta d}{W-\Delta} < \frac{1}{3}l \quad \text{then} \quad p'_f = \frac{2(W-\Delta)^2}{3(Wa-\Delta d)}$$

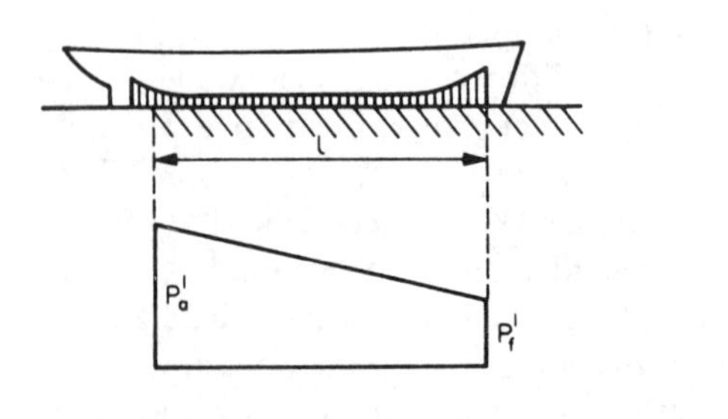

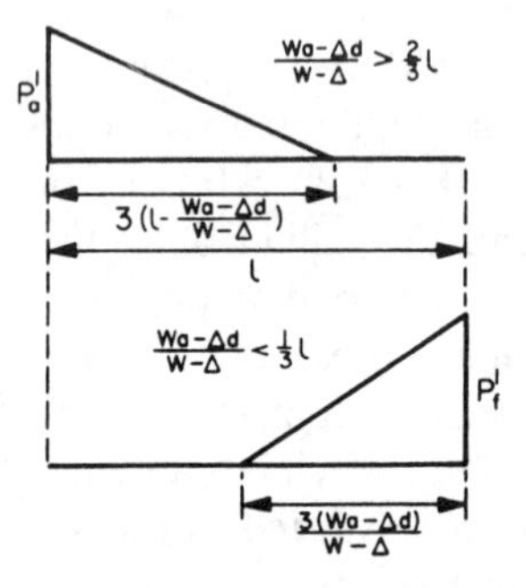

Fig. 8.6 Assumed load distribution

It is now possible to plot the maximum loads or pressures against travel down the slip (Fig. 8.7). Permissible grease pressures vary with grease, temperature and past experience. 5 tonf/ft² (55 tonnef/m²) is typical but figures of 10 or 15 are not unknown. The calculation helps determine poppet strength and where internal shoring is needed.

Apart from the assumption of linear variation in pressure the above method of calculation assumes the ship and slipway to be rigid. Reference 5 suggests

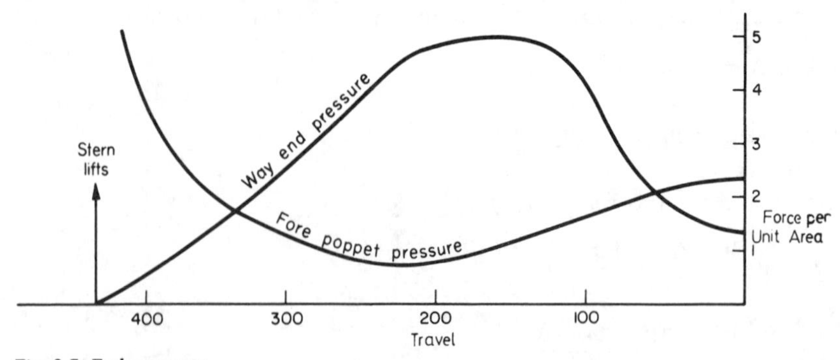

Fig. 8.7 End pressures

that the simple method predicts way end pressures which are unrealistically low and presents a more rigorous treatment allowing for ship elasticity, flexibility of the ways and the effects of bottom panel deflection. It shows that pressures may be reduced by champfering the ends of the ways and may be physically limited by using deformable packing or collapsible cushions.

THE DYNAMICS OF LAUNCHING

The force accelerating the ship down the groundways is, at any instant, approximately

$$(W-\Delta)\theta - \text{way friction} - \text{water resistance} - \text{drag forces}$$

θ is the slope of the ways at the centre of gravity of the ship. Way friction is $\mu(W-\Delta)$; the coefficient of friction μ is usually less than 0·02, although at the commencement it can be slightly higher and tests should be conducted to establish the figures for the particular lubricant over a range of temperatures. Water resistance is due to the hull friction, the creation of the stern wave and to the resistance of locked propellers, water brakes or masks, etc., where fitted; this resistance is expressed by $K\Delta^{2/3}V^2$ where V is the velocity of travel and K is a constant determined from similar ships with similar water braking devices (for *Queen Mary* it was 0·001). Retarding forces due to chain drags are found to follow a frictional law, $\mu'w$ where w is the weight of the chains and μ' is 0·40–0·80 depending upon the state of the slipway; this figure must be determined from trials or from previous launches.

For a particular ship the effects of entrained water can be expressed as a fraction z of the buoyancy. The equation of motion of the ship before it is waterborne is then

$$(W-\Delta)\theta - \mu(W-\Delta) - K\Delta^{\frac{2}{3}}\ V^2 - \mu' w = \text{nett force} = \frac{(W+z\Delta)}{g}\frac{dV}{dt}$$

This differential equation cannot, of course, be solved mathematically because of the presence of Δ.

Consideration of each of the factors at intervals of travel down the slip, however, enables a component force diagram to be built up as shown in Fig. 8.8 and a distance–time relationship estimated from it, by equating the nett force to the mass × acceleration. After the ship has become waterborne, the first two components of the expression become zero. Integration of each component force–distance curve gives the work done in overcoming that resistance. Velocity at any point of travel may therefore be checked by relating the kinetic energy at that point to the loss of potential energy minus work done in overcoming friction and resistance. Reference 1 describes a method of doing this.

STRENGTH AND STABILITY

Problems of local strength occur at the keel over the after end of the ways and at the fore poppet both for the ship and for the poppet structure. The difficulties

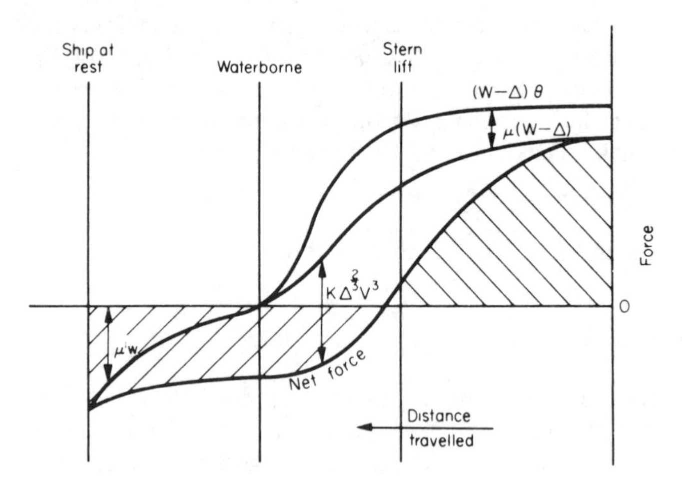

Fig. 8.8 Forces acting during launch

of containing the large forces at the fore poppet account for the increasing number of very large oil tankers and other bulk carriers being built in dry docks (productivity is of course high).

At the instant of stern lift, the hull girder undergoes a maximum sagging bending moment, the longitudinal strain from which in some large ships has been measured to give stresses of 5–6 tonf/in^2. The bending strain is easily found from the data already compiled to ensure a safe launch but strain measured in practice is usually somewhat below this, due probably to the effects of water resistance. Breakage is usually measured.

Stability is also at a minimum at the instant of stern lift due to the large fore poppet load at the keel. The $\overline{GM}$ is calculated by the methods described in Chapter 5 both for the ship with the concentrated load at the keel and fully afloat. While instability with the ship supported each side by the ways is unlikely, no naval architect would permit a ship to be launched with a negative virtual $\overline{GM}$ at the instant of stern lift. Any sudden sinking of the ways, for example, could be increased by an unstable ship. Ballasting is, in any case, often carried out to increase the longitudinal moment against tipping and presents no difficulty if needed for stability purposes.

SIDEWAYS LAUNCHING

When the ship is small or waterfront space is not at a great premium, ships may be built on an even keel broadside on to the water and consigned to the water sideways. There are three common methods of sideways launching:

(*a*) the ship slides down ways which are built well down under the water;
(*b*) the ship tips or drops off the end of the ways into the water, sometimes tipping a part of the ways too;
(*c*) the ship is built on piles which are made to collapse by a sideways push to allow the ship to fall into the water.

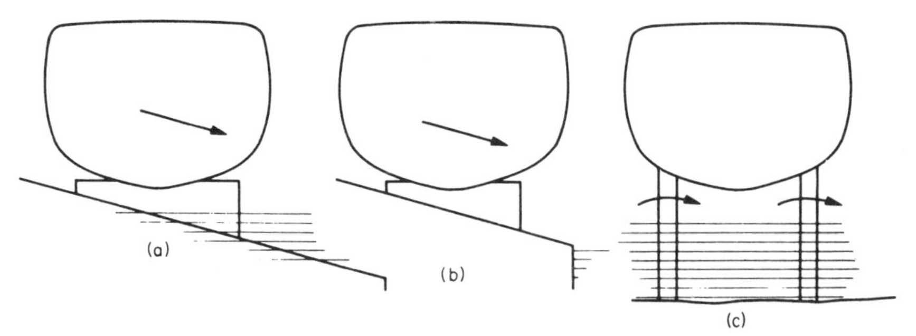

Fig. 8.9 Sideways launching

In all of these methods the ship takes to the water violently and may roll heavily—on entry, the ship may roll thirty degrees or more. Stability at large angles and watertightness are therefore important considerations. Waves may cause damage on adjacent shores.

Conventional calculations are not performed. Declivity of ways is usually of the order of 1 in 8 in order to give a high speed of launch to clear the end of the ways. Grease pressures permitted, 2·5–3·0 tonf/ft^2, are somewhat higher than with end launch.

Docking

From time to time, attention is needed to the outer bottom of a ship either for regular maintenance or to effect repairs. Access to the bottom may be obtained by careening, by hauling up a slipway or by placing in a dock from which the water is removed. There are limitations to what work can be done by careening, while the magnitude of the machinery needed to haul a ship up a slipway is such that this procedure is confined to ships up to a few hundred tonf in displacement.

The practical aspects of docks and docking are discussed fully in Refs. 2 and 6. This chapter is concerned with theory which assists such a successful operation and some design features of floating docks. Features of principal concern are

(*a*) load distribution between dock and ship;
(*b*) behaviour of blocks;
(*c*) strength of floating docks;
(*d*) stability.

This list by no means exhausts the problems associated with docking and constructing docks, which include design of caissons and pumping systems and the whole field of difficulties facing civil engineers in the construction of graving docks. It does, however, embrace the major concerns of the practising naval architect. Basically, docking is the placing of an elastic ship with an uneven weight distribution on to an elastic set of blocks supported in turn by an elastic floating dock or a relatively rigid graving dock. Blocks will not be of an even height and are subject to crush, creep and instability. In many aspects of ship design, the criteria of design are based on misuse or accident; with floating docks, there is little scope for permitting maloperation because if this were a major criterion of design, the lifting capacity of the dock would be unduly penalized.

LOAD DISTRIBUTION

A theoretical analysis of the behaviour of a ship in a floating dock is given in Ref. 3, which produces the longitudinal distribution of load on the blocks. A digital computer and the time to feed it data and program it may not always be available, however, and dockmasters may have to fall back on the practices which were current before this theory was deduced.

It has been assumed in the past that the load distribution along the blocks followed the weight distribution of the ship except at the after cut up and at any other gap in the blocks. Forward of the after cut up, the weight distribution is assumed augmented by the weight abaft the cut up spread over a length equal to twice the length of the overhang, and distributed according to a parabolic law such that the moment of weight is the same as the moment of overhang about the cut up. If the weight and moment of the overhang about the cut up are respectively W and M, the augment in load per unit length at the cut up, a, is

$$a = \frac{9W^2}{16M}$$

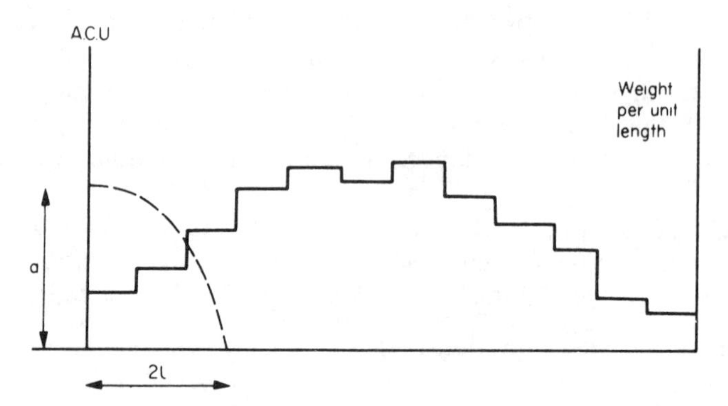

Fig. 8.10 Load distribution in dock

In a floating dock, the load distribution is further affected by the buoyancy distribution along the dock which is under the control of the dockmaster. For a given buoyancy distribution, the loading in the past has been determined by a method similar to the longitudinal strength calculation for a ship. The difference between the total buoyancy and ship's weight plus weight of dock and contained water gives the net loading. As explained in Chapter 6, shearing force on a girder is obtained by integrating the nett loading, and bending moment is obtained by integrating the shearing force along the length. In the case of the docked ship, the loading and the subsequent integrations apply to the combined ship and dock. Difficulty arises in trying to separate the two because their interaction is not known. In the past, bending moment has been divided up in the ratio of their respective second moments of area of section. This is not satisfactory and comprehensive analysis is needed.

Floating docks are classified by their lifting capacity. Lifting capacity is the buoyancy available when the dock is floating at its working freeboard. Working freeboard is often 6 in. (0·15 m) at the pontoon deck. Buoyancy available arises from the tanks which can be emptied to provide lift to the dock without undue strain.

BLOCK BEHAVIOUR

Depending as it does on defects, age, grain, moisture content and surface condition, it is not surprising to find wide variation in the behaviour of a stack of wooden dock blocks, as experiments have shown. Provided that the blocks are cribbed to prevent instability, it is common practice not to permit block pressures in excess of 40 tonf/ft², although the ultimate load intensity of a stack of English oak blocks may be as high as 100 tonf/ft².

Experiments have shown that the crippling pressure for a stack of blocks h high and a in width is given approximately by

$$p = 40 - \frac{h^2}{a^2} \text{ tonf/ft}^2$$

or

$$p = 11\left(40 - \frac{h^2}{a^2}\right)\text{tonnef/m}^2 = 1{\cdot}07\left(40 - \frac{h^2}{a^2}\right) \times 10^5 \text{ N/m}^2$$

where p is the load divided by the total plan area of a stack of blocks, irrespective of the half siding of keel. The formula is suitable for English oak and is obviously an approximation. For a 5 ft stack of 15 in. blocks it gives a crippling load of 24 tonf/ft²; for a 6 ft stack it is 17 tonf/ft². A 2 m stack of 40 cm blocks has a crippling load of 165 tonnef/m². In practice, 6 ft stacks are rare and pressures in excess of 20 tonf/ft² for a single uncribbed stack are not permitted. It is normal good practice to crib a certain number of blocks, which increases the crippling load as well as preventing tripping due to the trim of the docking ships.

The importance of exact alignment of blocks cannot be over-emphasized. Measurements of the loads in dock blocks have shown that a stack will take many times the load of its neighbour by being only a fraction of an inch higher. A lack of fit of half an inch between adjacent blocks is not unusual at present and is responsible for wild variation in dock block loading. It is responsible, too, for the departures of the block loading curve from the ship's weight distribution in graving docks.

Great care is needed to ensure that blocks are all of even stiffness. The number of blocks to ensure a mean deflection of stacks of x is

$$\text{Number of blocks} = \frac{2 \times \text{ship's weight}}{x \times \text{block stiffness}}$$

For a mean deflection of $\frac{1}{2}$ in. on blocks with a stiffness of 50 tonf/in., the minimum number of blocks required is $W/12{\cdot}5$ tonf. Clearly, this limitation is impractical for ships over about 2000 tonf in displacement and, in general, the larger the ship, the larger the block deflection that must be accepted.

STRENGTH OF FLOATING DOCKS

It is clear even from a cursory examination, that a dockmaster can place severe strains on a dock by a thoughtless selection of buoyancy tanks. In the transverse direction, if the dockmaster first selects the buoyancy tanks in the dock walls to begin raising the dock, a couple will be applied to the dock as shown in Fig. 8.11 which could cause damage. Similarly, in the longitudinal direction, if a ship has much weight amidships, the use of the buoyancy tanks at the ends of the dock could break it in two. It is good practice to select those buoyancy tanks immediately beneath the heaviest weights. Dock behaviour longitudinally, if the dock is made of steel, is checked continuously during lift by measuring the breakage. A strict limit to breakage is placed by the naval architect in the instructions for use of the dock. Reinforced concrete docks do not suffer much elastic distortion and breakage is not used as a measure of behaviour; instead, strict pumping patterns are imposed.

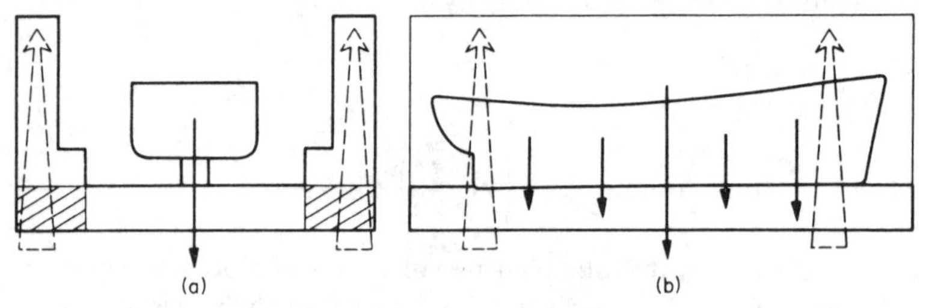

Fig. 8.11 Strains imposed by maloperation

Analysis of the strength of a floating dock was too difficult arithmetically until computers became available. The first reasonably precise analysis was published in 1966 (Ref. 3). This treated the pontoon deck as a grillage simply supported on its two long edges by the dock walls. The lumped weights of the ship were assumed supported by three rows of blocks to each stack of which was applied a lack of fit. Analysis was performed by reducing the problem to that of an unevenly loaded beam on an elastic foundation for which the differential equations can be readily formed, although the numerical solution is formidable without a computer.

The solution given by Ref. 3 is equally applicable to graving docks by applying very high figures to the dock rigidity. Standard comparative longitudinal strength calculations to cater for transit conditions are also performed on the floating dock structure itself as described for the ship in Chapter 6.

The conventional floating dock is capable of taking very large vessels. There are reports of a concrete floating dock capable of lifting a 350,000 dwt ship and a steel one of 400,000 dwt capacity.

STABILITY DURING DOCKING

Loss of stability due to grounding or docking has been discussed fully in Chapter 5. Calculation of this loss is a standard procedure before every docking, since a list developing before the ship is in contact with the blocks completely along the keel can be extremely dangerous and can dislodge the blocks.

A floating dock is trimmed approximately to the trim of the ship and large suing forces before the keel takes the blocks are thereby avoided. There is a critical stability condition when the ship is just clear of the water and the restoring waterplane for both ship and dock is provided only by the dock walls. In this condition, the dock designers normally demand a minimum $\overline{GM}$ of 5 ft (1·6 m). When the pontoon is lifted clear, the waterplane is, of course, much larger. Free surface effects in a floating dock are also large.

Large angle stability is important not only for ocean transit conditions but for the careening which is necessary during maintenance and repair operations.

SHIPLIFTS

Shiplifts can be used for launching or docking but the usual application is for docking. In a typical arrangement a vessel is floated above a cradle on a platform which can be raised to bring the cradle and ship to normal ground level. There is a system of rails which can then traverse the vessel sideways until it is in line with a shed into which it can be moved. In this way one lift can serve a number of docking sheds. Clearly care must be taken to avoid straining the ship or the platform during the lifting operation. There are a number of hydraulic hoisting points along the length of the platform and these can be fitted with load measuring devices to assist the operator, the data being fed into a computer to calculate load distribution and so assess trim, heel and draught before lowering into the water.

Platform elevator systems which lift vessels from the water and transfer them to the shore can take vessels of up to 25,000 tons. In other devices the vessel is floated over a cradle and both are winched up an inclined slipway. Often a transfer system is provided so that a single lift system can serve a number of refitting bays. Shiplifts are claimed to be significantly more economical than the corresponding floating dock system.

Problems

1. A vessel whose launching weight is 25 MN with CG 5 m abaft amidships has the fore poppet positioned 50 m forward of amidships. Find the force on the fore poppet and the travel when the stern lifts from the following data:

Travel down slip, m	50	60	70	80	90
Buoyancy, MN	10·1	12·3	15·3	19·4	24·6
CB abaft amidships, m	47	40	33	26	19

Sliding ways 70 cm wide each side are proposed, with a length of 95 m. What is the mean pressure on the lubricant before motion takes place?

2. In a certain ship, the length of sliding ways was 163 m and the breadth 1·63 m. The launching weight was 9755 tonnef and c.g. estimated at 75·50 m forward of the after end of the sliding ways. Calculate the mean pressure per square foot on the ways. Assuming the pressure to vary in a linear manner from forward to the after end of ways, what is the pressure at each end?

3. A ship of length 500 ft between perpendiculars is built on a slip with its keel at a slope of $\frac{5}{8}$ in. to one foot. It is launched on groundways which extend to the fore poppet only and have a camber of 2 ft over their length of 600 ft and a slope of chord of $\frac{3}{4}$ in. to one foot. If the underside of keel produced at the fore perpendicular is initially 18 ft above water level and the fore poppet is 60 ft aft of the fore perpendicular, what are the draughts at the fore perpendicular and at the after perpendicular when the ship has travelled 200 ft down the slip?

4. A ship is launched from ways 720 ft long, with a camber of 18 in. and a declivity of $\frac{9}{16}$ in./ft. The slope of the keel is $\frac{33}{64}$ in./ft. If the minimum height of stopping up is 2 ft 6 in., find the height of the fore poppet above the after end of the ways when the ship has travelled 200 ft down the ways. Assume the fore poppet initially at the fore end of ways.

5. Construct a set of launching curves from the following information:

launching weight	5230 tonnef
CG abaft midships	7 m
fore poppet before midships	69 m
after end of groundways from c.g. at rest	100 m

Travel down slip (m)	30	60	75	90	105	120	135
Buoyancy (tonnef)							
before stern lift	830	1960	2600	3350	4280	5600	
after stern lift					4050	4700	5550
Moment of buoyancy about fp before stern lift (1000 tonnef m)	132	219	282	356	450	585	
Moment of buoyancy about AEGW (1000 tonnef m)				42	76	160	316

(*a*) When does the stern lift?

(*b*) What is the maximum force at the fore poppet?

(*c*) What is the minimum moment against tipping?
(*d*) How does the ship leave the ways?

6. A ship weighs 28,500 tonnef at launch. The fore poppet is 72 m before the CG of the ship. The after end of the groundways is 155 m from the CG before the ship moves. The following table gives the data derived for the passage of the ship in full contact with the ways:

Travel, m	120	135	150	157·5	165
Buoyancy, tonnef	9000	15,600	24,400	29,200	35,000
CB abaft AEGW, m	—	7·5	21·5	28·0	33·0

Deduce a set of launching curves and pick off the important data.

7. A floating dock of rectangular bottom shape, 400 ft long and 90 ft wide floats, when empty, at a draught of 3·5 ft. It is used to dock a ship of 3600 tonf displacement, 360 ft long, which is symmetrically placed in the dock. The weight distribution of the ship is symmetrical about amidships and has the values shown in the table below:

Distance from midships, ft	0		20		40		60		80		100		120		140		160		180
Section from midships		1		2		3		4		5		6		7		8		9	
Weight per section, tonf		289		284		272		252		224		191		150		101		37	

Draw the curves of load, shearing force and bending moment for the dock and ship combination, in still water of 35 ft³/tonf.

The weight of the dock may be assumed equally distributed along its length.

8. A vessel has a launching weight of 5800 tonf, the CG being 26 ft abaft the mid-length and the fore poppet 230 ft before the mid-length. Construct a launching diagram from the following data:

Mid-length abaft AEGW, ft	0	20	40	60	80
Buoyancy, tonf	2560	3190	3840	4530	5330
CB abaft AEGW, ft	131	143	158	173	185

State:

(*a*) distance of mid-length abaft the after end of groundways when the stern lifts;
(*b*) force on fore poppets when the stern lifts;
(*c*) reserve moment against tipping.

References

1. McNeill, J. M. Launch of the quadruple screw turbine steamer, *Queen Mary*, *TINA*, 1935.
2. Todd, F. H. and Laws, E. Some model and full scale experiments on side launching, *Trans. N.E. Coast Inst.*, 1946–7.
3. Vaughan, H. Elastic analysis of a ship in a floating dock, *TRINA*, 1966.
4. Cameron, R. R. G. The sliding of large tanker on the slipway, *TRINA*, 1969.
5. Ratcliffe, A. T. Launchway pressure, *TRINA*, 1973.
6. Potvin, A. B., Hartz, B. and Nickum, G. C. Analysis of stresses in a floating dock due to a docked ship, *TSNAME*, 1969.
7. Lundgren, J., Price, W. G. and Yongshu, Wu. A hydroelastic investigation into the behaviour of a floating dry dock in waves, *TRINA*, 1989.

9 The ship environment and human factors

The naval architect, like the designer of any successful engineering product, must know the conditions under which his equipment is to exist and operate for its full life cycle, the attributes of those who will operate and maintain it and how those attributes vary with changing environment.

For convenience the environment can be divided into:

(*a*) the environment external to the ship which affects the ship as a whole and all exposed equipment. These conditions are caused by the sea and climate. Also the ship performance can be influenced by the depth or width of water present;
(*b*) the internal environment which affects the personnel and the internal equipment. To some extent, this environment is controllable, e.g. the temperature and humidity can be controlled by means of an air conditioning system.

The human element is covered by what is termed Human Factors (HF); a full study of which involves multi-discipline teams of physiologists, psychologists, engineers and scientists. The HF team can advise the naval architect on:

(i) how to design a system or equipment so that the man can most effectively play his part, so giving the greatest overall system efficiency. It is not necessarily, and in general is not, true that the maximum degree of automation is desirable. The blend of man and machine should build up the strengths of each and minimize the weaknesses;
(ii) in what way, and to what degree, the system efficiency will be reduced due to degradations in the man's performance due to the environment;
(iii) the levels of environmental parameters (e.g. noise, vibration) which should not be exceeded if a man's physical state is not to be temporarily or permanently harmed.

Clearly there is an interaction between (i) and (ii) above in that the initial design must allow for the likely in-service environment.

The external environment. The sea

WATER PROPERTIES

Certain physical properties of the sea are of considerable importance to the designer. They are:

(*a*) *Density*. The density of the water in which the ship floats affects her draught and trim (Chapter 3) and depends mainly upon the temperature and salinity. The standard values of density of fresh and salt water at various temperatures are given in Tables 9.1 and 9.2.

(*b*) *Kinematic viscosity*. This is particularly relevant to the frictional resistance experienced by a ship as it defines the Reynolds' number (Chapter 10). The standard values of the kinematic viscosity of fresh and salt water at various temperatures are given in Tables 9.3 and 9.4.

(*c*) *Salinity*. 'Standard' density and kinematic viscosity have been mentioned above. Values for actual samples of sea water will vary from area to area and will depend, amongst other things, upon the salinity. For instance, many objects will float in the Dead Sea, which has a very high salt content, which would sink in fresh water. For most standard calculations and tests a salinity of 3·5 per cent is assumed.

Table 9.1
Mass densities for fresh water
(last decimal figure is doubtful)

°C	ρ	°C	ρ	°C	ρ
0	999·79	10	999·59	20	998·12
1	999·79	11	999·49	21	997·92
2	999·89	12	999·40	22	997·72
3	999·89	13	999·30	23	997·43
4	999·89	14	999·10	24	997·24
5	999·89	15	999·00	25	996·94
6	999·89	16	998·91	26	996·75
7	999·79	17	998·71	27	996·45
8	999·79	18	998·51	28	996·16
9	999·69	19	998·32	29	995·87
				30	995·57

(*a*) Metric units; ρ in kg/m^3

°F	ρ	°F	ρ	°F	ρ
32	62·414	51	62·401	70	62·295
33	62·414	52	62·398	71	62·289
34	62·417	53	62·395	72	62·282
35	62·417	54	62·392	73	62·273
36	62·421	55	62·385	74	62·263
37	62·421	56	62·382	75	62·257
38	62·421	57	62·376	76	62·247
39	62·421	58	62·373	77	62·237
40	62·421	59	62·366	78	62·231
41	62·421	60	62·363	79	62·221
42	62·421	61	62·356	80	62·212
43	62·421	62	62·350	81	62·202
44	62·417	63	62·344	82	62·192
45	62·417	64	62·337	83	62·183
46	62·414	65	62·331	84	62·173
47	62·411	66	62·324	85	62·163
48	62·411	67	62·318	86	62·151
49	62·408	68	62·311		
50	62·405	69	62·305		

(*b*) British units; ρ in lb/ft^3

Table 9.2
Mass densities for salt water (salinity 3·5 per cent) (last decimal figure is doubtful)

°C	ρ	°C	ρ	°C	ρ
0	1028·03	10	1026·85	20	1024·70
1	1027·93	11	1026·66	21	1024·40
2	1027·83	12	1026·56	22	1024·11
3	1027·83	13	1026·27	23	1023·81
4	1027·74	14	1026·07	24	1023·52
5	1027·64	15	1025·87	25	1023·23
6	1027·44	16	1025·68	26	1022·93
7	1027·34	17	1025·38	27	1022·64
8	1027·15	18	1025·19	28	1022·25
9	1027·05	19	1024·99	29	1021·95
				30	1021·66

(*a*) Metric units; ρ in kg/m^3.

°F	ρ	°F	ρ	°F	ρ
32	64·177	51	64·100	70	63·949
33	64·174	52	64·094	71	63·939
34	64·174	53	64·087	72	63·930
35	64·171	54	64·081	73	63·920
36	64·168	55	64·071	74	63·910
37	64·165	56	64·065	75	63·901
38	64·161	57	64·058	76	63·891
39	64·158	58	64·052	77	63·878
40	64·155	59	64·042	78	63·869
41	64·152	60	64·036	79	63·859
42	64·145	61	64·029	80	63·846
43	64·142	62	64·020	81	63·836
44	64·136	63	64·010	82	63·824
45	64·132	64	64·004	83	63·814
46	64·126	65	63·994	84	63·801
47	64·123	66	63·988	85	63·791
48	64·116	67	63·978	86	63·779
49	64·110	68	63·968		
50	64·103	69	63·959		

(*b*) British units; ρ in lb/ft^3

THE SEA SURFACE

The sea presents an ever changing face to the observer. In the long term, the surface may be in any condition from a flat calm to extreme roughness. In the short term, the surface may present the appearance of a fairly steady level of roughness but the actual surface shape will be continuously varying.

Any observer is aware that a stream of air passing over a water surface causes ripples or waves to form, e.g. by blowing across the top of a cup of tea ripples are formed. The precise mechanism by which the transfer of energy takes place is not known. Nor is the minimum velocity of air necessary to generate waves known accurately, and figures between 0·6 and 6 m/s (2 and 21 ft/s) have been suggested by various workers in this field.

Table 9.3

Values of kinematic viscosity for fresh water, ν, in metric units of (m^2s^{-1}) × 10^6. *Temp. in degrees Celsius*

Deg. C	0·0	0·1	0·2	0·3	0·4	0·5	0·6	0·7	0·8	0·9
0	1·78667	1·78056	1·77450	1·76846	1·76246	1·75648	1·75054	1·74461	1·73871	1·73285
1	1·72701	1·72121	1·71545	1·70972	1·70403	1·69836	1·69272	1·68710	1·68151	1·67594
2	1·67040	1·66489	1·65940	1·65396	1·64855	1·64316	1·63780	1·63247	1·62717	1·62190
3	1·61665	1·61142	1·60622	1·60105	1·59591	1·59079	1·58570	1·58063	1·57558	1·57057
4	1·56557	1·56060	1·55566	1·55074	1·54585	1·54098	1·53613	1·53131	1·52651	1·52173
5	1·51698	1·51225	1·50754	1·50286	1·49820	1·49356	1·48894	1·48435	1·47978	1·47523
6	1·47070	1·46619	1·46172	1·45727	1·45285	1·44844	1·44405	1·43968	1·43533	1·43099
7	1·42667	1·42238	1·41810	1·41386	1·40964	1·40543	1·40125	1·39709	1·39294	1·38882
8	1·38471	1·38063	1·37656	1·37251	1·36848	1·36445	1·36045	1·35646	1·35249	1·34855
9	1·34463	1·34073	1·33684	1·33298	1·32913	1·32530	1·32149	1·31769	1·31391	1·31015
10	1·30641	1·30268	1·29897	1·29528	1·29160	1·28794	1·28430	1·28067	1·27706	1·27346
11	1·26988	1·26632	1·26277	1·25924	1·25573	1·25223	1·24874	1·24527	1·24182	1·23838
12	1·23495	1·23154	1·22815	1·22478	1·22143	1·21809	1·21477	1·21146	1·20816	1·20487
13	1·20159	1·19832	1·19508	1·19184	1·18863	1·18543	1·18225	1·17908	1·17592	1·17278
14	1·16964	1·16651	1·16340	1·16030	1·15721	1·15414	1·15109	1·14806	1·14503	1·14202
15	1·13902	1·13603	1·13304	1·13007	1·12711	1·12417	1·12124	1·11832	1·11542	1·11254
16	1·10966	1·10680	1·10395	1·10110	1·09828	1·09546	1·09265	1·08986	1·08708	1·08431
17	1·08155	1·07880	1·07606	1·07334	1·07062	1·06792	1·06523	1·06254	1·05987	1·05721
18	1·05456	1·05193	1·04930	1·04668	1·04407	1·04148	1·03889	1·03631	1·03375	1·03119
19	1·02865	1·02611	1·02359	1·02107	1·01857	1·01607	1·01359	1·01111	1·00865	1·00619
20	1·00374	1·00131	0·99888	0·99646	0·99405	0·99165	0·98927	0·98690	0·98454	0·98218
21	0·97984	0·97750	0·97517	0·97285	0·97053	0·96822	0·96592	0·96363	0·96135	0·95908
22	0·95682	0·95456	0·95231	0·95008	0·94786	0·94565	0·94345	0·94125	0·93906	0·93688
23	0·93471	0·93255	0·93040	0·92825	0·92611	0·92397	0·92184	0·91971	0·91760	0·91549
24	0·91340	0·91132	0·90924	0·90718	0·90512	0·90306	0·90102	0·89898	0·89695	0·89493
25	0·89292	0·89090	0·88889	0·88689	0·88490	0·88291	0·88094	0·87897	0·87702	0·87507
26	0·87313	0·87119	0·86926	0·86734	0·86543	0·86352	0·86162	0·85973	0·85784	0·85596
27	0·85409	0·85222	0·85036	0·84851	0·84666	0·84482	0·84298	0·84116	0·83934	0·83752
28	0·83572	0·83391	0·83212	0·83033	0·82855	0·82677	0·82500	0·82324	0·82148	0·81973
29	0·81798	0·81625	0·81451	0·81279	0·81106	0·80935	0·80765	0·80596	0·80427	0·80258
30	0·80091	0·79923	0·79755	0·79588	0·79422	0·79256	0·79090	0·78924	0·78757	0·78592

ν in English units of (ft^2s^{-1}) × 10^5. *Temp. in degrees Fahrenheit*

Deg. F	0·0	0·1	0·2	0·3	0·4	0·5	0·6	0·7	0·8	0·9
32	1·92314	1·91949	1·91585	1·91222	1·90860	1·90499	1·90139	1·89780	1·89422	1·89066
33	1·88710	1·88354	1·88000	1·87646	1·87294	1·86942	1·86592	1·86242	1·85894	1·85546
34	1·85200	1·84855	1·84512	1·84169	1·83828	1·83487	1·83148	1·82809	1·82471	1·82135
35	1·81799	1·81463	1·81129	1·80795	1·80462	1·80130	1·79799	1·79469	1·79140	1·78812
36	1·78485	1·78160	1·77835	1·77512	1·77189	1·76868	1·76547	1·76227	1·75908	1·75590
37	1·75273	1·74957	1·74642	1·74327	1·74014	1·73701	1·73389	1·73078	1·72768	1·72459
38	1·72150	1·71843	1·71536	1·71230	1·70926	1·70621	1·70318	1·70016	1·69714	1·69413
39	1·69114	1·68814	1·68516	1·68219	1·67922	1·67626	1·67331	1·67037	1·66744	1·66451
40	1·66160	1·65869	1·65578	1·65289	1·65000	1·64713	1·64426	1·64139	1·63854	1·63569
41	1·63285	1·63002	1·62720	1·62438	1·62157	1·61877	1·61598	1·61319	1·61042	1·60765
42	1·60488	1·60213	1·59938	1·59664	1·59390	1·59118	1·58846	1·58575	1·58304	1·58035
43	1·57766	1·57498	1·57231	1·56965	1·56700	1·56435	1·56171	1·55908	1·55646	1·55384
44	1·55123	1·54861	1·54601	1·54341	1·54082	1·53823	1·53565	1·53308	1·53052	1·52796
45	1·52541	1·52287	1·52034	1·51782	1·51530	1·51279	1·51029	1·50779	1·50530	1·50281
46	1·50034	1·49786	1·49540	1·49294	1·49049	1·48804	1·48560	1·48317	1·48074	1·47832
47	1·47591	1·47349	1·47108	1·46868	1·46628	1·46389	1·46151	1·45913	1·45675	1·45439
48	1·45203	1·44968	1·44734	1·44500	1·44268	1·44035	1·43803	1·43572	1·43342	1·43112
49	1·42882	1·42654	1·42425	1·42198	1·41971	1·41744	1·41518	1·41293	1·41068	1·40844
50	1·40620	1·40397	1·40175	1·39953	1·39731	1·39510	1·39290	1·39070	1·38851	1·38632
51	1·38414	1·38197	1·37980	1·37763	1·37547	1·37332	1·37117	1·36903	1·36689	1·36475
52	1·36263	1·36050	1·35839	1·35628	1·35417	1·35207	1·34997	1·34788	1·34579	1·34371
53	1·34164	1·33957	1·33750	1·33544	1·33338	1·33133	1·32929	1·32725	1·32521	1·32318
54	1·32116	1·31914	1·31713	1·31513	1·31313	1·31114	1·30915	1·30716	1·30518	1·30321
55	1·30124	1·29926	1·29730	1·29533	1·29337	1·29142	1·28947	1·28753	1·28559	1·28365
56	1·28172	1·27980	1·27789	1·27598	1·27408	1·27217	1·27028	1·26839	1·26650	1·26462
57	1·26274	1·26086	1·25898	1·25711	1·25524	1·25338	1·25152	1·24967	1·24782	1·24598
58	1·24414	1·24231	1·24048	1·23866	1·23684	1·23503	1·23322	1·23142	1·22962	1·22782
59	1·22603	1·22424	1·22245	1·22066	1·21888	1·21710	1·21533	1·21356	1·21180	1·21004
60	1·20828	1·20653	1·20479	1·20305	1·20132	1·19959	1·19786	1·19614	1·19442	1·19271
61	1·19100	1·18929	1·18759	1·18589	1·18420	1·18251	1·18082	1·17914	1·17746	1·17578
62	1·17411	1·17245	1·17078	1·16912	1·16747	1·16581	1·16416	1·16252	1·16088	1·15924
63	1·15761	1·15598	1·15435	1·15273	1·15111	1·14949	1·14788	1·14627	1·14467	1·14307
64	1·14147	1·13988	1·13829	1·13670	1·13512	1·13354	1·13196	1·13039	1·12882	1·12726
65	1·12570	1·12414	1·12258	1·12103	1·11948	1·11794	1·11640	1·11486	1·11333	1·11179
66	1·11027	1·10874	1·10722	1·10571	1·10419	1·10268	1·10117	1·09967	1·09817	1·09667
67	1·09518	1·09369	1·09220	1·09072	1·08923	1·08776	1·08628	1·08481	1·08334	1·08188
68	1·08042	1·07896	1·07750	1·07605	1·07460	1·07316	1·07171	1·07027	1·06884	1·06740
69	1·06597	1·06455	1·06313	1·06172	1·06031	1·05890	1·05749	1·05609	1·05469	1·05329
70	1·05190	1·05050	1·04910	1·04771	1·04633	1·04494	1·04356	1·04218	1·04081	1·03943
71	1·03806	1·03670	1·03533	1·03397	1·03261	1·03126	1·02990	1·02855	1·02721	1·02586
72	1·02452	1·02319	1·02186	1·02053	1·01921	1·01788	1·01657	1·01525	1·01394	1·01263
73	1·01132	1·01001	1·00871	1·00741	1·00611	1·00482	1·00353	1·00224	1·00095	0·99967
74	0·99839	0·99710	0·99582	0·99455	0·99327	0·99200	0·99073	0·98946	0·98820	0·98693
75	0·98567	0·98442	0·98317	0·98193	0·98068	0·97944	0·97820	0·97697	0·97573	0·97450
76	0·97327	0·97205	0·97082	0·96960	0·96838	0·96717	0·96595	0·96474	0·96353	0·96233
77	0·96112	0·95992	0·95871	0·95751	0·95631	0·95512	0·95392	0·95273	0·95154	0·95036
78	0·94917	0·94799	0·94682	0·94565	0·94448	0·94331	0·94214	0·94098	0·93982	0·93866
79	0·93751	0·93635	0·93520	0·93405	0·93291	0·93176	0·93062	0·92948	0·92834	0·92721
80	0·92608	0·92495	0·92382	0·92269	0·92157	0·92045	0·91933	0·91821	0·91709	0·91598
81	0·91487	0·91376	0·91266	0·91155	0·91045	0·90935	0·90825	0·90716	0·90607	0·90498
82	0·90389	0·90280	0·90172	0·90063	0·89955	0·89848	0·89740	0·89633	0·89525	0·89418
83	0·89312	0·89205	0·89099	0·88993	0·88887	0·88781	0·88676	0·88570	0·88465	0·88360
84	0·88256	0·88151	0·88047	0·87943	0·87839	0·87735	0·87632	0·87528	0·87425	0·87323
85	0·87220	0·87118	0·87016	0·86914	0·86813	0·86712	0·86611	0·86510	0·86409	0·86309
86	0·86208	0·86108	0·86008	0·85907	0·85807	0·85708	0·85608	0·85509	0·85410	0·85311

Table 9.4

Values of kinematic viscosity for salt water, ν, in metric units of $(m^2s^{-1}) \times 10^6$. *(Salinity 3·5 per cent.)*
Temp. in degrees Celsius

Deg. C	0·0	0·1	0·2	0·3	0·4	0·5	0·6	0·7	0·8	0·9
0	1·82844	1·82237	1·81633	1·81033	1·80436	1·79842	1·79251	1·78662	1·78077	1·77494
1	1·76915	1·76339	1·75767	1·75199	1·74634	1·74072	1·73513	1·72956	1·72403	1·71853
2	1·71306	1·70761	1·70220	1·69681	1·69145	1·68612	1·68082	1·67554	1·67030	1·66508
3	1·65988	1·65472	1·64958	1·64446	1·63938	1·63432	1·62928	1·62427	1·61929	1·61433
4	1·60940	1·60449	1·59961	1·59475	1·58992	1·58511	1·58032	1·57556	1·57082	1·56611
5	1·56142	1·55676	1·55213	1·54752	1·54294	1·53838	1·53383	1·52930	1·52479	1·52030
6	1·51584	1·51139	1·50698	1·50259	1·49823	1·49388	1·48956	1·48525	1·48095	1·47667
7	1·47242	1·46818	1·46397	1·45978	1·45562	1·45147	1·44735	1·44325	1·43916	1·43508
8	1·43102	1·42698	1·42296	1·41895	1·41498	1·41102	1·40709	1·40317	1·39927	1·39539
9	1·39152	1·38767	1·38385	1·38003	1·37624	1·37246	1·36870	1·36496	1·36123	1·35752
10	1·35383	1·35014	1·34647	1·34281	1·33917	1·33555	1·33195	1·32837	1·32481	1·32126
11	1·31773	1·31421	1·31071	1·30722	1·30375	1·30030	1·29685	1·29343	1·29002	1·28662
12	1·28324	1·27987	1·27652	1·27319	1·26988	1·26658	1·26330	1·26003	1·25677	1·25352
13	1·25028	1·24705	1·24384	1·24064	1·23745	1·23428	1·23112	1·22798	1·22484	1·22172
14	1·21862	1·21552	1·21244	1·20938	1·20632	1·20328	1·20027	1·19726	1·19426	1·19128
15	1·18831	1·18534	1·18239	1·17944	1·17651	1·17359	1·17068	1·16778	1·16490	1·16202
16	1·15916	1·15631	1·15348	1·15066	1·14786	1·14506	1·14228	1·13951	1·13674	1·13399
17	1·13125	1·12852	1·12581	1·12309	1·12038	1·11769	1·11500	1·11232	1·10966	1·10702
18	1·10438	1·10176	1·09914	1·09654	1·09394	1·09135	1·08876	1·08619	1·08363	1·08107
19	1·07854	1·07601	1·07350	1·07099	1·06850	1·06601	1·06353	1·06106	1·05861	1·05616
20	1·05372	1·05129	1·04886	1·04645	1·04405	1·04165	1·03927	1·03689	1·03452	1·03216
21	1·02981	1·02747	1·02514	1·02281	1·02050	1·01819	1·01589	1·01360	1·01132	1·00904
22	1·00678	1·00452	1·00227	1·00003	0·99780	0·99557	0·99336	0·99115	0·98895	0·98676
23	0·98457	0·98239	0·98023	0·97806	0·97591	0·97376	0·97163	0·96950	0·96737	0·96526
24	0·96315	0·96105	0·95896	0·95687	0·95479	0·95272	0·95067	0·94862	0·94658	0·94455
25	0·94252	0·94049	0·93847	0·93646	0·93445	0·93245	0·93046	0·92847	0·92649	0·92452
26	0·92255	0·92059	0·91865	0·91671	0·91478	0·91286	0·91094	0·90903	0·90711	0·90521
27	0·90331	0·90141	0·89953	0·89765	0·89579	0·89393	0·89207	0·89023	0·88838	0·88654
28	0·88470	0·88287	0·88105	0·87923	0·87742	0·87562	0·87383	0·87205	0·87027	0·86849
29	0·86671	0·86494	0·86318	0·86142	0·85966	0·85792	0·85619	0·85446	0·85274	0·85102
30	0·84931	0·84759	0·84588	0·84418	0·84248	0·84079	0·83910	0·83739	0·83570	0·83400

ν in English units of $(ft^2s^{-1}) \times 10^5$. *(Salinity 3·5 per cent.) Temp. in degrees Fahrenheit*

Deg. F	0·0	0·1	0·2	0·3	0·4	0·5	0·6	0·7	0·8	0·9
32	1·96810	1·96447	1·96085	1·95724	1·95364	1·95005	1·94647	1·94290	1·93934	1·93580
33	1·93226	1·92873	1·92520	1·92169	1·91819	1·91470	1·91122	1·90775	1·90429	1·90084
34	1·89740	1·89398	1·89057	1·88718	1·88379	1·88041	1·87704	1·87368	1·87034	1·86700
35	1·86367	1·86035	1·85705	1·85375	1·85046	1·84718	1·84391	1·84065	1·83740	1·83416
36	1·83093	1·82771	1·82450	1·82129	1·81810	1·81492	1·81174	1·80858	1·80542	1·80227
37	1·79914	1·79601	1·79289	1·78978	1·78667	1·78358	1·78050	1·77742	1·77436	1·77130
38	1·76825	1·76521	1·76218	1·75916	1·75614	1·75314	1·75014	1·74715	1·74417	1·74120
39	1·73824	1·73528	1·73234	1·72940	1·72647	1·72355	1·72063	1·71773	1·71483	1·71194
40	1·70906	1·70619	1·70332	1·70047	1·69762	1·69478	1·69194	1·68912	1·68630	1·68349
41	1·68069	1·67790	1·67512	1·67235	1·66959	1·66684	1·66409	1·66135	1·65861	1·65589
42	1·65317	1·65045	1·64774	1·64504	1·64234	1·63965	1·63697	1·63429	1·63163	1·62897
43	1·62631	1·62367	1·62104	1·61842	1·61580	1·61319	1·61059	1·60799	1·60540	1·60282
44	1·60025	1·59767	1·59510	1·59254	1·58998	1·58743	1·58489	1·58236	1·57983	1·57730
45	1·57479	1·57229	1·56979	1·56730	1·56482	1·56235	1·55988	1·55742	1·55496	1·55251
46	1·55007	1·54763	1·54519	1·54276	1·54033	1·53791	1·53550	1·53309	1·53069	1·52830
47	1·52591	1·52353	1·52117	1·51880	1·51645	1·51410	1·51175	1·50942	1·50708	1·50476
48	1·50244	1·50012	1·49782	1·49551	1·49322	1·49093	1·48864	1·48636	1·48409	1·48182
49	1·47956	1·47730	1·47505	1·47280	1·47057	1·46833	1·46610	1·46388	1·46166	1·45945
50	1·45724	1·45504	1·45283	1·45064	1·44845	1·44626	1·44408	1·44190	1·43973	1·43757
51	1·43541	1·43326	1·43112	1·42899	1·42685	1·42473	1·42261	1·42049	1·41838	1·41628
52	1·41418	1·41208	1·40999	1·40791	1·40583	1·40375	1·40169	1·39962	1·39756	1·39551
53	1·39346	1·39141	1·38937	1·38734	1·38531	1·38328	1·38126	1·37925	1·37724	1·37523
54	1·37323	1·37124	1·36926	1·36728	1·36530	1·36333	1·36137	1·35941	1·35745	1·35550
55	1·35355	1·35160	1·34966	1·34772	1·34578	1·34385	1·34192	1·34000	1·33808	1·33617
56	1·33426	1·33236	1·33046	1·32856	1·32667	1·32479	1·32290	1·32103	1·31915	1·31728
57	1·31542	1·31356	1·31170	1·30985	1·30801	1·30616	1·30432	1·30249	1·30066	1·29883
58	1·29701	1·29520	1·29339	1·29159	1·28979	1·28800	1·28621	1·28442	1·28264	1·28086
59	1·27908	1·27731	1·27553	1·27376	1·27200	1·27024	1·26848	1·26673	1·26498	1·26324
60	1·26150	1·25976	1·25802	1·25630	1·25457	1·25285	1·25113	1·24942	1·24771	1·24600
61	1·24430	1·24261	1·24092	1·23923	1·23755	1·23587	1·23420	1·23253	1·23086	1·22920
62	1·22754	1·22589	1·22424	1·22259	1·22094	1·21930	1·21767	1·21603	1·21440	1·21278
63	1·21115	1·20953	1·20791	1·20629	1·20467	1·20306	1·20146	1·19985	1·19825	1·19665
64	1·19506	1·19348	1·19189	1·19032	1·18874	1·18717	1·18561	1·18404	1·18248	1·18092
65	1·17937	1·17781	1·17626	1·17471	1·17316	1·17162	1·17008	1·16855	1·16701	1·16548
66	1·16396	1·16244	1·16092	1·15941	1·15790	1·15640	1·15490	1·15340	1·15191	1·15041
67	1·14892	1·14744	1·14596	1·14448	1·14300	1·14153	1·14006	1·13859	1·13713	1·13566
68	1·13421	1·13275	1·13130	1·12985	1·12841	1·12696	1·12552	1·12409	1·12265	1·12122
69	1·11979	1·11837	1·11695	1·11553	1·11411	1·11270	1·11129	1·10988	1·10848	1·10708
70	1·10568	1·10428	1·10289	1·10150	1·10011	1·09873	1·09735	1·09597	1·09459	1·09322
71	1·09185	1·09048	1·08912	1·08775	1·08639	1·08504	1·08368	1·08233	1·08098	1·07964
72	1·07830	1·07696	1·07562	1·07428	1·07295	1·07162	1·07030	1·06897	1·06765	1·06633
73	1·06502	1·06370	1·06239	1·06109	1·05978	1·05848	1·05718	1·05588	1·05458	1·05329
74	1·05200	1·05071	1·04943	1·04815	1·04687	1·04559	1·04432	1·04304	1·04177	1·04051
75	1·03924	1·03798	1·03672	1·03546	1·03421	1·03296	1·03171	1·03046	1·02922	1·02797
76	1·02673	1·02550	1·02427	1·02304	1·02182	1·02060	1·01937	1·01816	1·01694	1·01573
77	1·01452	1·01330	1·01209	1·01088	1·00968	1·00847	1·00727	1·00607	1·00487	1·00368
78	1·00248	1·00129	1·00011	0·99892	0·99774	0·99655	0·99538	0·99420	0·99302	0·99185
79	0·99068	0·98952	0·98836	0·98720	0·98604	0·98489	0·98374	0·98259	0·98144	0·98029
80	0·97915	0·97801	0·97686	0·97572	0·97458	0·97344	0·97231	0·97118	0·97004	0·96892
81	0·96779	0·96667	0·96555	0·96444	0·96332	0·96221	0·96110	0·96000	0·95889	0·95779
82	0·95669	0·95558	0·95448	0·95338	0·95228	0·95119	0·95009	0·94900	0·94791	0·94683
83	0·94574	0·94466	0·94359	0·94251	0·94144	0·94037	0·93930	0·93823	0·93717	0·93611
84	0·93504	0·93398	0·93292	0·93186	0·93080	0·92975	0·92869	0·92764	0·92659	0·92554
85	0·92450	0·92346	0·92242	0·92138	0·92035	0·91932	0·91829	0·91726	0·91623	0·91521
86	0·91418	0·91316	0·91213	0·91111	0·91009	0·90907	0·90806	0·90704	0·90603	0·90502

The surface, then, is disturbed by the wind, the extent of the disturbance depending upon the strength of the wind, the time for which it acts and the length of the water surface over which it acts. These three qualities are referred to as the *strength, duration* and *fetch* of the wind, respectively. The disturbance also depends on tide, depth of water and local land contours. Wave characteristics become practically independent of fetch when the fetch is greater than about 300 miles.

Once a wave has been generated it will move away from the position at which it was generated until all its energy is spent. Waves generated by local winds are termed *sea* and those which have travelled out of their area of generation are termed *swell*. Sea waves are characterized by relatively peaky crests and the crest length seldom exceeds some two or three times the wave-length. Swell waves are generally lower with more rounded tops. The crest length is typically six or seven times the wave-length. In a swell, the variation in height between successive waves is less than is the case for sea waves.

The sea surface presents a very confused picture which defied for many years any attempts at mathematical definition. Developments in the theory of random variables since about 1940 have now made it possible to construct a mathematical model of the sea surface. The essential feature of these theories was, however, foreseen by R. E. Froude when, in 1905, (Ref. 1) he stated:

> . . . irregular wave systems are only a compound of a number of regular systems (individually of comparatively small amplitude) of various periods, ranging through the whole gamut (so to speak) represented by our diagrams and more. And the effect of such a compound wave series on the models would be more or less a compound of the effects proper to the individual units composing it.

Unfortunately, at that time the mathematics needed to develop his ideas into a practical tool for the naval architect had not been developed.

Before proceeding to consider the irregular sea surface, it is first of all necessary to consider a regular wave system as this is the basic building brick from which the picture of any irregular system is built up. Such a system with crests extending to infinity in a direction normal to the direction of propagation of the wave has the appearance of a large sheet of corrugated iron. The section of the wave is regular and two cases are of particular significance. They are: (i) the trochoidal wave, and (ii) the sinusoidal wave.

The theory of these two wave forms is developed in the following paragraphs which can be omitted by the student wishing to have only a general appreciation of wave systems.

Waves

TROCHOIDAL WAVES

The trochoidal wave theory has been adopted for certain standard calculations, e.g. that for longitudinal strength. By observation, the theory appears to reflect many actual ocean wave phenomena although it can be regarded as only an approximation to the complex wave form actually existing.

The section of the trochoidal wave surface is defined mathematically as the path traced by a point fixed within a circle when that circle is rolled along and below a straight line as shown in Fig. 9.1.

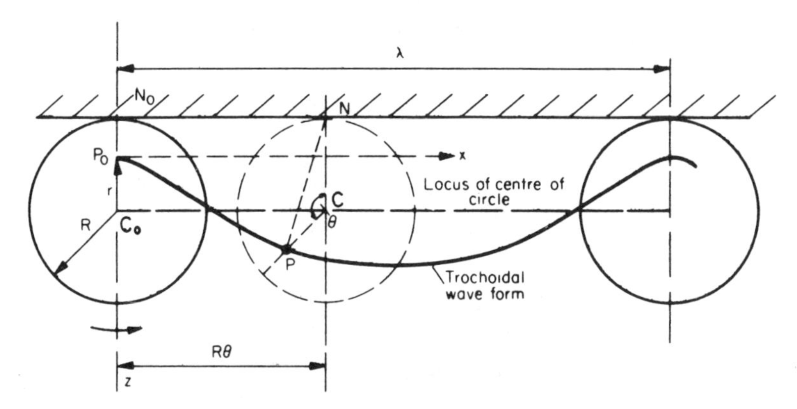

Fig. 9.1 Generation of trochoidal wave form

Suppose the height and length of the wave to be generated are h_w and λ respectively: then the radius of the generating circle is R, say, where

$$\lambda = 2\pi R$$

and the distance of the generating point from the centre of the circle is r, say, where

$$h_w = 2r$$

Let the centre of the generating circle start at position C_0 and let it have turned through an angle θ by the time it reaches C. Then

$$C_0C = N_0N = R\theta$$

Relative to the x-, z-axes shown in Fig. 9.1, the co-ordinates of P are

$$x = R\theta - r\sin\theta$$
$$= \frac{\lambda}{2\pi}\theta - \frac{h_w}{2}\sin\theta$$

$$z = r - r\cos\theta = \frac{h_w}{2}(1-\cos\theta)$$

As the circle rolls it turns, instantaneously, about its point of contact, N, with the straight line. Hence $\overline{NP}$ is the normal to the trochoidal surface and the instantaneous velocity of P is

$$\overline{NP}\frac{d\theta}{dt} \text{ normal to } \overline{NP}$$

Surfaces of equal pressure below the water surface will also be trochoidal. Crests and troughs will lie vertically below those of the surface trochoid, i.e. the length of the trochoidal wave is the same in all cases and they are all generated by a rolling circle of the same diameter. At great depths, any surface disturbance is not felt so that the radius of the point generating the trochoidal surface must reduce with increasing depth.

Consider two trochoidal sub-surfaces close to each other as shown in Fig. 9.2.

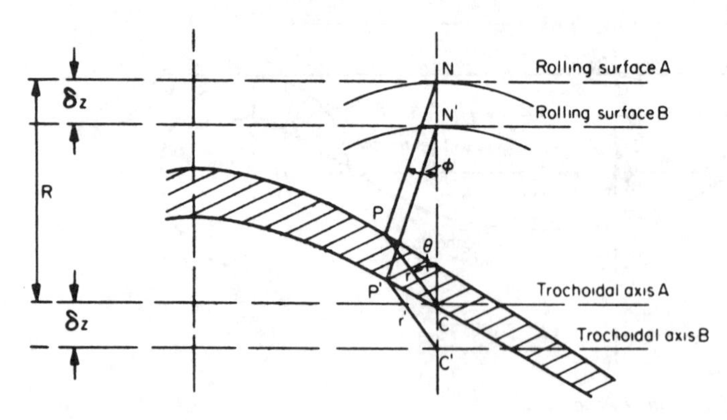

Fig. 9.2 Sub-surface trochoids

By definition, no particles of water pass through a trochoidal surface. Hence, for continuity the space shown shaded in Fig. 9.2 must remain filled by the same volume of water.

Let t = thickness of layer at P (measured normal to the trochoidal surface).

$$\text{Velocity at P} = \overline{\text{PN}}\frac{\mathrm{d}\theta}{\mathrm{d}t}$$

Hence the condition of continuity implies

$$t\overline{\text{PN}}\frac{\mathrm{d}\theta}{\mathrm{d}t} = \text{constant for all values of } \theta$$

i.e.

$$t\overline{\text{PN}} = \text{constant for all values of } \theta$$

Now $\overline{\text{CP}}$ and $\overline{\text{C'P}}'$ are parallel and, provided δz is small, $\overline{\text{PN}}$ and $\overline{\text{P'N}}'$ are very nearly parallel. Resolving along PN

$$\overline{\text{P'N}}' + \delta z \cos\phi = \overline{\text{PN}} + t$$

$$\therefore \quad t = \delta z\cos\phi + \delta\overline{\text{PN}}, \quad \text{where} \quad \delta\overline{\text{PN}} = \overline{\text{P'N}}' - \overline{\text{PN}}$$

$$= \delta z\left(\frac{R - r\cos\theta}{\overline{\text{PN}}}\right) + \delta\overline{\text{PN}}$$

$$\therefore \quad t\overline{\text{PN}} = \delta z(R - r\cos\theta) + \overline{\text{PN}}\delta\overline{\text{PN}}$$

Also

$$\overline{\text{PN}}^2 = R^2 + r^2 - 2Rr\cos\theta$$

Differentiating and remembering that R and θ are constant

$$\overline{2\mathrm{PN}\delta\mathrm{PN}} = 2r\delta r - 2R\cos\theta\,\delta r$$

$$\therefore\quad t\overline{\mathrm{PN}} = \delta z(R - r\cos\theta) + r\,\delta r - R\cos\theta\,\delta r$$

$$= R\,\delta z + r\,\delta r - \cos\theta\,(r\,\delta z + R\,\delta r)$$

Since $t\overline{\mathrm{PN}}$ is constant for all values of θ

$$r\,\delta z + R\,\delta r = 0$$

$\therefore$ in the limit

$$\frac{1}{r}\,\mathrm{d}r = -\frac{1}{R}\mathrm{d}z$$

Integrating

$$\log_e r = c - \frac{z}{R} \text{ where } c = \text{constant}$$

If $\quad r = r_0$ at $z = 0$, $c = \log_e r_0$

$$\therefore\quad \log_e r = \log_e r_0 - \frac{z}{R}$$

i.e.

$$r = r_0 \exp\left(-\frac{z}{R}\right)$$

That is to say, the radius of the point tracing out the sub-surface trochoid decreases exponentially with increasing depth. This exponential decay is often met within natural phenomena, and is very rapid as is illustrated in Table 9.5 for a surface wave 150 m long and 15 m high, i.e. a wave for which

$$R = \frac{\lambda}{2\pi} = \frac{75}{\pi} = 23{\cdot}87\,\mathrm{m}$$

$$r_0 = \frac{15}{2} = 7{\cdot}5\,\mathrm{m}$$

Table 9.5
Decay of orbital radius with depth

z (m)	r (m)	z (m)	r (m)
0	7·50	25	2·57
5	6·06	50	0·87
10	4·88	100	0·10
23·87	2·76		

Although it has been convenient to study the shape of the trochoidal wave using the artifice of a point within a rolling circle, a particle in the water surface itself does not trace out a trochoid. The absolute motion is circular and neglecting bodily movement of masses of water such as those due to tides, is obtained by superimposing on Fig. 9.1 an overall velocity of C from right to left such that

$$C = \frac{\lambda}{T}$$

This superposition of a constant velocity in no way invalidates the results obtained above or those which follow. The use of the 'stationary' trochoid as in Fig. 9.1 is merely one of mathematical convenience.

The magnitude of C can be expressed in terms of λ by considering the forces acting on a particle at P in Fig. 9.1. Such a particle will suffer a downward force due to gravity and a 'centrifugal force' in the line $\overline{\text{CP}}$.

If the mass of the particle is m, the gravitational force is mg and the centrifugal force $mr\omega^2$ where $\omega = \mathrm{d}\theta/\mathrm{d}t$.

Since the surface is one of equal pressure, the resultant of these two forces must be normal to the surface, i.e. it must lie in the direction of $\overline{\text{NP}}$.

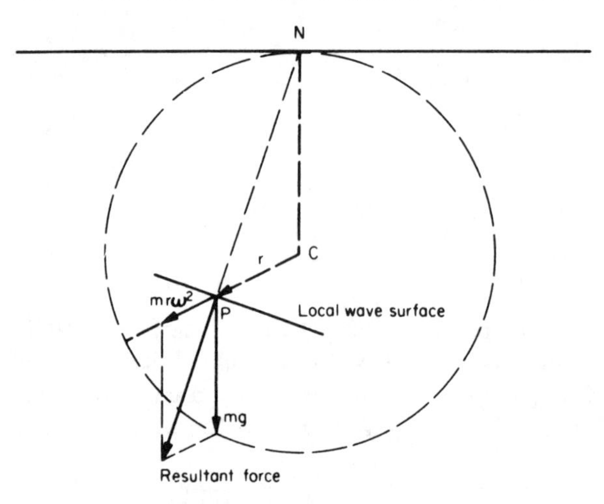

Fig. 9.3 Forces on particle in wave surface

Referring to Fig. 9.3,

$$\frac{mg}{\overline{\text{NC}}} = \frac{mr\omega^2}{\overline{\text{CP}}}$$

i.e.

$$\frac{mg}{R} = \frac{mr\omega^2}{r}$$

Whence

$$\omega^2 = \frac{g}{R}$$

Now $C = \lambda/T$ where T, the wave period $= 2\pi/\omega$.

$$\therefore \quad C = \lambda\omega/2\pi = \omega R$$

Hence

$$C^2 = \omega^2 R^2 = gR = \frac{g\lambda}{2\pi}$$

Relationship between line of orbit centres and the undisturbed surface

In Fig. 9.4, the trochoidal wave form is superimposed on the still water surface. It is clear that the volume of water in a crest, as measured above the

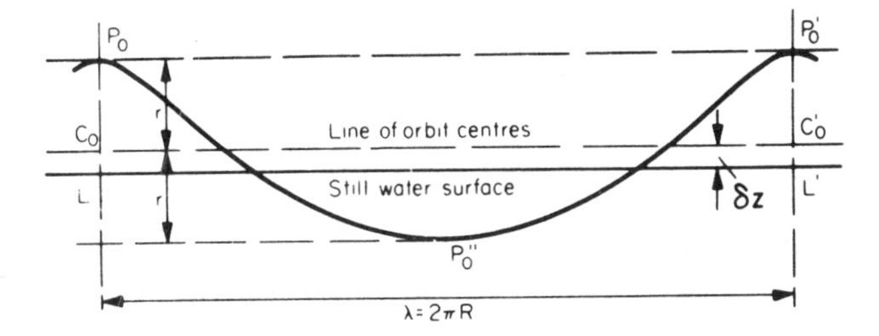

Fig. 9.4 Line of orbit centres in relation to the still water surface

still water level, must be equal to the volume of a trough measured below this level. Put another way, the area $P_0\ P_0'\ P_0''$ must equal the area of the rectangle $P_0\ P_0'\ L'\ L$ or, for simplification of the integration

$$\int_0^{x=\pi R} z\,dx = \pi R\,(r+\delta z)$$

where δz = distance of still water surface below the line of orbit centres. Since $x = R\theta - r\sin\theta$ and $z = r - r\cos\theta$, we obtain

$$\int_{\theta=0}^{\theta=\pi} (r - r\cos\theta)(R - r\cos\theta)\,d\theta = \pi R(r+\delta z)$$

$$\int_0^{\pi} (rR - rR\cos\theta - r^2\cos\theta + r^2\cos^2\theta)\,d\theta = \pi R(r+\delta z)$$

i.e.

$$\pi Rr + \frac{\pi}{2}r^2 = \pi R(r+\delta z)$$

$$\therefore \quad \delta z = \frac{r^2}{2R}$$

This property has been used in the past as a wave pressure correction in the longitudinal strength calculations.

SINUSOIDAL WAVES

Mathematically, the trochoidal wave is not easy to manipulate and the basic units from which an irregular sea is assumed to be built up are sinusoidal in profile.

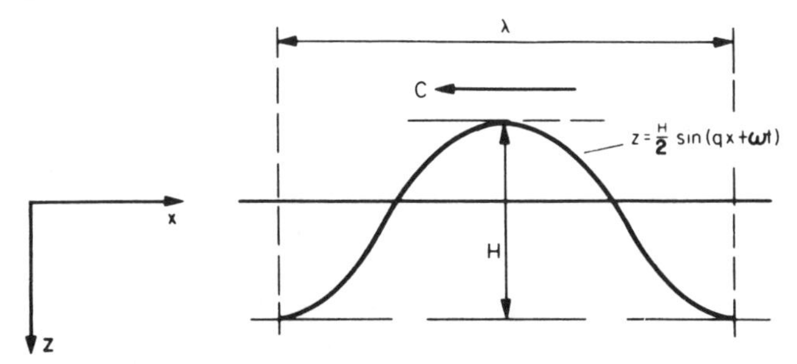

Fig. 9.5 Profile of sinusoidal wave

For a wave travelling with velocity C in the direction of decreasing x, Fig. 9.5, the profile of the wave can be represented by the equation

$$z = \frac{H}{2}\sin(qx+\omega t)$$

In this expression, $q = 2\pi/\lambda$ is known as the *wave number* and $\omega = 2\pi/T$ is known as the *wave frequency*. The wave velocity as in the case of the trochoidal wave, is given by

$$C = \frac{\lambda}{T} = \frac{\omega}{q}$$

Other significant features of the wave are

$$T^2 = \frac{2\pi\lambda}{g}$$

$$\omega^2 = \frac{2\pi g}{\lambda}$$

$$C^2 = \frac{g\lambda}{2\pi}$$

Water particles within the waves move in orbits which are circular, the radii of which decrease with depth in accordance with the expression

$$r = \frac{H}{2}\exp(-qz) \text{ for depth } z$$

From this it can be deduced that at a depth $z = +\lambda/2$, the orbit radius is only $0{\cdot}02H$ so that for all intents and purposes motion is negligibly small at depths equal to or greater than half the wavelength. (See Table 9.5.) The proof of this relationship is similar to that adopted for the trochoidal wave.

The hydrodynamic pressure at any point in the wave system is given by

$$p = \rho g\frac{H}{2}\exp(-qz)\sin(qx+\omega t)$$

The average potential and kinetic energies per unit area of the wave system are equal, and the average total energy per unit area is

$$\rho g\frac{H^2}{8}$$

The energy of the wave system is transmitted at half the velocity of advance of the waves. Thus, when a train of regular waves enters calm water, the front of the waves advances at the velocity of energy transmission, and the individual waves travel at twice this velocity and 'disappear' through the front.

Two of the above relationships are presented graphically in Fig. 9.6.

A fuller treatment of sinusoidal waves will be found in Ref. 2. As mentioned above, this wave form is assumed in building up irregular wave systems. It is also used in studying the response of a ship to regular waves. Differences in

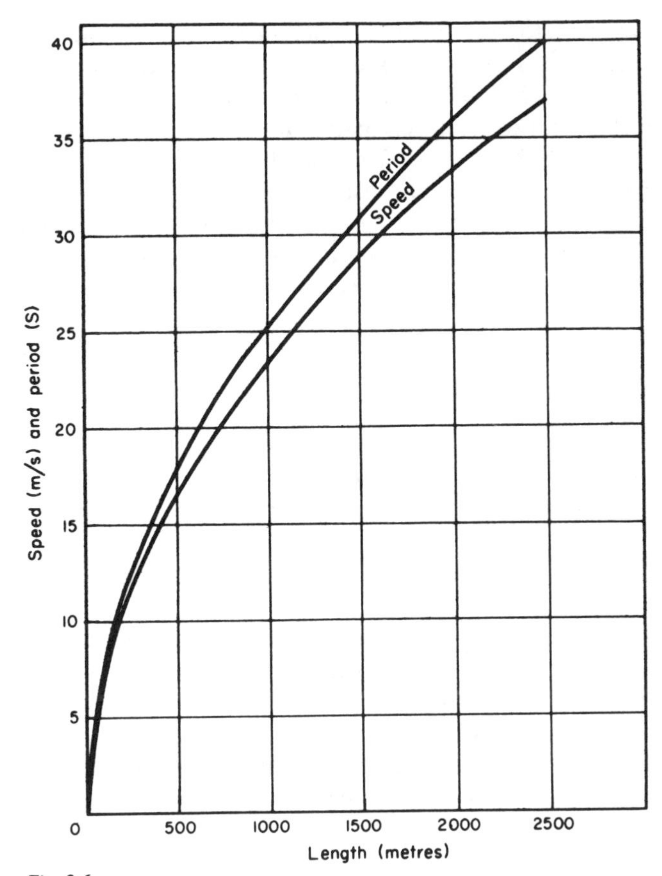

Fig. 9.6

response to a trochoidal wave would be small if the waves are of the same height and length.

IRREGULAR WAVE PATTERNS

Behind the apparent confusion of the sea there is statistical order and the sea surface may be regarded as the result of superimposing a large number of regular sinusoidal waves of different, but small, heights and various lengths in random phase. An important parameter is the range of lengths, and hence frequencies, present in the system.

If all these wave components travel in the same direction, the irregular pattern will exhibit a series of straight crests extending to infinity in a direction normal to the direction of wave travel. Such a system is termed a *long-crested irregular wave system* and is referred to as *one-dimensional* (frequency). In the more general case, the individual wave components travel in different directions and the resultant wave pattern does not exhibit long crests but rather a series of humps and hollows. Such a system is termed an *irregular wave system* or *two-dimensional system* (frequency and direction). The remainder of this section is devoted mainly to the long-crested system.

How is any irregular system to be defined? It is possible to measure the time interval between successive crests passing a fixed point and the heights between successive troughs and crests. These vary continuously and can be misleading in representing a particular sea. For instance, a part of the system in which many component waves cancelled each other out because of their particular phase relationships would appear to be less severe than was in fact the case.

Provided the record of surface elevation against time is treated carefully, however, it can yield some very useful information. If $\bar{\lambda}$ is the average distance between crests and $\bar{T}$ the average time interval in seconds, it can be shown that, approximately

$$\bar{\lambda} = \frac{2}{3}\frac{g\bar{T}^2}{2\pi} = 3{\cdot}41\bar{T}^2,\ \text{ft}$$

$$= 1{\cdot}04\bar{T}^2,\ \text{m}$$

This relationship between $\bar{\lambda}$ and $\bar{T}$ is represented in Fig. 9.7. Both can be related to wind speed as defined in Ref. 3 by the expression:

Average period $= 0{\cdot}285\ V_w$, s

where V_w = wind speed in knots.

Since an observer is more likely to record $\bar{T}$ using a stop watch, applying the classical relationship $\lambda = gT^2/2\pi$ would lead to an error.

In general, in this chapter record is used to refer to a record of the variation of sea surface elevation, relative to its mean level, with time. A complete record would be too difficult to analyse and a sample length is usually taken from time to time, the duration of each being such as to ensure a reasonable statistical representation of the sea surface.

The values of wave height in a sample can be arranged in descending order of magnitude and the mean height of the first third of the values obtained. This mean height of the highest one-third of the waves is termed the *significant wave height*. The height of the highest wave occurring in a given record can also be found. If a large number of wave records is so analysed for a given ocean area, it is possible to plot the probability of exceeding a given significant wave height. Results, generally based on Ref. 20, are presented in Fig. 9.8 for the North Atlantic, Northern North Atlantic and world wide. It will be noted that the Northern North Atlantic is the most severe area as would be expected from general experience.

SEA STATE CODE

A generally accepted description of the sea state appropriate to various significant wave heights is afforded by the sea state code.

Sea state code

Code	Description of sea	Significant wave height (m)
0	Calm (glassy)	0
1	Calm (rippled)	0–0·1
2	Smooth (wavelets)	0·1–0·5

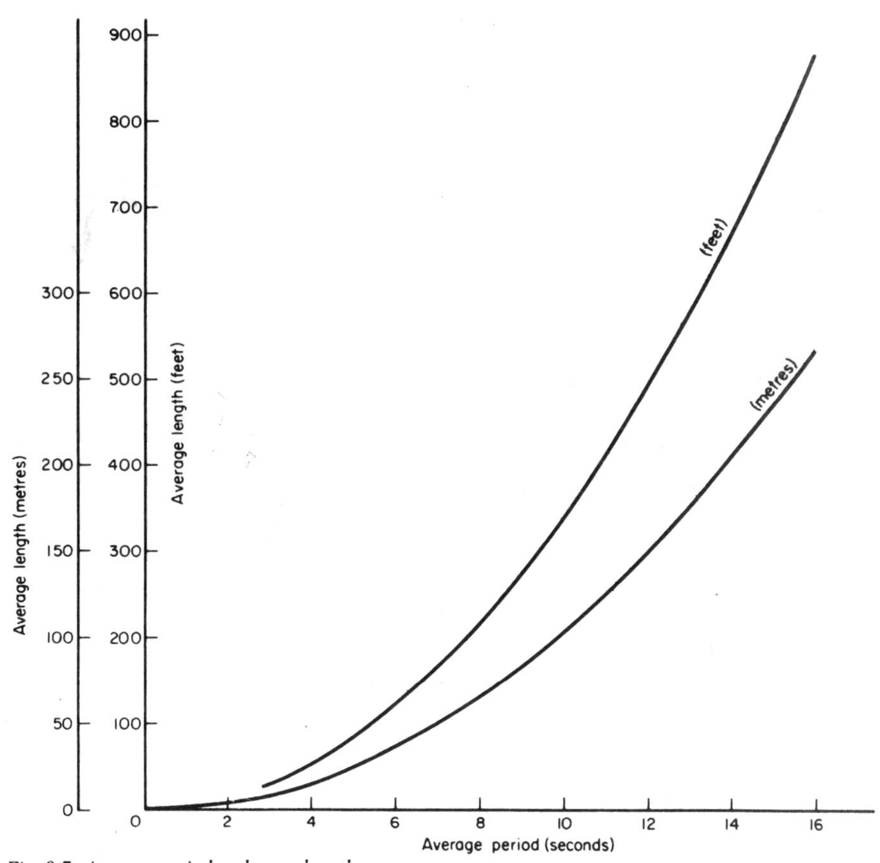

Fig. 9.7 Average period and wave-length

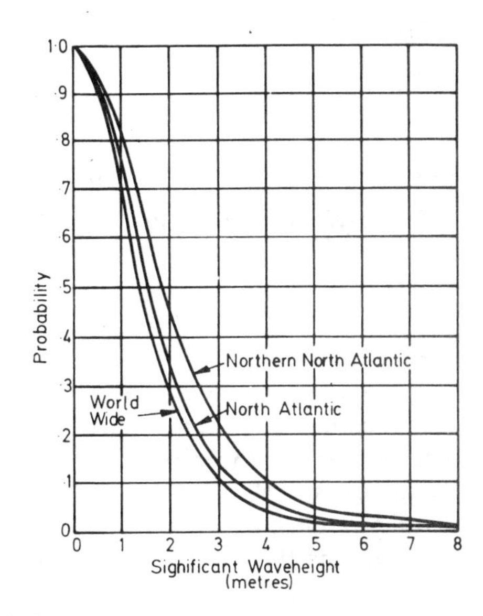

Fig. 9.8 Waves at OWS station India

Code	Description of sea	Significant wave height (m)
3	Slight	0·5–1·25
4	Moderate	1·25–2·50
5	Rough	2·50–4·00
6	Very rough	4–6
7	High	6–9
8	Very high	9–14
9	Phenomenal	Over 14

HISTOGRAMS AND PROBABILITY DISTRIBUTIONS

The wave heights can also be arranged in groups so that the number of waves with heights falling into various intervals can be counted. These can be plotted as a histogram as in Fig. 9.9 in which the area of each rectangle represents the number of waves in that interval of wave height. In practice, it is usual to arrange scales such that the total area under the histogram is unity. The histogram shows the distribution of heights in the wave sample, but in many applications subsequent analysis is easier if a mathematical curve can be fitted to the results. Three theoretical distributions are of particular importance.

The Normal or Gaussian distribution which is defined by the equation

$$p(x) = \frac{1}{\sigma\sqrt{(2\pi)}}\exp\left\{-\frac{1}{2}\frac{(x-\mu)^2}{\sigma^2}\right\}$$

where μ = mean of values of x, σ^2 = variance, and σ = standard deviation.

The Logarithmically Normal or Log-Normal Distribution which is defined by the equation

$$p\,(\log x) \doteq \frac{1}{\sigma_1\sqrt{(2\pi)}}\exp\left\{-\frac{1}{2}\frac{(\log x-u)^2}{\sigma_1^2}\right\}$$

where u = mean value of $\log x$, σ_1^2 = variance, and σ_1 = standard deviation.

The Rayleigh distribution which is defined by the equation

$$p(x) = \frac{x}{a}\exp\left\{-\frac{x^2}{2a}\right\}, \qquad x > 0$$

where $2a$ is the mean value of x^2.

In these expressions, $p(x)$ is a probability density. Plotted to a base of x the total area under the curve must be unity, expressing the fact that it is certain that the variable under examination will take some value of x. The area under the curve in a small interval of x, $\mathrm{d}x$, is $p(x)\,\mathrm{d}x$ and this represents the probability that the variable will take a value in the interval $\mathrm{d}x$.

Integrating the curve leads to a *cumulative probability distribution* in which the ordinate at a given value of x represents the area under the probability distribution between 0 and x. The ordinate thus represents the probability that the variable will have a value less than, or equal to, x.

It will be noted that the normal and log-normal distributions as functions of x are defined by two variables, μ and σ or u and σ_1, and the Rayleigh distribution by one variable a. Hence, it is necessary to determine the values of these quantities which most closely fit the observed wave data.

EXAMPLE 1. A record contains 1000 waves with heights up to 10 m, the numbers of readings falling into various 1-m groups are

Height (m)	0–1	1–2	2–3	3–4	4–5	5–6	6–7	7–8	8–9	9–10
Number of waves	6	29	88	180	247	260	133	42	10	5

Plot a histogram for these data and derive the corresponding normal distribution.

Solution: The histogram is obtained by erecting a series of rectangles on each height band such that the area of each rectangle is proportional to the number of waves in that height band. For ease of comparison with the probability curve, the total area of all rectangles is made unity by dividing by 1000 and the histogram is plotted as in Fig. 9.9.

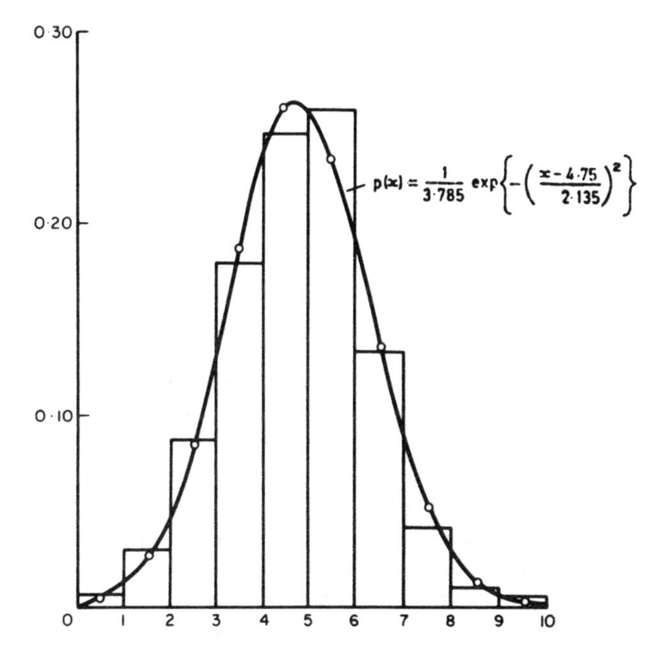

Fig. 9.9 Histogram and normal probability curve superimposed

The mean, μ, of the wave heights is found by summating the products of the average height in each band (x_r say) by its frequency of occurrence (F_r) and the variance follows as in the table. Using this table (Table 9.6),

$$\mu = \frac{4750}{1000} = 4{\cdot}75$$

$$\sigma^2 = \frac{2282}{1000} = 2{\cdot}282$$

$$\therefore \quad \sigma = 1{\cdot}510$$

Hence the equation to the normal curve is

$$p(x) = \frac{1}{1{\cdot}510\sqrt{(2\pi)}} \exp\left\{-\tfrac{1}{2}\left(\frac{x-4{\cdot}75}{1{\cdot}51}\right)^2\right\}$$

$$= \frac{1}{3{\cdot}785} \exp\left\{-\left(\frac{x-4{\cdot}75}{2{\cdot}135}\right)^2\right\}$$

Corresponding values of x and $p(x)$ are given in the tables (Table 9.6) and these have been plotted in Fig. 9.9 superimposed on the histogram.

Table 9.6

x_r	F_r	$x_r F_r$	$x_r - \mu$	$(x_r - \mu)^2$	$F_r(x_r - \mu)^2$
0·5	6	3	−4·25	18·06	108·5
1·5	29	43·5	−3·25	10·56	306·2
2·5	88	220	−2·25	5·06	445·5
3·5	180	630	−1·25	1·56	281·3
4·5	247	1111·5	−0·25	0·06	15·5
5·5	260	1430	+0·75	0·56	146·2
6·5	133	864·5	+1·75	3·06	407·3
7·5	42	315	+2·75	7·56	317·7
8·5	10	85	+3·75	14·06	140·8
9·5	5	47·5	+4·75	22·56	112·7
	1000	4750			2281·7

x	0·5	1·5	2·5	3·5	4·5	5·5	6·5	7·5	8·5	9·5
$p(x)$	0·0050	0·0260	0·0870	0·1874	0·2606	0·2340	0·1350	0·0504	0·0120	0·0018

In applying the above techniques, it must be realized that particular sets of records are only *samples* of all possible records, known in statistics as the *population*, e.g. the wave data reproduced above for weather station India were made in one area only and for limited periods of time. When an analytical expression is fitted to the observed data, it is desirable to demonstrate that the expression is reasonably valid. The reliability of the hypothesis can be evaluated by applying statistical tests of significance which enable estimates to be made of the expected variation of the measured data from the analytically defined data if the hypothesis is true.

WAVE SPECTRA

A very useful method of presenting data on the wave system is in the form of a wave spectrum. When there are n regular sine wave components present and

n is not too large, a *line spectrum* can be drawn as Fig. 9.10 where the height of each ordinate represents the component wave amplitude or half the square of the component wave amplitude. In this latter case the ordinate is proportional to the component wave energy.

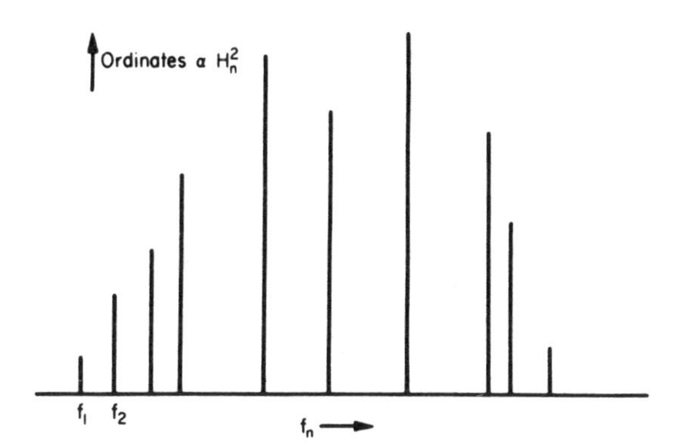

Fig. 9.10 Line spectrum

In practice, n is usually so large and the amplitude of each component so small, that the concept of a *continuous spectrum* is more convenient. Taking the energy spectrum form the area between frequencies f and $f + \delta f$ represents

$$\sum_{f}^{f+\delta f} \tfrac{1}{2}\zeta_A^2$$

That is, the area under the curve between frequencies f and $f + \delta f$ represents the sum of the energy of all component waves having frequencies in the band. In the limit, as δf tends to zero, the curve becomes smooth. This more general curve can be represented by a plot of $S(\omega)$ against ω as in Fig. 9.11.

Area under spectrum $= m_0 =$ mean square of surface elevation

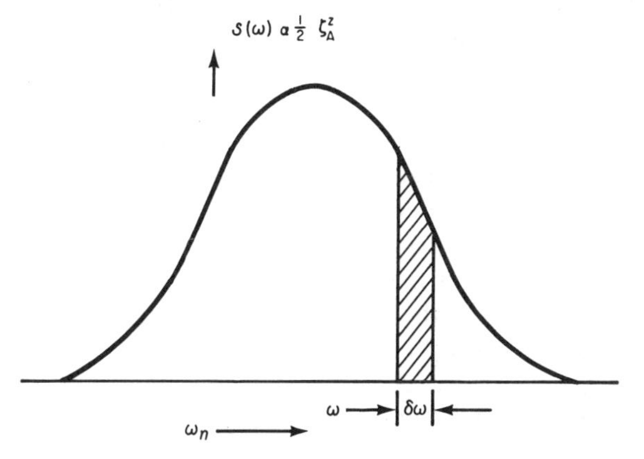

Fig. 9.11 Continuous spectrum

The area under the energy spectrum is given by

$$\int_0^\infty S(\omega)\mathrm{d}\omega$$

In more general terms the nth moment of the curve about the axis $\omega = 0$ is

$$m_n = \int_0^\infty \omega^n S(\omega)\mathrm{d}\omega$$

It can be shown that

m_0 = area under spectrum = mean square of wave surface elevation

Referring to Fig. 9.12, m_0 can be obtained by measuring the surface elevation ζ, relative to the mean level, at equal time intervals. If n such measurements are taken:

$$m_0 = \frac{1}{2n}\Sigma\zeta^2 = \frac{1}{2n}(\zeta_1^2 + \zeta_2^2 + \ldots + \zeta_n^2)$$

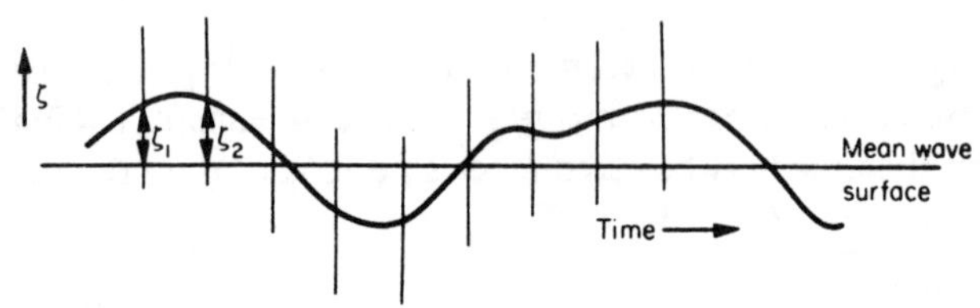

Fig. 9.12

For reasonable accuracy, the time interval chosen should not be greater than half the minimum time interval between successive crests and the number of readings must be large.

WAVE CHARACTERISTICS

If the distribution of wave amplitude is Gaussian then the probability that at a random instant of time the magnitude of the wave amplitude exceeds some value ζ is given (for a normal distribution) by

$$P(\zeta) = 1 - \mathrm{erf}\,[\zeta/\surd(2m_0)]$$

The error function (erf) is tabulated in standard mathematical tables.

The percentages of wave amplitudes exceeding certain values are shown in Fig. 9.13. It will be seen that more than 60 per cent of waves have amplitudes of $\surd m_0$ or less.

However, observations at sea show that many statistical properties of the wave surface can be closely represented by the Rayleigh distribution. It can be shown that for waves represented by a Rayleigh distribution:

Most frequent wave amplitude $= 0{\cdot}707\surd(2m_0) = \surd m_0$
Average wave amplitude $= 0{\cdot}886 \quad \surd(2m_0) = 1{\cdot}25\surd m_0$
Average amplitude of $\frac{1}{3}$ highest waves $= 1{\cdot}416 \quad \surd(2m_0) = 2\surd m_0$
Average amplitude of $\frac{1}{10}$ highest waves $= 1{\cdot}800 \quad \surd(2m_0) = 2{\cdot}55\surd m_0$

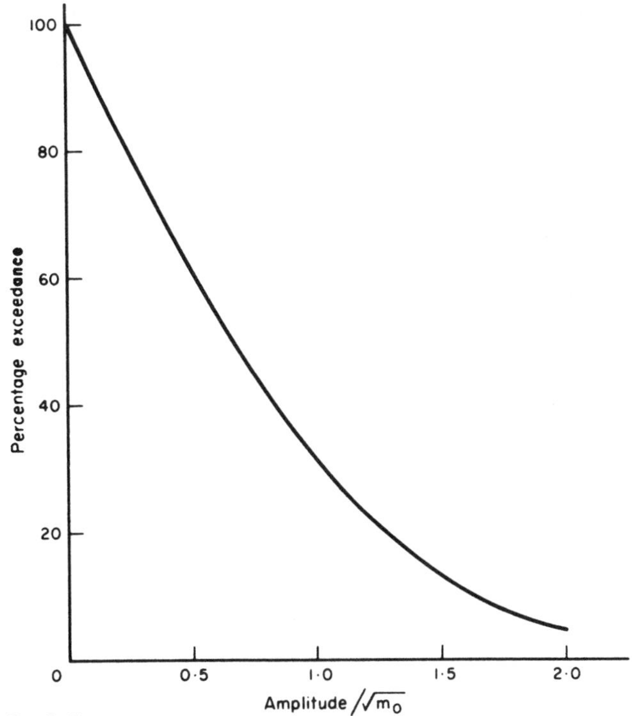

Fig. 9.13

Note: Some authorities plot ζ^2 against f as the spectrum. In this case, the area under the spectrum is $2m_0$ where m_0 is defined as above.

The average amplitude of $\frac{1}{3}$ highest waves is denoted by $\zeta_{\frac{1}{3}}$ and is termed the *significant wave amplitude*.

More generally if the values of ζ were to be arranged in descending order of magnitude, the mean value of the first cn of these is given by $k_1\ \zeta_S$ where corresponding values of c and k_1 are given below:

Table 9.7
Corresponding values of c and k_1

c	k_1	c	k_1
0·01	2·359	0·4	1·347
0·05	1·986	0·5	1·256
0·1	1·800	0·6	1·176
0·2	1·591	0·7	1·102
0·25	1·517	0·8	1·031
0·3	1·454	0·9	0·961
0·3333	1·416	1·0	0·886

It also follows that:

(*a*) the probability that ζ should exceed a certain value r is given by

$$\int_r^\infty p(r)\,\mathrm{d}r = \exp(-r^2/\bar{\zeta}^2)$$

(*b*) the expected value of the maximum amplitude in a sample of n values is $k_2\zeta_S$ where corresponding values of n and k_2 are given in Table 9.8.

Table 9.8
Corresponding values of n and k_2

n	k_2	n	k_2
1	0·707	500	2·509
2	1·030	1000	2·642
5	1·366	2000	2·769
10	1·583	5000	2·929
20	1·778	10,000	3·044
50	2·010	20,000	3·155
100	2·172	50,000	3·296
200	2·323	100,000	3·400

(*c*) in such a sample the probability that ζ will be greater than r is

$$1-\exp[-\exp\{-(r^2-r_0^2)/\bar{\zeta}^2\}]$$

where $r_0^2/\bar{\zeta}^2 = \log n$

(*d*) The number of zero up-crossings (Fig. 9.14) per unit time is:

$$\frac{1}{2\pi}\sqrt{\frac{m_2}{m_0}};\ \text{Average period } \bar{T}_0 = 2\pi\sqrt{\frac{m_0}{m_2}}$$

(*e*) The number of maxima per unit time is:

$$\frac{1}{2\pi}\sqrt{\frac{m_4}{m_2}};\ \text{Average period } \bar{T}_m = 2\pi\sqrt{\frac{m_2}{m_4}}$$

(*f*) Average wavelengths are:

$$\bar{\lambda}_0 = 2\pi g\sqrt{\frac{m_0}{m_4}}\ \text{based on zero up-crossings}$$

$$\bar{\lambda}_m = 2\pi g\sqrt{\frac{m_4}{m_8}}\ \text{based on maxima}$$

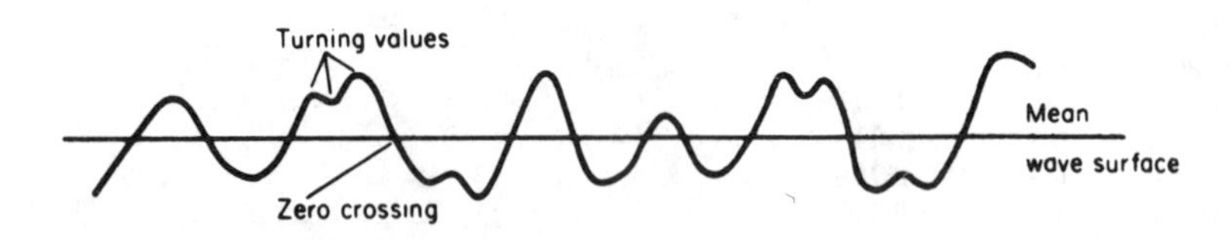

Fig. 9.14 Wave elevation recorded at a fixed point

In the more general case when the spectrum is not narrow its width can be represented by a paramter ε. The extreme values of ε are 0 and 1 representing a Rayleigh and normal distribution of wave amplitude respectively.

Then, if m_0 is the mean square value of the wave surface relative to the mean level, the variance as previously defined, it can be shown that the

$$\text{mean square crest height} = (2-\varepsilon^2)m_0$$

By putting $\varepsilon = 0$, it follows that

$$\bar{\zeta}^2 = 2m_0$$

On the other hand in a very broad spectrum with $\varepsilon = 1$,

$$\text{mean square crest height} = m_0$$

Also, mean crest to trough height is:

$$2\{\frac{\pi}{2}(1 - \varepsilon^2)\, m_0\}^{1/2}$$

and

$$\varepsilon^2 = 1 - \frac{m_2^2}{m_0 m_4} = 1 - \frac{\overline{T}_\lambda^{\,2}}{\overline{T}_0^{\,2}}$$

$$\text{hence } (1 - \varepsilon^2)^{1/2} = \frac{\text{number of zero crossings}}{\text{number of turning points}}$$

Thus, referring to Fig. 9.14 and remembering that, in practice, a much longer record would be used

Number of zero crossings = 11; Number of turning values = 19

Hence

$$(1-\varepsilon^2)^{\frac{1}{2}} = \tfrac{11}{19} = 0{\cdot}58$$

i.e.

$$\varepsilon = 0{\cdot}815$$

FORM OF WAVE SPECTRA

There has been much debate concerning the form of representative wave spectra. The dependence of the wave system upon the strength, duration and fetch of the wind has already been discussed. Some of the differences between the spectra advanced by various authorities are probably due, therefore, to the difficulty of deciding when a sea is fully developed because a uniform wind seldom blows long enough for this to be determined. When a wind begins to blow, short, low amplitude waves are formed at first. As the wind continues to blow, larger and longer waves are formed. As the wind dies down the longer waves move out of the area by virtue of their greater velocity, leaving the shorter waves. Thus the growth and decay of wave spectra is as shown in Figs. 9.15 and 9.16.

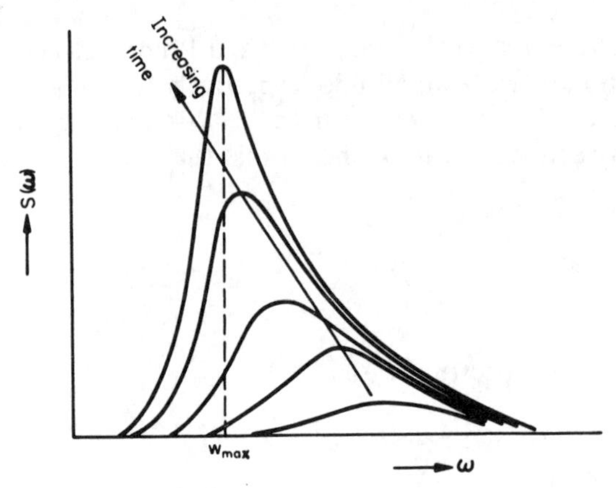

Fig. 9.15 Growth of spectra

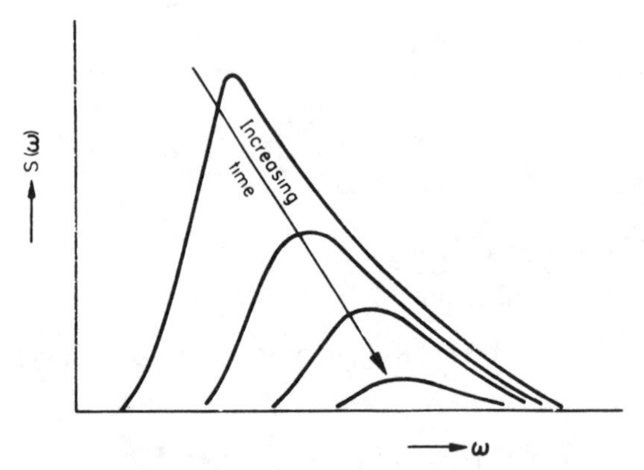

Fig. 9.16 Decay of spectra

Because of the dependence of a fully developed sea on wind strength earlier attempts to define wave spectra used wind speed as the main parameter. One such was that due to Pierson and Moskowitz, viz.:

$$S(\omega) = \frac{8{\cdot}1 \times 10^{-3}\, g^2}{\omega^5} \exp\left\{-0{\cdot}74 \left(\frac{g}{V\omega}\right)^4\right\} m^2 s$$

g in ms^{-2}
V in ms^{-1}

Typical plots are given in Fig. 9.17. These show how rapidly the area under the curves increases with increasing wind speed, while the peak frequency decreases.

The spectrum now most widely adopted is that recommended by the International Towing Tank Conference:

$$S(\omega) = \frac{A}{\omega^5} \exp\left(-B/\omega^4\right) \quad \text{(the Bretschneider spectrum)}$$

ω is the circular frequency of the waves (s^{-1}) and A and B are constants.

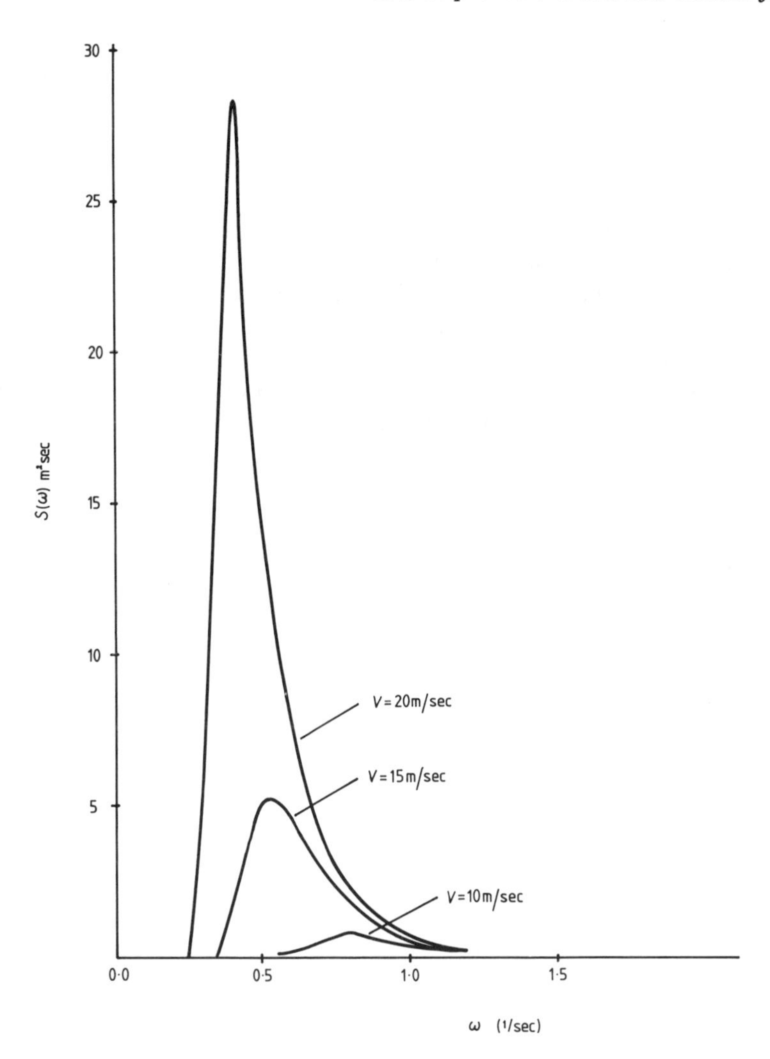

Fig. 9.17

It will be noted that this is of the same general form as that due to Pierson and Moskowitz. For this form

$$\text{Maximum } S(\omega) = A(0\cdot 8B)^{-5/4}\, e^{-5/4} \text{ at } \omega = (0\cdot 8B)^{1/4}$$

When both significant wave height and the characteristic wave period, T_1, are known,

$$A = \frac{173\,\zeta^2_{\frac{1}{3}}}{T_1{}^4} \qquad \zeta_{\frac{1}{3}} = \text{significant wave height in m}$$

$$B = 691/T_1^4 \qquad = 4\sqrt{m_0}$$

$$T_1 = 2\pi\frac{m_0}{m_1}$$

When only the significant wave height is known, an approximation to $S(\omega)$ is given using

$$A = 8 \cdot 10 \times 10^{-3} g^2$$
$$B = 3 \cdot 11/\zeta^2_{\frac{1}{3}}$$

A suggested relationship for significant wave height in terms of wind speed for a fully developed sea is shown in Fig. 9.23. A typical family of spectra for long crested seas using the single parameter formula is shown in Fig. 9.18. With this simple version the ratio between the frequency or wave length at peak energy and the significant wave height is constant for all spectra the value of the ratio being 39.2. This is not true for all ocean spectra although many have values in the range 30 to 50. Hence for general purposes, such as prediction of

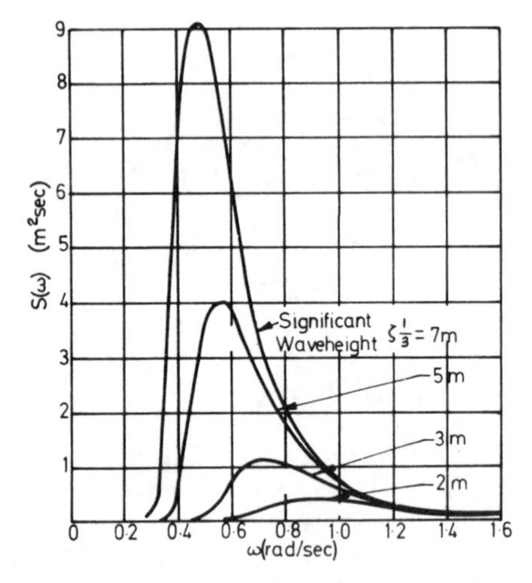

Fig. 9.18

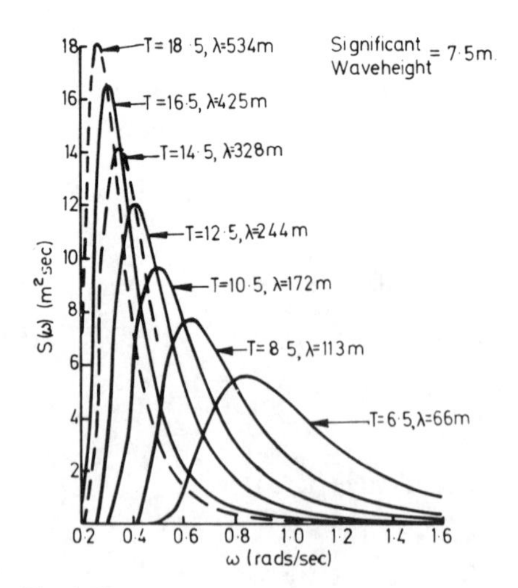

Fig. 9.19

ship motion likely to be experienced by a new design, the use of the simple formula is acceptable. Where it is necessary to vary the ratio the more elaborate formula can be used. Figure 9.19 presents a family of spectra with significant wave height of 7·5 metres.

For limited fetch conditions, the ITTC recommend

$$S_J(\omega) = 0{\cdot}658\ S(\omega)(3{\cdot}3)\exp\left(-\frac{0{\cdot}206\omega T_1 - 1}{\sqrt{2}\sigma}\right)^2 m^2S$$

where $\sigma = 0{\cdot}07$ for $\omega \leqslant 4{\cdot}85T_1$
$\sigma = 0{\cdot}09$ for $\omega > 4{\cdot}85T_1$

For a multi-directional sea the ITTC recommend applying a spreading function as follows

$$S(\omega, \mu) = K\cos^n \mu S(\omega) \qquad -\frac{\pi}{2} < \mu < \frac{\pi}{2}$$

where μ = direction of wave components relative to the predominant wave direction.

The 1978 ITTC recommended use of $K = 2/\pi$ and $n = 2$ in the absence of better data.

EXTREME WAVE AMPLITUDES

The naval architect is often concerned with the extreme waves that can occur. Suppose a number of records, say M, are taken with the same significant wave amplitude and N values of wave amplitude are recorded from each record. If the greatest wave amplitude in each record is noted there will be M such maxima. Then Table 9.9 expresses the maximum wave amplitudes in terms of the significant amplitude $\zeta_{\frac{1}{3}}$ for various values of N.

Table 9.9
Exceptionally high waves

N	5% lower	Most frequent	Average	5% greater
20	0·99	1·22	1·32	1·73
50	1·19	1·40	1·50	1·86
100	1·33	1·52	1·61	1·94
200	1·45	1·63	1·72	2·03
500	1·60	1·76	1·84	2·14
1000	1·70	1·86	1·93	2·22

Thus, if each record has 200 wave amplitudes then, of the M maximum readings,

5 per cent can be expected to be lower than $1{\cdot}45\zeta_{\frac{1}{3}}$
The most frequent value = $1{\cdot}63\zeta_{\frac{1}{3}}$
The average value = $1{\cdot}72\zeta_{\frac{1}{3}}$
5 per cent can be expected to be greater than $2{\cdot}03\zeta_{\frac{1}{3}}$

Extreme value statistics were studied by Gumbel (Ref. 4) and are based on a probability distribution function with a pronounced skew towards the higher values of the variate. Type I extreme value distributions represent the numerical value of probability for the largest values of a reduced variate y as

$$\Phi(x) = \exp[-\exp(-y)]$$

where
$$y = \alpha(x-u)$$
$$x = \text{extreme variate}$$
$$\alpha = \text{a scale parameter}$$
$$u = \text{the mode of the distribution}$$

The probability distribution for the smallest values is

$$\Phi(x) = 1 - \exp[-\exp(y)]$$

In type II distributions the logarithm of the variate is used. An example of this type of distribution applied to a naval gunfire problem is given in Ref. 5.

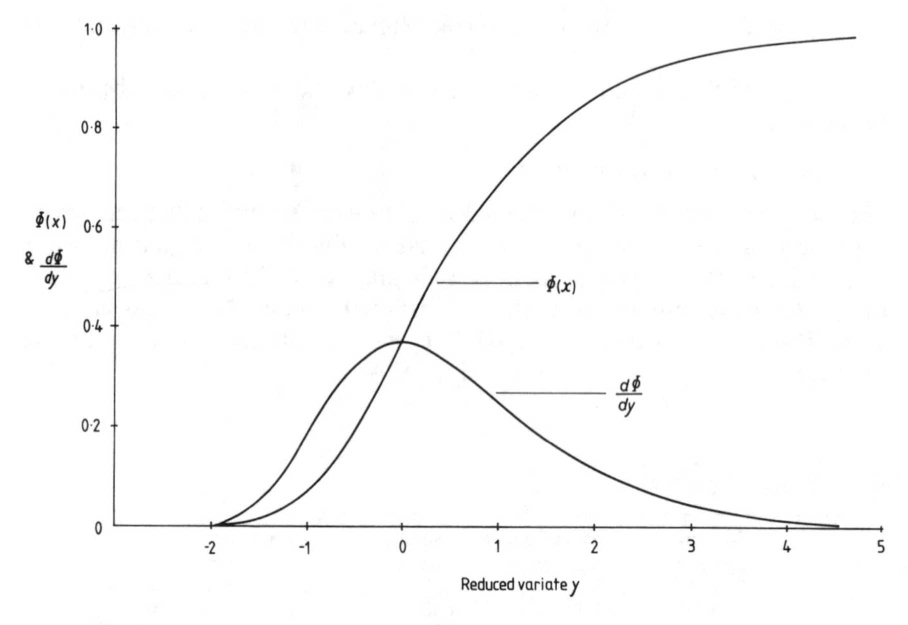

Fig. 9.20 Gumbel Type I Extreme value probability and cumulative probability distribution (after Ref. 20)

Plots of $\Phi(x)$ and $d\Phi/dy$ are presented in Fig. 9.20. In this distribution the mode occurs at $y = 0$, the median at $y = 0{\cdot}367$ and the mean at $y = 0{\cdot}577$. A special paper is available for plotting data for extreme value statistics (see Fig. 9.21). The reduced variate and ordinate scales are linear. The frequency scale follows from the relationship of Fig. 9.20. The 'return period' scale at the top derives its name from an early application of extreme value statistics to flooding problems.

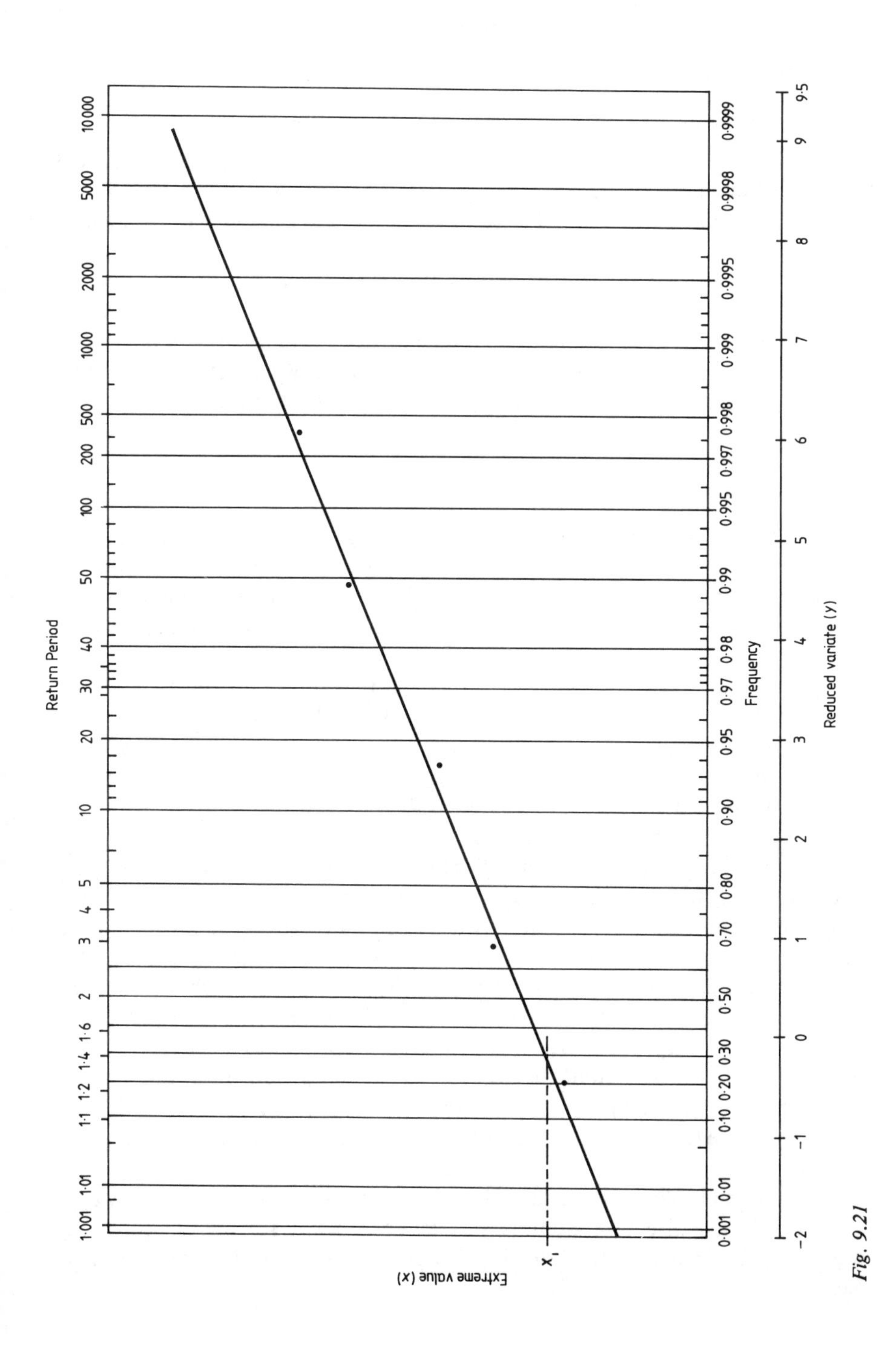
Return Period
1·001
1·01
1·1
1·2
1·4
1·6
2
3
4
5
10
20
30
40
50
100
200
500
1000
2000
5000
10 000
Frequency
0·001
0·01
0·10
0·20
0·30
0·50
0·70
0·80
0·90
0·95
0·97
0·98
0·99
0·995
0·997
0·998
0·999
0·9995
0·9998
0·9999
Reduced variate (y)
−2
−1
0
1
2
3
4
5
6
7
8
9
9·5
Extreme value (x)
x_1

Fig. 9.21

If observed data are in accord with the Gumbel distribution they will lie close to a straight line when plotted on the special paper, the line being

$$x = u + y/\alpha$$

Such a plot can provide a lot of useful information, e.g.

(*a*) If x_1 is the ordinate of the straight line corresponding to a probability of 0·30 then only 30 per cent of the results will be x_1 or less in value. Conversely, the probability of a result being less than any particular value of x can be found.

(*b*) The most probable extreme value is obtained by noting the ordinate corresponding to zero on the reduced variate scale. The mean and median values follows from the corresponding reduced variate values quoted above.

(*c*) To discover how many results are needed to give a 50 per cent probability of observing an extreme value of x_1 the return period corresponding to this ordinate on the straight line is read off.

OCEAN WAVE STATISTICS

The ITTC formulation defines the wave spectrum in terms of the characteristic wave period and the significant wave height. If a designer wishes to ascertain the motions of a design in waves with specific values of these parameters he can do so provided he knows the response of the design to a range of regular waves. This aspect of ship motions is dealt with in Volume 2.

More generally the designer is concerned with the motions likely to be experienced during a given voyage or during the life of the ship. For this he needs statistics on the probability of occurrence of various sea conditions at different times of the year on the likely trade routes together with the predominant wave directions. Such statistics are presented in Ref. 20 (see also Ref. 19), which has some 3,000 tables based on a million sets of observations of wave heights, periods and directions. Although individually the observations may be suspect, taken in aggregate the overall statistics can provide an accurate picture. There is evidence to show that the characteristic period and significant wave height of the ITTC formula correspond closely to the observed period and height (Ref. 19).

The data are presented for 50 sea areas (Fig. 9.20) for which a reasonable number of observations are available and in which conditions are fairly homogeneous. Thus they cover the well-frequented sea routes. Where observers reported both sea and swell the tables present the group of waves with the greater height or, if heights are equal, that with the longer period. The heights and periods are those obtained from the larger well-formed waves of the system observed.

The detailed tables of wave period against wave height are arranged by area, season and wave direction. Summary tables covering all seasons and directions are provided. They are illustrated by Tables 9.10 to 9.12 which are taken from Ref. 20.

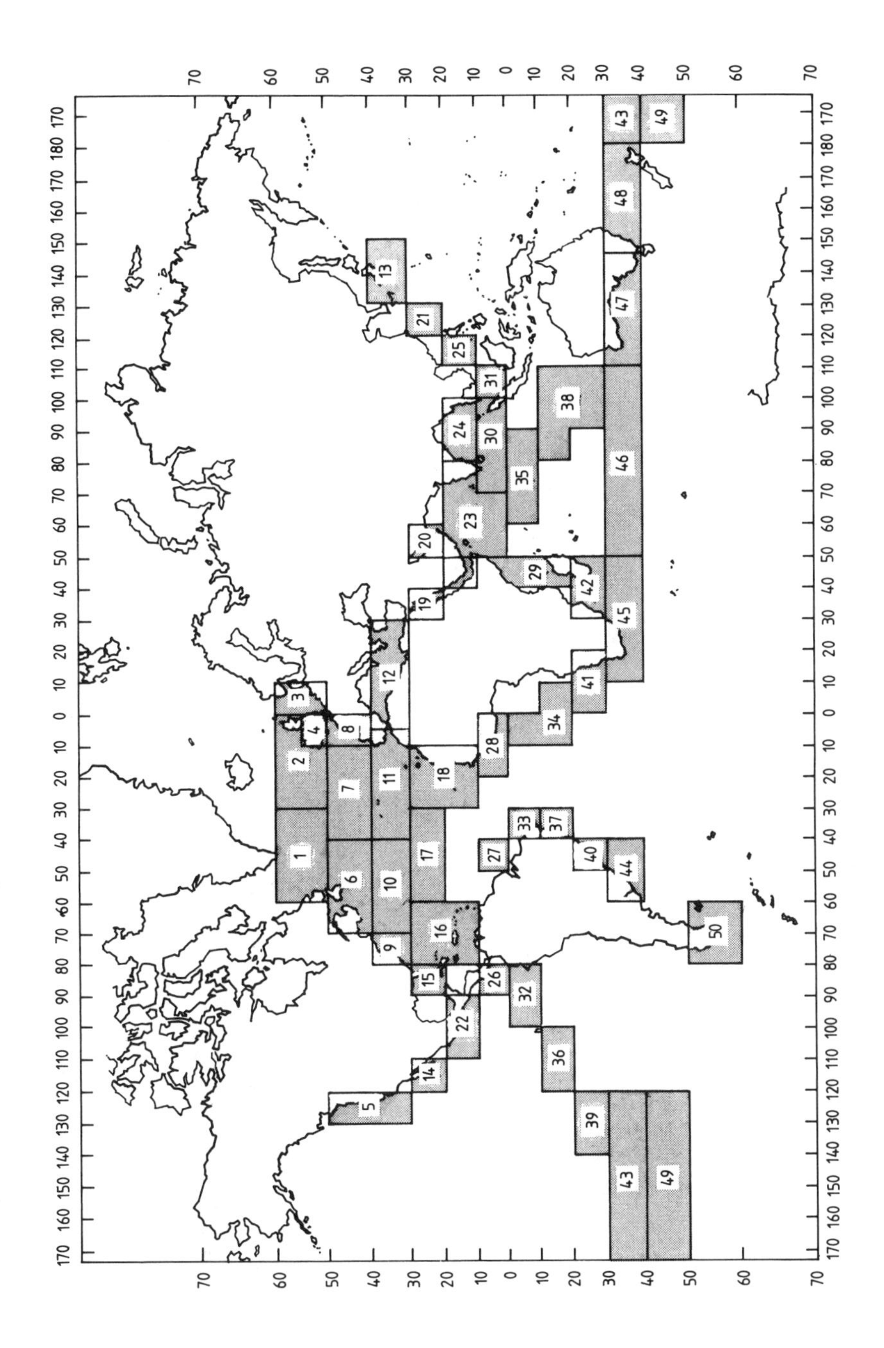
170 160 150 140 130 120 110 100 90 80 70 60 50 40 30 20 10 0 10 20 30 40 50 60 70 80 90 100 110 120 130 140 150 160 170 180 170
70 60 50 40 30 20 10 0 10 20 30 40 50 60 70
1
2
3
4
5
6
7
8
9
10
11
12
13
14
15
16
17
18
19
20
21
22
23
24
25
26
27
28
29
30
31
32
33
34
35
36
37
38
39
40
41
42
43
44
45
46
47
48
49
50

Fig. 9.22

The data can be combined and presented in a number of ways. Table 9.13 presents the data for areas 1, 2, 6, 7 and 8 for all seasons and all directions in terms of the expected number of days occurrence in the year. Table 9.14 presents the same data in terms of percentage frequency of occurrence with the data reported under 'calm or period undetermined' spread over the range of periods reported in proportion to the observations recorded. Tables 9.15 and 9.16 give similar percentage frequencies for areas 1, 2, 6–11, 16–18 and world wide.

Table 9.10
Area 2. All seasons. All directions (after Ref. 20)

	WAVE PERIOD CODE											
	x	2	3	4	5	6	7	8	9	0	1	TOTALS
00	524	514	24	18	11	4		2	1	7	4	1109
01	73	1558	179	46	20	8	8	4	6	5	63	1970
02	186	2800	1448	321	123	42	6	3	6	5	48	4988
03	289	1565	2860	1083	275	85	35	13	6	3	4	6218
04	232	476	2049	1460	388	119	30	10	3			4767
05	190	229	1325	1455	620	185	65	13	6	1	2	4091
06	118	96	681	1015	584	224	48	13	10	2		2791
07	97	48	435	769	555	253	75	13	5			2250
08	52	25	204	480	411	195	58	28	7	2	1	1463
09	69	25	193	421	397	241	69	28	17	1		1461
10	14	3	34	71	80	25	14	2		1	1	245
11	15		21	79	65	56	24	5	1		1	267
12	12	3	34	97	120	78	33	15	3		2	397
13	10	6	40	68	103	71	24	13	2		3	340
14	4	5	21	35	34	31	18	8	1			157
15	4	1	16	45	41	30	14	7	3			161
16	10	2	10	31	35	40	26	9	2		3	168
17	4		1	15	13	21	13	4				71
18	3		4	7	25	15	7	3				64
19	8	2	8	26	32	47	23	12	3			161
91					1				1			2
92					1	2						3
95					1	1						2
TOTALS	1914	7358	9587	7542	3935	1773	590	205	83	27	132	33146

WAVE PERIOD CODE	WAVE PERIOD SECONDS
x	CALM OR PERIOD UNDETERMINED
2	5 OR LESS
3	6 OR 7
4	8 OR 9
5	10 OR 11
6	12 OR 13
7	14 OR 15
8	16 OR 17
9	18 OR 19
0	20 OR 21
1	OVER 21

Wave height code number in left hand column is twice wave height in metres except:

00 = 0.25 m
91 = 11 m
92 = 12 m
95 = 15 m

Table 9.11
Area 2. All seasons (after Ref. 20)

			260°		270°		280°				
	2	3	4	5	6	7	8	9	0	1	
00	50		2	3	1				1		57
01	199	16	7	3	1	2	1			6	235
02	381	252	46	16	5	1	1	1		8	711
03	248	498	191	46	20	4	3			1	1011
04	72	382	276	76	22	7	5	1			841
05	37	285	317	159	46	12	2	3			861
06	19	149	220	131	54	10	1	4	1		589
07	8	84	197	150	67	22	4	2			534
08	6	49	136	105	54	17	14	2	1		384
09	7	48	120	115	82	19	15	5	1		412
10	1	11	29	32	10	3	1			1	88
11		3	27	19	12	5	1				67
12		11	21	42	28	8	3	1			114
13	1	14	18	35	29	5	4	1			107
14		8	16	10	7	7	1				49
15		5	16	18	11	7	3	3			63
16	2	2	16	9	11	13	5	1		3	62
17			5	4	11	7	2				29
18		2	5	13	6	1	1				28
19		1	6	8	17	3	1	1			37
91				1				1			2
92				1	2						3
95					1						1
	1031	1820	1671	996	497	153	68	26	4	19	6285

Table 9.12
Area 2. December to February (after Ref. 20)

			260°		270°		280°				
	2	3	4	5	6	7	8	9	0	1	
00	5		1								6
01	28			1							29
02	39	23	4	2	1					1	70
03	43	53	21	4	2						123
04	12	56	42	9	6	2	3	1			131
05	5	52	39	19	5	2		1			123
06	2	23	45	22	10	1		1			104
07	2	19	42	44	13	4	2				126
08	1	10	24	23	17	5	2				82
09	4	15	38	37	21	7	4				126
10		2	2	13	1		1			1	20
11			12	6	6	2					26
12		2	7	17	6	1	2				35
13	1	4	3	9	8	1	2				28
14			4		5	4	1				14
15			2	6	5	1	2	2			18
16	1		6	8	4	6	2			2	29
17			3	3	7	4	1				18
18			3	7	5						15
19			1	3	10	3	1				18
91								1			1
92					1						1
95					1						1
	143	259	299	233	134	43	23	6		4	1144

Table 9.13
Wave statistics. All seasons. All directions. Days per year northern North Atlantic

Wave height (metres)	Wave Period T_1 (seconds)											Total
	*	⩽ 5	6–7	8–9	10–11	12–13	14–15	16–17	18–19	20–21	> 21	
0·25	6·127	7·163	0·264	0·151	0·106	0·036	0·014	0·022	0·012	0·156	0·102	14·153
0·5	0·706	18·977	2·224	0·612	0·295	0·088	0·055	0·028	0·024	0·043	0·863	23·915
1·0	1·747	36·768	17·697	4·029	1·342	0·485	0·146	0·061	0·040	0·071	0·395	62·781
1·5	2·136	19·507	34·577	12·423	3·249	1·071	0·380	0·140	0·045	0·029	0·066	73·623
2·0	1·721	5·638	23·020	17·059	5·355	1·487	0·430	0·133	0·050	0·014	0·014	54·921
2·5	1·272	2·349	13·101	14·968	7·236	2·207	0·725	0·191	0·061	0·021	0·017	42·148
3·0	0·966	0·940	6·340	10·187	6·529	2·437	0·747	0·194	0·055	0·009	0·009	28·413
3·5	0·744	0·555	3·771	7·080	5·353	2·543	0·860	0·225	0·080	0·007	0·007	21·225
4·0	0·437	0·238	1·973	4·128	3·748	2·018	0·690	0·277	0·087	0·012	0·005	13·613
4·5	0·541	0·222	1·642	3·580	3·367	1·992	0·829	0·347	0·106	0·016	0·012	12·654
5·0	0·128	0·038	0·236	0·624	0·648	0·362	0·158	0·040	0·005	0·007	0·005	2·251
5·5	0·097	0·045	0·220	0·576	0·603	0·397	0·212	0·062	0·007	0·005	0·003	2·227
6·0	0·102	0·055	0·342	0·844	0·967	0·586	0·258	0·102	0·028	0·002	0·007	3·293
6·5	0·076	0·057	0·293	0·687	0·927	0·584	0·276	0·107	0·023	0·003	0·007	3·040
7·0	0·047	0·026	0·120	0·347	0·350	0·234	0·113	0·040	0·007	0·000	0·002	1·286
7·5	0·059	0·009	0·149	0·350	0·437	0·298	0·158	0·068	0·016	0·002	0·005	1·551
8·0	0·061	0·016	0·081	0·272	0·317	0·316	0·123	0·073	0·010	0·009	0·012	1·290
8·5	0·029	0·003	0·057	0·120	0·189	0·163	0·102	0·049	0·007	0·005	0·003	0·727
9·0	0·023	0·003	0·031	0·087	0·179	0·140	0·062	0·042	0·007	0·000	0·005	0·579
9·5	0·054	0·010	0·055	0·192	0·321	0·274	0·161	0·085	0·057	0·028	0·014	1·251
> 10	0·000	0·000	0·004	0·007	0·015	0·024	0·005	0·000	0·002	0·002	0·000	0·059
Totals	17·073	92·619	106·197	78·323	41·533	17·742	6·504	2·286	0·729	0·441	1·553	365·000

*Calm or period undetermined

Table 9.14
Percentage frequency of occurrence of sea states, northern North Atlantic

Wave height (metres)	Wave Period T_1 (seconds)										Total
	⩽ 5	6–7	8–9	10–11	12–13	14–15	16–17	18–19	20–21	> 21	
Up to 0·50	5·1870	0·3864	0·1364	0·0785	0·0254	0·0125	0·0122	0·0077	0·0655	0·1500	6·0616
0·5–1·25	13·9308	5·4071	1·2508	0·4336	0·1534	0·0515	0·0223	0·0158	0·0282	0·2738	21·5673
1·25–2·5	7·4338	18·1186	10·4412	3·4524	1·0344	0·3308	0·1043	0·0356	0·0153	0·0251	40·9915
2·5–4	0·7898	5·0003	7·5983	4·9232	2·0100	0·6557	0·1850	0·0592	0·0091	0·0077	21·2383
4–6	0·1290	0·9294	2·0748	1·9919	1·1577	0·4774	0·1824	0·0500	0·0099	0·0076	7·0101
6–9	0·0398	0·2517	0·6354	0·7925	0·5554	0·2644	0·1158	0·0227	0·0055	0·0099	2·6931
9–14	0·0033	0·0209	0·0691	0·1201	0·1039	0·0564	0·0302	0·0178	0·0082	0·0047	0·4346
14+	—	0·0005	0·0002	0·0014	0·0012	—	—	—	0·0002	—	0·0035
Totals	27·5135	30·1149	22·2062	11·7936	5·0414	1·8487	0·6522	0·2088	0·1419	0·4788	100·0000

Table 9.15
Percentage frequency of occurrence of sea states, North Atlantic

Wave height (metres)	Wave Period T_1 (seconds) ≤ 5	6–7	8–9	10–11	12–13	14–15	16–17	18–19	20–21	> 21	Total
Up to 0·50	7·1465	0·5284	0·1608	0·0792	0·0318	0·0147	0·0140	0·0083	0·0876	0·2390	8·3103
0·5–1·25	18·6916	6·8744	1·4616	0·4616	0·1798	0·0616	0·0355	0·0158	0·0392	0·3785	28·1996
1·25–2·5	8·6495	19·1881	9·7609	3·0633	0·9056	0·2969	0·0967	0·0269	0·0104	0·0290	42·0273
2·5–4	0·6379	3·7961	5·4827	3·4465	1·4291	0·4597	0·1368	0·0420	0·0067	0·0060	15·4435
4–6	0·0863	0·5646	1·2574	1·2199	0·7040	0·3005	0·1143	0·0338	0·0076	0·0054	4·2938
6–9	0·0258	0·1468	0·3486	0·4417	0·3052	0·1451	0·0614	0·0126	0·0039	0·0057	1·4968
9–14	0·0018	0·0114	0·0362	0·0614	0·0520	0·0320	0·0163	0·0082	0·0046	0·0024	0·2263
14+	—	0·0002	0·0001	0·0006	0·0006	—	—	—	0·0001	—	0·0016
Totals	35·2394	31·1100	18·5083	8·7742	3·6081	1·3105	0·4750	0·1476	0·1601	0·6660	99·9992

Table 9.16
Percentage frequency of occurrence of sea states, world wide

Wave height (metres)	Wave Period T_1 (seconds) ≤ 5	6–7	8–9	10–11	12–13	14–15	16–17	18–19	20–21	> 21	Total
Up to 0·50	10·0687	0·5583	0·1544	0·0717	0·0272	0·0117	0·0087	0·0076	0·0959	0·2444	11·2486
0·5–1·25	22·2180	6·9233	1·4223	0·4355	0·1650	0·0546	0·0268	0·0123	0·0394	0·3878	31·6850
1·25–2·5	8·6839	18·1608	9·0939	2·9389	0·8933	0·2816	0·0889	0·0222	0·0079	0·0230	40·1944
2·5–4	0·5444	3·0436	4·4913	2·9163	1·2398	0·4010	0·1220	0·0322	0·0050	0·0049	12·8005
4–6	0·0691	0·3883	0·8437	0·8611	0·5214	0·2244	0·0866	0·0229	0·0045	0·0033	3·0253
6–9	0·0223	0·0928	0·2164	0·2632	0·1842	0·0908	0·0419	0·0081	0·0032	0·0034	0·9263
9–14	0·0014	0·0067	0·0186	0·0316	0·0258	0·0177	0·0094	0·0044	0·0023	0·0011	0·1190
14+	—	0·0001	0·0001	0·0003	0·0003	—	—	—	0·0001	—	0·0009
Totals	41·6078	29·1739	16·2408	7·5187	3·0571	1·0814	0·3844	0·1098	0·1583	0·6679	100·0001

Climate

THE WIND

The main influence of the wind is felt indirectly through the waves it generates on the surface of the sea. As stated above, the severity of these waves will depend upon the strength (i.e. velocity) of the wind, the time it acts (i.e. its duration) and the distance over which it acts (i.e. the fetch).

The strength of the wind is broadly classified by the *Beaufort Scale*. The numbers 0 to 12 were introduced by Admiral Sir Francis Beaufort in 1806, 0 referring to a calm and force 12 to a wind of hurricane force. Originally, there were no specific wind speeds associated with these numbers but the values shown in Table 9.17 have now been adopted internationally. The figures relate to an anemometer at a height of 6 m above the sea surface.

No fixed relation exists between the spectra of a sea and the speed of wind generating it. Figure 9.23 shows a relationship agreed by the ITTC but it applies only to fully developed seas where duration and fetch are large. Figure 9.24 shows the probability of exceeding a given wind speed in the North Atlantic based on Ref. 20.

Table 9.17
The Beaufort Scale

	Limits of speed		
	(Kts)	(m/s)	(ft/s)
0 Calm	1	0·3	2
1 Light air	1–3	0·3–1·5	2–5
2 Light breeze	4–6	1·6–3·3	6–11
3 Gentle breeze	7–10	3·4–5·4	12–18
4 Moderate breeze	11–16	5·5–7·9	19–27
5 Fresh breeze	17–21	8·0–10·7	28–36
6 Strong breeze	22–27	10·8–13·8	37–46
7 Near gale	28–33	13·9–17·1	47–56
8 Gale	34–40	17·2–20·7	57–68
9 Strong gale	41–47	20·8–24·4	69–80
10 Storm	48–55	24·5–28·4	81–93
11 Violent Storm	56–63	28·5–32·6	94–106
12 Hurricane	64 & over	32·7 & over	107 & over

The wind velocity varies with height. At the surface of the water, the relative velocity is zero due to a boundary layer effect. Reference 8 defines a nominal wind at 33 ft above the waterline, and the variation in wind velocity with height is assumed to be in accord with Fig. 9.25 which is based on that reference. It will be noted, that these nominal velocities will be about 6 per cent less than those defined by the Beaufort Scale at a height of 6 m above the surface. Also given in Fig. 9.25 is a curve based on this nominal height.

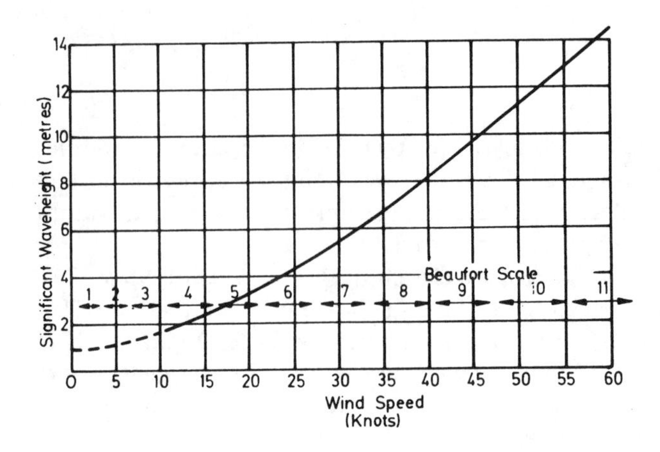

Fig. 9.23

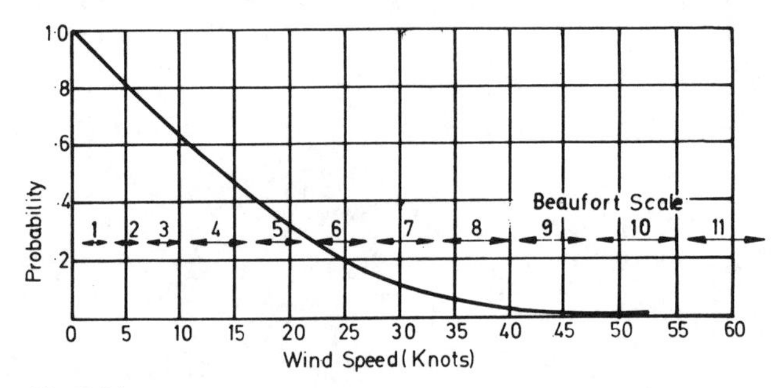

Fig. 9.24

For nominal velocities other than 100 knots, the true velocities can be scaled in direct proportion to the nominal velocities.

More direct influences of the wind are felt by the ship as forces acting on the above water portions of the hull and the superstructure. The fore and aft components of such forces will act either as a resistive or propulsive force. The lateral force will act as a heeling moment. Thus, standards of stability adopted in a given design must reflect the possible magnitude of this heeling moment, e.g. large windage area ships will require a greater stability standard as discussed in Chapter 4.

Another direct influence of the wind, including the effect of the ship's own speed, is to cause local wind velocities past the superstructure which may entrain the funnel gases and bring them down on to the after decks, or which may produce high winds across games decks to the discomfort of passengers. These effects are often studied by means of model experiments conducted in a wind tunnel, in order to obtain a suitable design of funnel and superstructure.

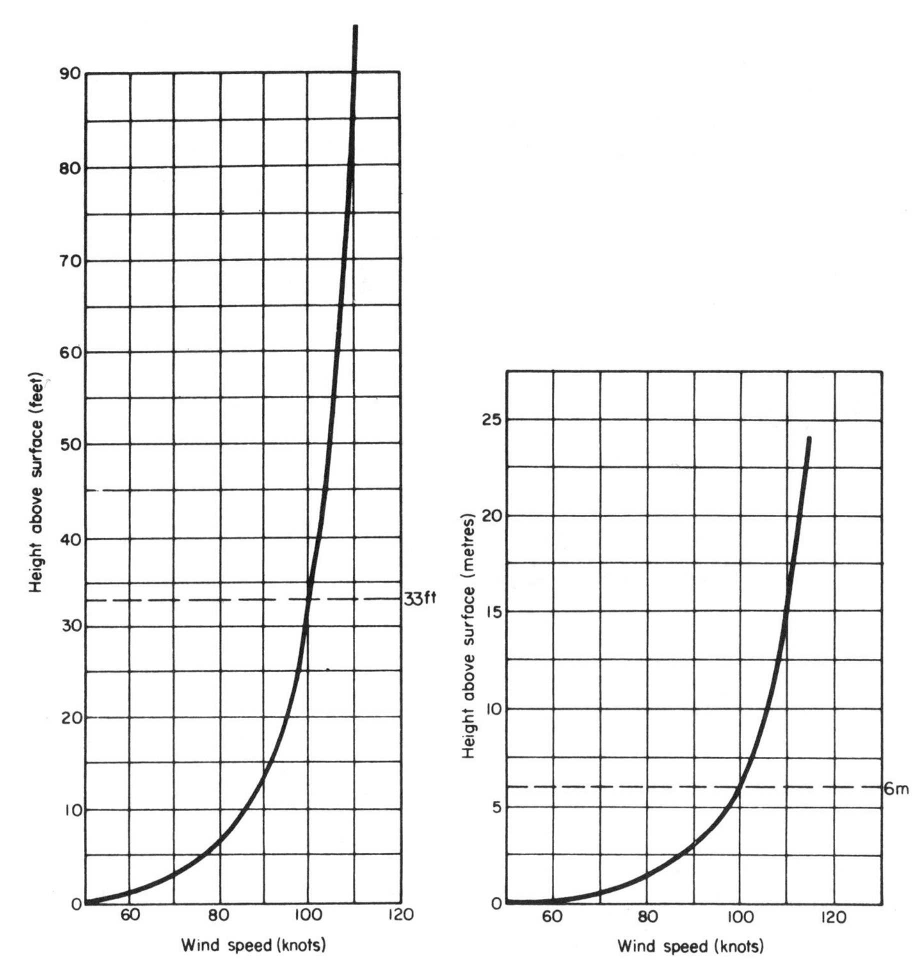

Fig. 9.25 Variation of wind speed with height above sea surface (nominal wind speed 100 knots)

AMBIENT AIR

A ship is completely immersed in two fluids—air for the upper portion and water for the lower portion. Because the specific heat of water is much greater than that for air, the former suffers a much slower variation of temperature due to climatic changes. In defining the temperature environment of the ship, it is necessary to quote temperatures for both the air and water. Further, it is common experience that the humidity of air is important in determining the degree of comfort in hot climates. This idea is elaborated later in this chapter, but for the present, it is enough to know that for air it is necessary to specify the temperature that would be recorded on both a 'wet bulb' and a 'dry bulb' thermometer.

Standard temperatures considered by the Ministry of Defence as being appropriate to different regions of the world at various times of the year are given in Table 9.18.

Table 9.18
Standard climatic figures, Ministry of Defence

Region	Average max. Summer temperature (°C)			Average min. Winter temperature (°C)		
	Air		Deep sea	Air		Deep sea
	Dry bulb	Wet bulb	Water temp. near surface	Dry bulb	Wet bulb	Water temp. near surface
Extreme Tropics	34.5	30	33			
Tropics	31	27	30			
Temperate	30	24	29			
Temperate Winter				−4	—	2
Sub-Arctic Winter				−10	—	1
Arctic and Antarctic Winter				−29	—	−2

Reference 9 suggests the conditions given in Table 9.19; figures in parentheses are the approximate values in degrees Celsius.

Table 9.19
Recommended values from Ref. 9

	Dry bulb (°F)(°C)	Wet bulb (°F)(°C)	Relative humidity (%)
Red Sea passenger vessels	90 (32)	86 (30)	85
Other passenger vessels	88 (31)	80 (27)	70
Extreme Tropical cargo vessels and tankers	90 (32)	84 (29)	78

Table 9.20
Extreme recorded air temperatures

	Harbour	At sea
Tropics	125°F (52°C)	100°F (38°C)
Arctic	−40°F (−40°C)	−22°F (−30°C)

These temperatures are important to the ship design in a number of ways. The air temperature, for instance, influences:

(*a*) the amount of heating or cooling required for air entering the ship to maintain a given internal temperature;
(*b*) the amount of thermal insulation which can be economically justified;
(*c*) efficiency of some machinery;
(*d*) the amount of icing likely to occur in cold climates and thus the standard of stability required to combat this (Chapter 4), and the need to provide special de-icing facilities.

The water temperature influences:

(*a*) density of water and thus the draught;
(*b*) the efficiency of, and hence the power output from, machinery, e.g. steam machinery may suffer a loss of about 3 per cent of power in the tropics compared with temperate conditions. The effect on other machinery may be greater, but all types are affected to some degree including auxiliaries such as electrical generators, pumps, and air conditioning plant;
(*c*) whether ice is likely to be present and if so how thick and, thus, whether special strengthening of the ship is necessary.

Both air and water temperatures can cause expansion or contraction of materials leading to stresses where two dissimilar materials are joined together. Temperature differences between the air and water can lead to relative expansion or contraction of the upper deck relative to the keel causing the ship to hog or sag. This leads to stresses in the main hull girder and can upset the relative alignment of items some distance apart longitudinally. Similarly, if one side of the ship is in strong sunlight the main hull will tend to bend in a horizontal plane.

The above temperatures are used in calculations for assessing the capacity required of ventilation or air conditioning system or for the degradation in machinery power generation.

Equipment should be tested in accordance with standard specifications:

(*a*) in locations where the ambient temperature rises and remains at a relatively constant level, e.g. in machinery spaces where the temperature is mainly influenced by the heat generated by machinery;
(*b*) where the adjacent air temperatures are indirectly affected by solar heating, e.g. equipment in the open but not fully exposed;
(*c*) fully exposed to solar radiation. In this case not only surface heating is important but also the degradation of materials caused by the ultra-violet radiation;
(*d*) in low temperature environments.

The tests may have to cater for high or low humidity conditions and may need to allow for diurnal variations.

CLIMATIC EXTREMES

The designer must allow for extreme climatic conditions although these may only be met on a few occasions. These are usually specified by an owner. The MOD use Naval Engineering Specifications from which many of the figures in the following section are drawn. Others are British Standards Specifications and the MIL Specifications of the USA. Care is necessary to ensure that the latest specifications are used.

Dust and sand. These cause abrasion of surfaces and penetrate into joints where movements due to vibration, and to changes in temperature, may bring about serious wear. Movements of particles can also set up static electrical charges. Although arising from land masses, dust and sand can be carried considerable distances by high winds. Specifications define test conditions, e.g. grade of dust or sand, concentration, air velocity and duration.

Rain. Equipments should be able to operate without degradation of performance in a maximum rainfall intensity of 0·8 mm/min at 24°C and a wind speed of 8 m/s for periods of 10 min with peaks of 3 mm/min for periods of 2 min. The test condition is 180 mm/h of rain for 1 hour. Driving sea spray can be more corrosive than rain.

Icing. Equipments fitted in the mast head region must remain operational and safe with an ice accretion rate of 6·4 mm/h with a total loading of 24 kg/m^2 Corresponding figures for equipment on exposed upper decks are 25 mm/h and 70 kg/m^2. Such equipments must be able to survive a total loading of 120 kg/m^2. Equipments fitted with de-icing facilities should be able to shed 25 mm ice per hour.

Hail. Equipment should be able to withstand without damage or degradation of performance a hailstorm with stones of 6 to 25 mm, striking velocity 14 to 25 m/s, duration 7 min.

High winds. Exposed equipments should remain secure and capable of their design performance in a relative steady wind to ship speed of 30 m/s with gusts of 1 min duration of 40 m/s. They must remain secure, albeit with some degradation of performance in wind to ship speed of 36 m/s with gusts of 1 min duration of 54 m/s.

Green sea loading. With heavy seas breaking over a ship, exposed equipment and structure are designed to withstand a green sea frontal pulse loading of 70 kPa acting for 350 ms with transients to 140 kPa for 15 ms.

Solar radiation. Direct sunlight in a tropical summer will result in average surface temperatures of up to 50°C for a wooden deck or vertical metal surfaces and 60°C for metal decks. Equipment should be designed to withstand the maximum thermal emission from solar radiation which is equivalent to a heat flux of 1120 W/m^2 acting for 4 hours causing a temperature rise of about 20°C on exposed surfaces.

Mould growth. Mould fungi feed by breaking down and absorbing certain organic compounds. Growth generally occurs where the relative humidity (rh) is greater than 65 per cent and the temperature is in the range 0 to 50°C. Most rapid growth occurs with rh greater than 95 per cent and temperature 20 to 35°C and the atmosphere is stagnant. Equipments should generally be able to withstand exposure to a mould growth environment for 28 days (84 for some critical items) without degradation of performance. Moulds can damage or degrade performance by:

(*a*) direct attack on the material, breaking it down. Natural products such as wood, cordage, rubber, greases, etc., are most vulnerable;
(*b*) indirectly by association with other surface deposits which can lead to acid formation;
(*c*) physical effects, e.g. wet growth may form conducting paths across insulating materials or change impedance characteristics of electronic circuits.

Physical limitations

There are various physical limitations, either natural or man-made, which must be considered by the designer. They are:

(*a*) *Depth of water available*

The draught of the ship including any projections below keel level must not exceed the minimum depth of water available at any time during service. This depth may be dictated by the need to negotiate rivers, canals, harbour or dock entrances. If the vessel has to operate for any distance in shallow water, the increased resistance which occurs in these conditions must be taken into account. Various empirical formulae exist for estimating the depth of water needed to avoid loss of speed. One is:

$$\text{Depth in fathoms} > \frac{5}{3} \times \frac{VT}{(L)^{\frac{1}{2}}}$$

where T and L are in ft and V is in knots.

If T and L are in metres the expression becomes:

$$\text{Depth in fathoms} > 3{\cdot}02\frac{VT}{(L)^{\frac{1}{2}}}$$

Allowance must be made for the sinkage and trim that will occur. For example, a large ship drawing say 25 ft of water may strike bottom in 28 ft of water if it attempts to negotiate the shallows too quickly.

(*b*) *Width of water available*

This is usually limited by the need to negotiate canals, harbour or dock entrances. As with shallow water the ship is acted upon by additional interference forces when negotiating narrow stretches at speed. It may be necessary to adopt special manoeuvring devices for ships habitually operating in confined waters.

(*c*) *Height available above the water line*

Most ships have, at one time or another, to pass below bridges in entering ports. Unless the bridge itself can be opened, the ship must either keep masts and funnels below the level of the bridge or the uppermost sections must be capable of being lowered. Tidal variations must be catered for.

(*d*) *Length*

The length of a ship may be limited by the building slips, fitting-out berths and docks available or by the length of canal lock or harbour dimensions.

The internal environment

A full study must embrace all those aspects of the environment which affect the efficiency of equipment and/or the crew. Thus it must include

(*a*) Movement. The ship is an elastic body acted on by a series of external forces. A thorough study (Ref. 15) should deal with the vital ship response in an integrated way. In practice it is usual to deal separately with the response of the ship as a rigid body (ship motions), as an elastic body (vibrations) and under impulsive loading (wave impact, collision or enemy action). The last item is dealt with in Chapter 5.

(*b*) Ambient conditions including quality of air (temperature, humidity, freshness), noise levels (from machinery, sea, wind) and lighting levels. In some spaces the levels of the ambient conditions will be more critical than in others. This may dictate layout or special environmental control systems.

Motions

The study of ship motions *per se* is considered in Chapter 12. As far as the ship itself is concerned, the designer is concerned with the effects on the structure, equipment and crew.

For structure and equipment it is often sufficient to consider certain limiting conditions of amplitude and frequency for design purposes when associated with a factor of safety based upon previous successful design. More strictly, any amplitudes of motion should be associated with the probability of their occurrence during the life of a ship.

To avoid overdesign, it is usual to consider two sets of motion figures; those under which equipment should be capable of meeting its specification fully, and a second set in which the equipment must be able to function albeit with reduced performance. Again, certain equipment needs only to function in certain limited sea states, e.g. that associated with transfer of stores between two ships at sea.

Typical figures for which equipment must remain secure and able to operate without degradation are given in Table 9.21 for a medium size warship. Figures for acceleration are taken to be 1·5 times those calculated assuming simple harmonic motion. They are based on Sea State 7.

Table 9.21

	Roll	Pitch	Yaw	Heave
Period (s)	10	5–6·5		7
Amplitude (±)	18°	8°		3·5 m
Acceleration (°/s²)			1·75	

As regards effects on the crew, most people have felt nausea when subject to large motion amplitudes, having earlier experienced a reduction in mental alertness and concentration. Whilst the underlying cause is reasonably clear surprisingly little is known about the degree of degradation suffered and the motions which contribute most to this degradation. The subject assumes greater significance for modern ships with smaller complements and more complex systems requiring greater mental capability. For warships the problems are compounded by the desire to go to smaller ships which, for conventional hull forms at least, implies larger motion amplitudes.

It is the effect that motions have on the vestibula and labyrinthine systems of the human body that causes the feelings of nausea. In one interesting experiment carried out by the US Coastguard (Ref. 6), a number of 'labyrinthine defectives' were subject to very severe motion conditions. None of these subjects vomitted although normal subjects in the same situation all did. All the stimulants believed to be relevant to inducing nausea in subjects were present including fear for own safety and witnessing others being sick.

What is not clear is which motions (roll, pitch) and which aspects of those motions (amplitude, velocity, acceleration) have most effect. It appears that linear or angular accelerations provide the best correlation between motion and sickness, the linear acceleration being caused mainly by a combination of pitch and heave with a smaller contribution from roll and the most significant angular acceleration being that due to roll. Frequency of motion is a vital factor (see Fig. 9.26 from Ref. 16).

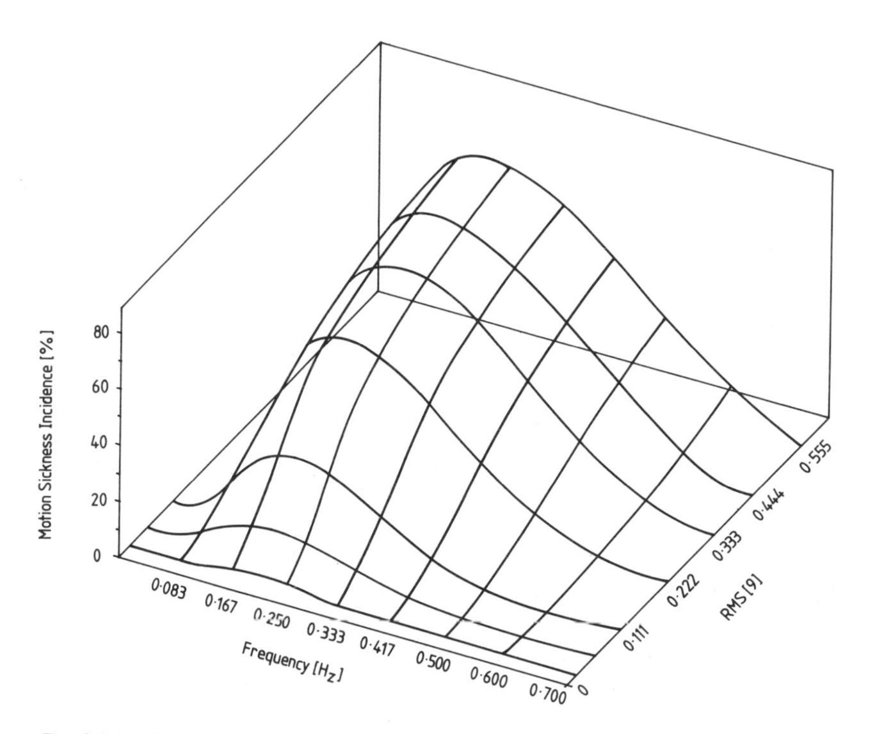

Fig. 9.26 Motion sickness incidence (Ref. 16)

A number of steps can be taken to mitigate the effects of motion. Drugs can suppress feelings of nausea although it is not clear whether the side effects, such as drowsiness, mean the person is any better on balance in carrying out some complex mental task. The ship designer can arrange for vital human operations to be arranged in an area of minimum motion or for the operators to be in a prone position. The designer could even provide a stabilized platform but all these imply some offsetting disadvantage in the overall design and he therefore needs some objective means of judging the 'trade-offs'. These are not yet available but the problem is receiving increasing attention.

Motions may cause not only a fall off in human abilities but also make a task itself physically more difficult. Thus moving heavy weights about the ship, during a replenishment at sea operation, say, is made more difficult and dangerous. The designer is normally expected to design for such operations up to some limiting sea state. See also Chapter 12.

The air

If no action were taken to modify the air in a closed compartment within the ship, what would happen? Consider the case where people and heat producing equipment are both present. Then:

(*a*) due to the people and equipment the temperature within the space would rise steadily until the heat passing through the compartment boundaries matched the heat input from within;
(*b*) due to the moisture breathed out by the people the moisture content of the air would increase;
(*c*) due to the people breathing, the oxygen content would decrease and the carbon dioxide content increase;
(*d*) due to the movement of the people dust would be created;
(*e*) smells would be produced.

Put briefly, the temperature, humidity, chemical purity and physical purity are changed. Complete air conditioning therefore involves control of all these factors. Only in special cases, such as the submarine, is the chemical purity of the air controlled by using special means to produce new oxygen and absorb the carbon dioxide, although in all surface ships a certain minimum quantity of fresh air is introduced to partly control the chemical balance. This is typically about 0·3 m^3 (10 ft^3) of fresh air per man per minute. In general, the physical purity is only controlled within broad limits by filters in air distribution systems. Exceptions to these may be special 'clean' workshops or operating theatres. With only one device, namely cooling, to effect control of both temperature and humidity some compromise between the two is necessary. More precise control of both can be effected by cooling in conjunction with after-warming. This is discussed more fully in Chapter 14.

Temperature is measured by means of a thermometer in either *degrees Celsius* (*Centigrade*) (°C) or *degrees Fahrenheit* (°F). Pure water freezes at 0°C or 32°F and boils at 100°C or 212°F at sea level. To convert from one system to the other,

$$\text{Temperature in °C} = \tfrac{5}{9}(\text{temperature in °F} - 32)$$

It is a common experience, that the sunny side of a house feels warmer than the side in shade. This is because the body feels not only the warmth of the surrounding air but also the warmth of the sun's radiation. To define temperature precisely it is necessary to distinguish between these two. The *dry bulb temperature* is that measured in still air by a thermometer which is unaffected by radiated heat. To measure the effect of radiated heat a special device known as a globe thermometer is used. In this, the conventional thermometer is enclosed in a black sphere. This device records the *globe temperature*.

Another common experience is that one feels chilly in wet clothing. This is because the water, as it evaporates, absorbs heat from the body. This effect is measured by the *wet bulb temperature* which is that recorded by a thermometer with its bulb covered by wet muslin and subject to a moving air stream. It follows, that the wet bulb temperature can never be higher than the dry bulb and will, in general, be lower. Air can only hold a certain quantity of water vapour at a given temperature. The higher its moisture content the slower the rate of water evaporation from a wet body and the closer the wet bulb temperature will approach the dry bulb measurement.

Thus, the amount of water vapour present is important in relation to the maximum amount the air can contain at that temperature. This ratio is known as the *relative humidity*. The ability of air to hold water vapour increases with increasing temperature. If the temperature of a given sample of air is lowered, there comes a time when the air becomes *saturated* and further reduction leads to condensation. The temperature at which this happens is known as the *dew point* for that sample of air. If a cold water pipe has a temperature below the dew point of the air in the compartment, water will condense out on the pipe. This is why chilled waterpipes are lagged if dripping cannot be tolerated.

Since a human being's comfort depends upon temperature, humidity and air movement, it is difficult to define comfort in terms of a single parameter. For a given air state, an *effective temperature* is defined as the temperature of still, saturated air which produces the same state of comfort.

When a kettle of water is heated, the temperature rises at first and then remains constant (at 100°C) while the water is turned into steam. To distinguish between these two conditions, heat which causes only a temperature change is called *sensible heat* (i.e. it is clearly detectable) and heat which causes only a change of state is called *latent heat* (i.e. hidden heat). The sum of these two heats is known as the *total heat*.

The *British Thermal Unit*, Btu, which is the heat required to raise one pound of pure water through one degree Fahrenheit, and the *calorie*, which is the heat required to raise one gramme of pure water through one degree Celsius were the standard measurements but the basic SI unit is the joule. Some useful equivalents are:

1 Btu	=	1·05506 kJ
1 hp hr	=	2·68452 MJ
1 calorie*	=	4·1868 joules (J)
1 kilowatt	=	56·869 Btu/min
	=	3412 Btu/hr
1 horsepower	=	42·407 Btu/min
	=	2544 Btu/hr
1 Btu	=	778·17 ft lbf

*Note that the 'calorie' used loosely in food energy measurement is, in fact, a kilocalorie.

This section covers briefly the basic definitions associated with air conditioning comfort. These are put to use in the design of an air conditioning system in Chapter 14.

Lighting

There is a clear need for certain compartments to have a minimum level of illumination in order that work in them can be carried out efficiently.

Light is an electromagnetic radiation and the eye responds to radiation in that part of the spectrum between 7600 Å and 3800 Å. In this, Å stands for angstrom which is a wave-length of 10^{-10} m. The energy of the radiation can be expressed in terms of watts per unit area but the eye does not respond equally to all frequencies. Response is greater to greens and yellows than to reds and blues. Initially, wax candles were used as standards for illumination. A *lumen* (lm) is the amount of luminous energy falling per unit area per second on a sphere of unit radius from a point source of one candle-power situated at the centre of the sphere. The intensity of illumination is defined by either:

$$1\ \text{lm/ft}^2 = 1\ \text{foot-candle}$$

or

$$1\ \text{lm/m}^2 = 1\ \text{lux} = 1\ \text{metre-candle}$$

Reference 7 suggests standards of illumination for merchant ships and these are given in Table 9.22. Figures in parentheses are the approximate equivalents in lux.

Table 9.22
Suggested standards of illumination

Space	lm/ft^2	Space	lm/ft^2
Passenger cabins	7 (75)	Nursery	10 (108)
Dining rooms	10–15 (108–161)	Engine rooms	15–20 (161–215)
Lounges	7–10 (75–108)	Boiler rooms	10 (108)
Passageways	2–5 (22–54)	Galleys	15 (161)
Toilets	7 (75)	Laundries	15 (161)
Shops	20 (215)	Store-rooms	7 (75)

Standards used within the Ministry of Defence are given in Table 9.23—local figures relate to desk tops, instrument dials, etc. Special fittings are used in dangerous areas such as magazines, paint shops, etc., and an emergency system of red lighting is also provided. In darkened ship conditions, only this red lighting is used over much of the ship so as not to impair night vision.

In some ships, lighting may be regarded as merely a necessary service but in large ships and particularly in passenger liners, lighting can have a considerable influence on atmosphere. This can be especially important in the public rooms, and shipowners often enlist the services of lighting engineers and designers. Light intensity and colour is important and also the design of the light fittings themselves and their arrangement, e.g. it is essential to avoid glare.

Table 9.23
Lighting standards, RN ships

Space	Level of illumination (lux) General	Local
Accommodation spaces	150	300
Machinery spaces	100	300
Galley, bakery	100	200
Workshops	150	400
Magazines	150	300
Stores, general	100	300
Switchboards	150	300
Bathrooms	100	—
Dental clinic	200	400

Reference 7 states that, economic considerations lead to a desire to rationalize light fittings throughout the major part of the ship and suggests that a 230 volt, 60 hertz unearthed system will become standard for merchant ships. The Ministry of Defence (Navy Department) favours a three wire 115/0/115 volt, 60 hertz, unearthed system which provides a very flexible system with the choice of 115 or 230 volts to suit electrical demands of all sorts, including lighting.

In deciding upon the type of light fitting, it is necessary to consider the efficiency of the appliance, its probable life and the effect it will have on the general lighting system. Efficiency is important in air-conditioned spaces as less heat is generated in the production of a given light intensity. Fluorescent lighting is fitted in all important, regularly used, compartments in RN ships. The levels of illumination achieved can be expected to drop by 30 per cent over time due to the slow deterioration of the reflectance of bulkheads etc. and within the light fitting itself.

Table 9.24
Efficiency and life of light fittings. From Ref. 7

Type of fitting	Efficiency (lm/W)	Life (hr)
Single coil tungsten	8–20	1000
Tungsten–quartz iodine	22	2000
Fluorescent	24–66	5000
Cold cathode	10–40	30,000

Vibration and noise

Vibration theory is dealt with in many textbooks. In the present context, it is sufficient to point out that the ship is an elastic structure containing a number of discreet masses and, as such, it will vibrate when subject to a periodic force.

VIBRATION

Ships should be designed so as to provide a suitable environment for continuous and efficient working of equipment and one in which the crew can perform comfortably, efficiently and safely. References 10 and 14 are two international documents setting out criteria which have been the subject of much debate. Reference 11 discusses the application of these.

EXCITATION

Periodic forces causing excitation can arise from:

The propulsion system. Misalignment of shafts and propeller imbalance can cause forces at a frequency equal to the shaft revolutions. Forces should be small with modern production methods. Because it operates in a non-uniform flow the propeller is subject to forces at blade rate frequency — shaft revs × number of blades. These are unlikely to be of concern unless there is resonance with the shafting system or ship structure. Even in a uniform flow a propulsor induces pressure variations in the surrounding water and on the ship's hull in the vicinity. The variations are more pronounced in non-uniform flow particularly if cavitation occurs. If cavitation is fairly stable over a relatively large arc it represents in effect an increase in blade thickness and the blade rate pressures increase accordingly. If cavitation is unstable pressure amplitudes may be many times greater. Whilst the number of blades is important to the frequency it has little effect on pressure amplitude. The probability of vibration problems in single screw ships can be reduced by using bulbous or U- rather than V-sections in the after body, avoiding near horizontal buttock lines above the propeller, providing good tip clearance between propeller and hull, avoiding shallow immersion of the propeller tips to reduce the possibility of air drawing and avoiding low cavitation numbers. Generally the wake distribution in twin screw ships is less likely to cause vibration problems. If A-brackets are used, the angle between their arms must not be the same as that between the propeller blades.

Wave forces. A ship in waves is subject to varying pressures around its hull. The ship's rigid body responses are dealt with in Chapter 12. Because the hull is elastic some of the wave energy is transferred to the hull causing main hull and local vibrations. They are usually classified as springing or whipping vibrations. The former is a fairly continuous and steady vibration in the fundamental hull mode due to the general pressure field acting on the hull. The latter is a transient vibration caused by slamming or shipping green seas. Generally vertical vibrations are most important because the vertical components of wave forces are dominant. However, horizontal and torsional vibrations can be important in ships with large openings or of relatively light scantlings, e.g. container ships or light carriers. The additional bending stresses due to vibration may be significant in fatigue because of the frequency, and the stresses caused by whipping can be of the same order of magnitude as the wave bending stresses.

Machinery. Rotating machinery such as turbines and electric motors generally produce forces which are of low magnitude and relatively high frequency.

Reciprocating machinery on the other hand produces larger magnitude forces of lower frequency. Large diesels are likely to pose the most serious problems particularly where, probably for economic reasons, 4 or 5 cylinder engines are chosen with their large unbalance forces at frequencies equal to the product of the running speed and number of cylinders. Auxiliary diesels are a source of local vibrations. Vibration forces transmitted to the ship's structure can be much reduced by flexible mounting systems. In more critical cases vibration neutralizers can be fitted in the form of sprung and damped weights which absorb energy, or active systems can be used which generate forces equal but anti-phase to the disturbing forces.

RESPONSES

As with any vibratory phenomemon, the response of the ship, or part of the ship, to an exciting force depends upon the frequency of the excitation compared with the natural frequency of the structure and the damping present as indicated in Fig. 9.27.

In this figure, the magnification Q is defined as

$$Q = \frac{\text{dynamic response amplitude}}{\text{static response amplitude}}$$

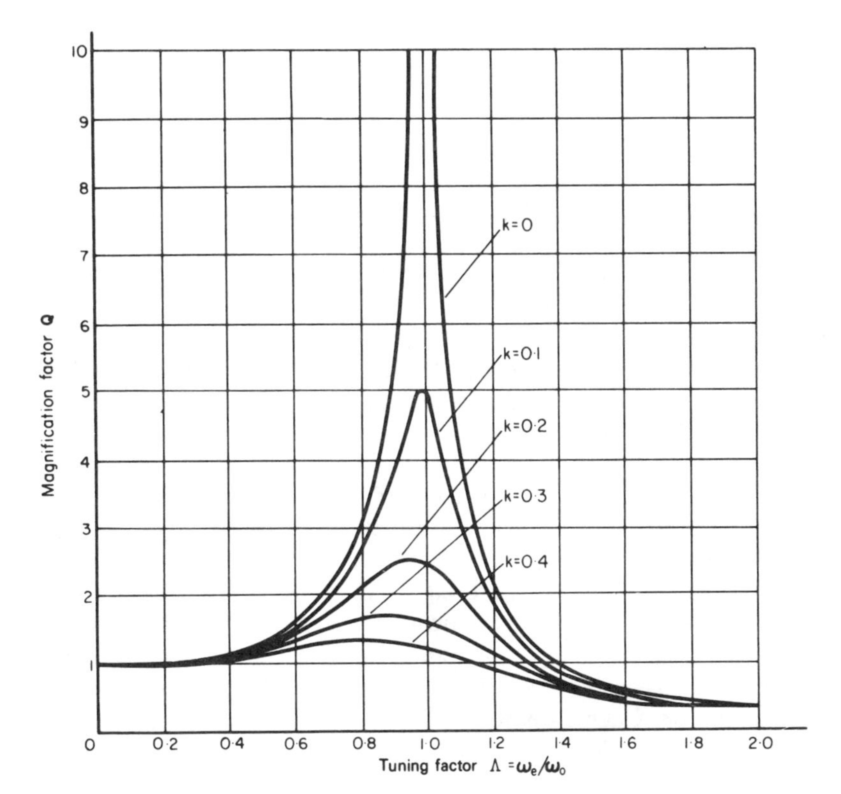

Fig. 9.27 ***Magnification factor***

and ω/ω_0 is the ratio of the frequency of the applied disturbance to the natural frequency of the structure. It should be noted, that the most serious vibrations occur when the natural frequency of the structure is close to that of the applied force, i.e. at resonance.

The response by the ship may be as a whole or in a local area or piece of structure. In the former case the ship responds to the exciting forces by vibrating as a free-free beam. In this type of vibration, certain points along the length suffer no displacement and these points are called *nodes*. The term *anti-nodes* is used for the points of maximum displacement between nodes. The hull can vibrate in different ways, or *modes*, involving 1, 2, 3, 4 or more nodes (Fig. 9.28), although the single node mode applies only to torsional vibration. The natural frequency of the vibration increases as the number of nodes increases as is discussed later.

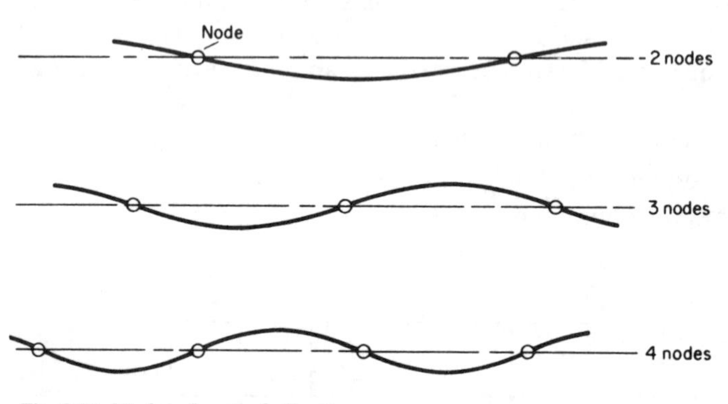

Fig. 9.28 ***Modes of vertical vibration***

There is nothing a designer can do to prevent this free-free vibration, and there is little he can do to alter the frequencies at which resonances occur. He can only recognize that they exist, calculate the critical frequencies and try to avoid any exciting forces at these values.

The figures given in Fig. 9.29 (From Ref. 10) and Table 9.25 relate specifically to the vertical vibration. Figure 9.29 is intended to be used for evaluating hull and superstructure vibration indicating where adverse comment is to expected. It is applicable to turbine and diesel driven merchant ships 100 m long and longer. It is not intended to establish vibration criteria for acceptance or testing of machinery or equipment. The figures in the third column of Table 9.25 are used to evaluate the responses of equipments and detect resonances which the designer will endeavour to design out. Resonances are considered significant when the dynamic magnification factor, Q, exceeds 3. The endurance tests are then conducted at the fixed frequencies shown in the fourth column plus any frequencies, determined by the response tests, giving rise to significant resonances which the designer was unable to eliminate. Transverse vibration will generally be rather less, but for design purposes is usually assumed equal to it. Fore and aft vibration amplitudes are generally insignificant except at the masthead position, but even here they are low compared with the vertical and transverse levels.

Table 9.25
Vibration response and endurance test levels for surface warships

Ship type	Region	Standard test level Peak values and frequency range	Endurance tests
Minesweeper size and above	Masthead	1·25 mm, 5 to 14 Hz 0·3 mm, 14 to 23 Hz 0·125 mm, 23 to 33 Hz	1·25 mm, 14 Hz 0·3 mm, 23 Hz 0·125 mm, 33 Hz Each 1 hour
	Main	0·125 mm, 5 to 33 Hz	0·125 mm, 33 Hz For 3 hours
Smaller than minesweeper	Masthead and main	0·2 mm or a velocity of 63 mm/s whichever is less 7 to 300 Hz	0·2 mm, 50 Hz For 3 hours
	Aftermost 1/8 of ship length	0·4 mm or a velocity of 60 mm/s whichever is less 7 to 300 Hz	0·4 mm, 24 Hz For 3 hours

Notes:
1. The masthead region is that part of the ship above the main hull and superstructure.
2. The main hull includes the upper deck, internal compartments and the hull.

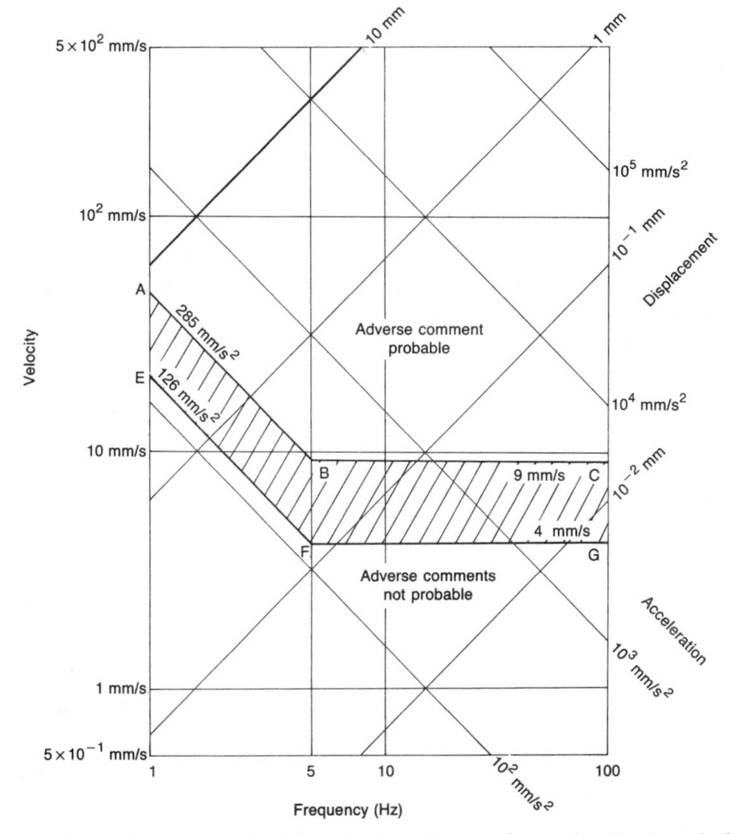

Fig. 9.29 (After Ref. 10) Guidelines for the evaluation of vertical and horizontal vibration in merchant ships (peak values)

The remaining vibratory mode—the torsional— is not common and there is not a lot of available evidence on its magnitude.

The local response of the ship, due either to ship girder vibration or direct resonance with the disturbing forces, is curable and is therefore of less vital interest. A deck, bulkhead, fitting or panel of plating may vibrate. Considering all the possible vibratory systems in local structures, it is impossible to avoid some resonances with the exciting forces. Equally, it is impossible to calculate all the frequencies likely to be present. All the designer can do is select for calculation those areas where vibration would be particularly obnoxious; for the rest, troubles will be shown up on trial and can be cured by local stiffening, although this is an inconveniently late stage.

BODY RESPONSE

The human body responds to the acceleration rather than the amplitude and frequency of vibration imposed upon it. Humans can also be upset by vibrations of objects in their field of view, e.g. by a VDU which they need to monitor. Reference 14 and Fig. 9.30 derived from it, are concerned with vibrations transmitted to the body through a supporting surface such as feet or buttocks. The limits shown in Fig. 9.30 relate to vibration at the point of contact with the man. Thus for a seated man they refer to the vibration transmitted to him through the cushion. In working from the basic ship structure vibration levels allowance must be made for the transmission properties of anything, e.g. the chair, between the structure and the man. The curves are presented as r.m.s. acceleration against frequency and refer to exposure times beyond which there is likely to be a decrease in proficiency particularly in carrying out tasks where fatigue is likely to be a significant factor. The limiting accelerations at which humans are likely to experience some discomfort can be obtained by dividing the accelerations from Fig. 9.30 by 3·15 (i.e. they are 10 dB lower) and those for maximum safe exposure are double those in Fig. 9.30 (i.e. 6 dB higher). The figures apply to each 24 hour period.

The accelerations are related to orthogonal axes with origin at the heart. The 'longitudinal' acceleration is in the line from foot (or buttock) to head. 'Transverse' accelerations are those from the chest to the back and from right to left side. In using the limiting curve, Ref. 14 recommends

(*a*) If vibration occurs simultaneously at more than one discrete frequency or in more than one direction the r.m.s. acceleration of each component shall be evaluated separately with reference to the appropriate limits.

(*b*) For vibration concentrated in a third-octave band or less the r.m.s. acceleration within the band shall be evaluated with reference to the appropriate limit at the centre frequency of the band.

(*c*) For broad band distributed vibration occurring in more than one third-octave band, the r.m.s. acceleration in each such band shall be evaluated separately with respect to the appropriate limit.

(*d*) The effective total daily exposure to an interrupted constant intensity vibration is obtained by summating the individual exposure times, i.e. no allowance is made for human recovery which is likely to occur during pauses.

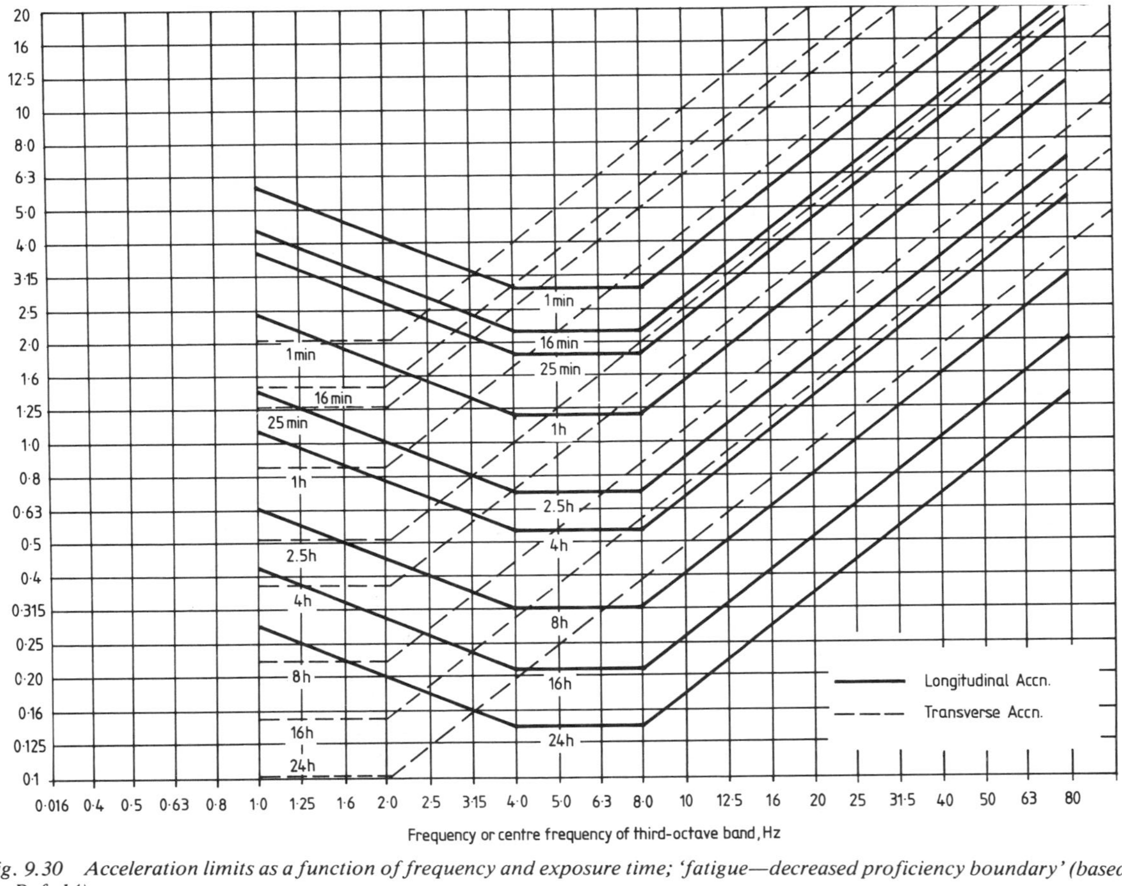

Fig. 9.30 Acceleration limits as a function of frequency and exposure time; 'fatigue—decreased proficiency boundary' (based on Ref. 14)

(*e*) Where intensity varies significantly with time the total exposure is divided into a number of exposures t_i at level A_i. A convenient notional A^1 is chosen in the range of values A_i. If T_i is the permissible time at A_i then the equivalent time for exposure at A^1 is

$$t_i \frac{T^1}{T_i} \text{ where } T^1 = \text{permissible exposure at } A^1$$

The equivalent total exposure time at the notional level A^1 is then given by

$$T^1 \sum \frac{t_i}{T_i}$$

CALCULATIONS

Reference 12 discusses a design procedure capable of dealing with vibration problems other than those arising from wave excitation. Because of the complicated mathematics, cost and time may limit what can be achieved in practice. In the early design stages of a ship, it is more likely that use will be made of one of the empirical formulae available.

One of the earliest empirical formulae was proposed by *Schlick*, viz.

$$\text{Frequency} = \phi\left(\frac{I}{\Delta L^3}\right)^{\frac{1}{2}} \text{c.p.m.}$$

where Δ = ship displacement in tonf, L = length of ship in ft, and I = moment of inertia of the midship section including all continuous members in $\text{in}^2\,\text{ft}^2$. ϕ is a coefficient which is best calculated from data for a ship similar to that being designed. Typical values of ϕ for the 2-node vertical vibration are given:

Large tankers	130,000 (282,000)
Small trunk deck tankers	100,000 (217,000)
Cargo ships at about 60 per cent load displacement	112,000 (243,000)

If I is in m^4, L is in m and Δ in MN, then the values of the constants become as shown in parentheses. For warships, with their finer lines, values of about 160,000 (347,000) are more appropriate. If Δ is in tonnef and other parameters in the metric system, the constants are approximately ten times those in parentheses.

For merchant ships the fundamental frequencies of the hull are generally larger than the encounter frequency with the waves. However, increasing size accompanied by increased flexibility results in lowering the hull frequencies. Taken with higher ship speeds resonances become more likely. To illustrate the point the value of $(I/\Delta L^3)^{1/2}$ in Schlick's formula for a 300,000 DWT tanker may be only a third of that of a 100,000 DWT ship.

The Schlick formula is useful for preliminary calculations but it does ignore the effects of entrained water and is therefore only likely to give good results where a similar ship is available. It also involves a knowledge of I which may not be available during the early design stages. This is overcome in the *Todd* formula which can be expressed as:

$$\text{Frequency} = \beta\left(\frac{BH^3}{\Delta L^3}\right)^{\frac{1}{2}} \text{ c.p.m.}$$

where Δ is measured in tonf and B, H and L in ft.

Values of β were found to be

Large tankers	61,000 (11,000)
Small trunk deck tankers	45,000 (8150)
Cargo ships at about 60 per cent load displacement	51,000 (9200)

If Δ is expressed in MN and linear dimensions in metres the values of the constants are as in parentheses. If Δ is expressed in tonnef and linear dimensions in metres the constants, within the accuracy of the formula, can be taken as ten times those in parentheses.

To account for the entrained water, a virtual weight Δ_1 can be used instead of the displacement Δ, where

$$\Delta_1 = \Delta\left(1{\cdot}2+\frac{1}{3}\frac{B}{T}\right)$$

Having estimated the frequency of the 2-node vibration, it is necessary to estimate the 3-node period and so on. The theory of a vibrating uniform beam predicts that the 3-node and 4-node vibrational frequencies should be 2·76 and 5·40 times those for the 2-node mode. It is found that in ships the ratios are lower and for very fine ships may be 2, 3, 4, etc.

Some typical natural frequencies of hull for various ship types are given in Table 9.26.

Table 9.26
Natural frequencies of hull for several types of ship

Type of ship	Length (m)	Condition of loading	Frequency of vibration						
			Vertical				Horizontal		
			2-node	3-node	4-node	5-node	2-node	3-node	4-node
Cargo ship	85	Light	150	290			230		
		Loaded	135	283			200		
Cargo ship	130	Light	106	210			180	353	
		Loaded	85	168			135	262	
Passenger ship	136		104	177			155	341	
Tanker	227	Light	59	121	188	248	103	198	297
		Loaded	52	108	166	220	83	159	238
Destroyer	160	Average action	85	180	240		120	200	
G.M. Cruiser	220	Average action	70	130	200		100	180	270

NOISE

Noise is of growing importance both for merchant ships and warships. This book can only introduce the subject briefly. The internationally agreed unit for sound intensity is 10^{-16} watts/cm^2 and at 1000 hertz this is close to the threshold of hearing. Due to the large range of intensity to which a human ear is sensitive a logarithmic scale is used to denote the intensity of sound and the usual unit in which noise levels are expressed is the *decibel*. If two noise intensities are w_1 and w_2 then the number of decibels, n, denoting their ratio is

$$n = 10 \log_{10} \frac{w_1}{w_2} \text{ dB}$$

As a reference level, $w_2 = 10^{-16}$ watts/cm^2 is used.

Instruments measuring noise levels in air record sound pressure so that

$$n = 10 \log_{10} \frac{w_1}{w_2} = 10 \log_{10} \left(\frac{p_1}{p_2}\right)^2 = 20 \log_{10} \frac{p_1}{p_2}$$

where p_2 is the pressure in dynes/cm^2 corresponding to the threshold of hearing. This becomes $p_2 = 2 \times 10^{-5}$ N/m^2.

Thus a sound pressure level of 0·1 N/m^2 (1 dyne/cm^2) is represented by

Level in dB = 20 log 1/0·0002 = 74 dB

In the open, the sound intensity varies inversely as the square of the distance from the source. Hence halving the distance from the noise source increases the level in dB by 10 log 4 = 6 dB and doubling the distance reduces the level by 6 dB. The addition of two equal sounds results in an increase of 3 dB and each 10 dB increase is roughly equivalent to a doubling of the subjective loudness.

A noise typically contains many components of different frequency and these varying frequencies can cause different reactions in the human being. In order to define fully a given noise it is necessary, therefore, to define the level at all frequencies and this is usually achieved by plotting results as a noise spectrum. The alternative is to express noise levels in dB(A). The A weighted decibel is a measure of the total sound pressure modified by weighting factors, varying with frequency, reflecting a human's subjective response to noise. People are more sensitive to high (1000 Hz+) than low (250 Hz and less) frequencies.

Primary sources of noise are machinery, propulsors, pumps and fans. Secondary sources are fluids in systems, electrical transformers and ship motions. Noise from a source may be transmitted through the air surrounding the source or through the structure to which it is attached. The actions are complex since airborne noise may excite structure on which it impacts and directly excited structure will radiate noise to the air. Taking a propulsor as an example, much of the noise will be transmitted into the water. That represented by pressure fluctuations on the adjacent hull will cause the structure to vibrate transmitting noise both into the ship and back into the water. Other transmissions will be through the shaft and its support system.

A designer will be concerned to limit noise for two reasons:

(*a*) noise transmitted into the water can betray the presence of the ship, provide a signal on which weapons can home and reduce the effectiveness of the ship's own sensors;

(*b*) internal noise levels may have effects on crew and passengers.

It is the latter effects with which we are primarily concerned here. Noise may:

(*a*) annoy;
(*b*) disturb sleep;
(*c*) interfere with conversation. Conversational speech at a distance of 1 m represents a noise level of about 60 dB. The greatest interference is created by noises in the frequency range 300–5000 hertz;
(*d*) damage hearing. Noise levels of 130–140 dB can cause pain in the ear and higher levels can cause physical damage;
(*e*) upset the normal senses, e.g. it can lead to a temporary disturbance of vision.

Thus noise effects can vary from mere annoyance, itself very important in passenger vessels, to interference with the human being's ability to function efficiently and eventually can cause physical damage to the body.

Reference 13 outlines basic acoustical calculations a designer can use and their application to ship noise estimation and the design of noise control systems.

Various methods are available for reducing the effect of noise, the particular method depending upon the noise source and why it is undesirable. For instance, the noise level within a communications office with many teleprinters can be reduced by treating the boundary with noise absorption material. A heavy, vibrating, machine transmitting noise through the structure can be mounted on special noise isolating mounts, care being taken to ensure that the mounts are not 'short-circuited' by pipes connected to the machine. Such pipes must include a flexible section. Where a piece of equipment produces a high level of air-borne noise in a compartment adjacent to an accommodation space the common boundary can be treated with noise insulation material. For the best results, the whole boundary of the offending compartment should be treated to reduce the level of structure borne noise which can arise from the impact of the pressure waves on, say, deck and deckhead.

In recent years active noise cancellation techniques have been developing. The principle used is to generate a noise of equivalent frequency content and volume but in anti-phase to the noise to be cancelled. Thus to cancel the noise of a funnel exhaust a loudspeaker could be placed at the exhaust outlet. For structure borne noise from a machine force generators could be used at the mounting.

ICE

When the temperature falls sufficiently, ice forms on the rigging and upperworks of ships constituting a hazard, as discussed in Chapter 4. Remedial action may also be difficult because the deck machinery itself may have been rendered unusable by ice. As well as reducing stability, ice may hazard the whole ship by structural damage caused by sea ice or by the icebergs which originate in glaciers. Reference 17 discusses the formation of sea ice. While fresh water has its maximum density at 4°C sea water of salinity greater than 24·7 per cent increases in density right down to its freezing point (which is — 1·33°C at 24·7 per cent salinity). Thus freezing does not occur at the surface until the water below is also at the freezing point. Initially individual ice crystals are formed, creating a slush. Movement of the water surface leads to the formation of flat circular discs 30 to 100 cm in diameter with upturned rims, called pancake ice. As the temperature drops further the discs become cemented together to form a continuous sheet which in a typical Arctic winter reaches an average thickness of about 3 metres. In way of pressure ridges the thickness may be much greater.

The majority of icebergs in the North Atlantic originate at the Greenland glaciers. They vary a great deal in size but whatever the size represent a formidable threat to shipping, as witnessed by the *Titanic*. Due to their porosity, their effective density may be appreciably less than that of pure ice and the depth below water can vary between three and eight times their height above water.

Problems

1. Discuss the process by which sea waves are thought to be formed and the factors on which the wave characteristics depend. What do you understand by the term *swell?*

Describe the type of sea you would expect to be generated by a wind blowing for some time in one direction. How would you describe, quantitatively, the resulting surface characteristics?

2. For a regular sinusoidal wave deduce expressions for speed of propagation and period.

Calculate the speed and period of a wave 100 m long. If the surface wave height is 3 m calculate the depth at which the height of the sub-surface profile is 0·5 m.

3. A wave spectrum is defined by the following table:

ω (1/s)	0·20	0·40	0·60	0·80	1·00	1·20	1·40
$S_\zeta(\omega)$ (m^2 s)	0·00	7·94	11·68	5·56	2·30	0·99	0·45

ω (1/s)	1·60	1·80	2·00	2·20
$S_\zeta(\omega)$ (m^2 s)	0·23	0·12	0·07	0·00

Calculate the significant wave height, average wave height and mean of the tenth highest waves.

4. A large number of wave heights are recorded and on analysis it is found that the numbers falling in various height bands are:

wave height band (m)	0–$\frac{1}{2}$	$\frac{1}{2}$–1	1–$1\frac{1}{2}$	$1\frac{1}{2}$–2	2–$2\frac{1}{2}$	$2\frac{1}{2}$–3	3–$3\frac{1}{2}$	$3\frac{1}{2}$–4
No. of waves	5	10	20	40	55	40	25	5

Plot these as a histogram record and deduce the normal and Rayleigh distributions which most closely fit the measured data.

5. A 20 MN ship presents a profile as below

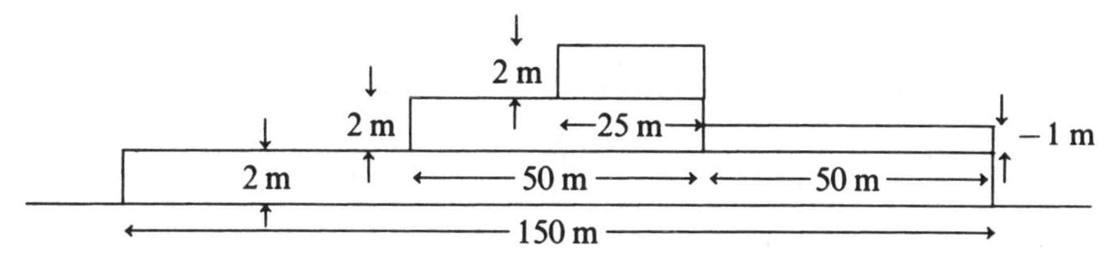

Fig. 9.31

Assuming that the force acting on area A m^2 due to a wind of velocity V knots is given by

$$\text{Force} = 0{\cdot}19V^2A \text{ newtons}$$

calculate the force due to a nominal wind of 50 knots (i.e. as measured 6 m above the water surface) allowing for the wind gradient as defined in Fig. 9.25. Deduce the effective average wind speed.

Assuming the ship has a mean draught of 6 m and a metacentric height of 1·5 m calculate the approximate angle of heel due to the wind.

6. In a given wave sample, the number of waves of various heights are:

Wave height (m)	0·5	1·0	1·5	2·0	2·5	3·0	3·5	4·0
No. of waves	10	39	50	35	16	5	4	1

Calculate the average and significant wave height and the average of the $\frac{1}{10}$th highest waves.

7. Calculate the wave-lengths in metres and wave velocities in metres per second corresponding to wave periods of 1, 2, 4, 6, 8, 10, 14, 18 and 20 s.

8. The successive crests of the wave profile along the side of a ship at speed in calm water are observed to be about 100 m apart. What is the approximate speed of the ship in knots? If the ship speed is reduced to 10 knots what will the distance apart of the crests become?

9. Check the following statement:

The orbits and velocities of the particles of water are diminished by approximately one-half for each additional depth below the mid-height of the surface trochoid equal to one-ninth of a wave-length. For example:

Depth in fraction of wave-length	0	$\frac{1}{9}$	$\frac{2}{9}$	$\frac{3}{9}$	$\frac{4}{9}$
Proposed velocities and diameters	1	$\frac{1}{2}$	$\frac{1}{4}$	$\frac{1}{8}$	$\frac{1}{16}$

10. Using empirical formulae, calculate the 2-node frequency of vibration of the hull of a cargo ship to which the following data relates:

Displacement	7800 tonf
Length	420 ft
Beam	60 ft
Depth	30 ft
Draught	20 ft

Allowance should be made for the entrained water.

References

1. Froude, R. E. Model experiments on hollow versus straight lines in still water and among artificial waves, *TINA*, 1905.
2. Milne-Thomson, L. M. *Theoretical hydrodynamics*, Macmillan, 1949.
3. *Practical methods for observing and forecasting ocean waves*, HO Pub. No. 603. US Navy hydrographic office, 1960.
4. Gumbel, E. J. *Statistics of extremes*, Columbia U.P., 1958.
5. Campbell, L. M., Some applications of extreme value and non-parametric statistics to naval engineering, *Naval Engineers Jnl*, Dec. 1977.
6. Kennedy, R. S., Graybiel, A., McDonough, R. C. and Beckwith, Fr. D. Symptomatology under storm conditions in the North Atlantic in control subjects and in persons with bilateral labyrinthine defects, *Acta oto-laryngologica* 66, 533–40 (1968).
7. Carter, G. and Fothergill, A. E. Lighting in ships, *TRINA*, 1963.
8. Sarchin, T. H. and Goldberg, L. L. Stability and buoyancy criteria for US surface ships, *TSNAME*, 1962.
9. Jones, S. J. and MacVicar, J. K. W. The development of air conditioning in ships, *TINA*, 1959.
10. BS 6634:1985. ISO 6954-1984. Overall evaluation of vibration in merchant ships.
11. Ward, G. *Vibration of ship structures with particular reference to warships*, Advances in marine structures, Elsevier Applied Science Publications, 1986.
12. Ward, G. The application of current vibration technology to routine ship design work, *TRINA*, 1982.
13. Morrow, R. T. Noise reduction methods for ships, *TRINA*, 1989.
14. ISO 2631, Guide for the evaluation of human exposure to whole-body vibration, 1978.
15. Bishop, R. E. D. and Price, W. G. *Hydroelasticity of ships*, Cambridge U.P., 1979.
16. McCauley, M. E. et al. *Motion sickness incidence; exploratory studies of habituation, pitch and roll, and the refinement of a mathematical model*, Human Factors Research Inc., 1976.
17. Bowden, K. F. The marine environment—some features of concern to navel architects, *TRINA*, 1970.
18. Warnaka, G. E. Active attentuation of noise—the state of the art. *Noise Control Engineering*, 1982.
19. Hogben, N. *Ocean wave statistics—five minutes slow after six years*, National Physical Laboratory Report Ship 180, Aug., 1974.
20. Hogben, N. and Lumb, F. E. *Ocean Wave Statistics*, HMSO, 1967.
21. Hogben, N., Dacunha, N. M. C. and Oliver, G. F. *Global wave statistics*. British Maritime Technology Ltd, 1986.

Answers to problems

Chapter 2

1. 0·73, 0·80, 0·97, 3·70 m.
2. 8771 tonnef, $86{\cdot}1 \times 10^6$ newtons, 103 m^3, 0·722.
3. 0·785, 0·333, 0·500.
4. 146·25 m^2, 145·69 m^2, 1·0039.
5. 2048 ft^2.
6. 1·2 per cent low, 0·46 per cent low.
7. 9·33 m^3, 0·705 MN.
8. 5286 m^2, 1·65 m abaft 6 ord.
9. 2306 tonf, 16·92 ft above 10 ord.
10. 403 tonnes.
11. 36,717 m^3, 4·85 m, 5·14 m.
12. 243,500 m^4, 23,400 m^4.
13. $182{\cdot}2 \times 10^6$ ft^4.
14. 39·42 ft, 48·17 ft.
15. 2·48 tonf, 8·50 ft.
16. 26·3 ft.
17. (*a*) 539·3 (Simpson 2), 561·0 (trap).
    (*b*) 278·7 (Simpson 58–1), 290·0 (trap).
18. 0·41 per cent.
19. $M = \frac{1}{6}h^2(y_1 + 6y_2 + 12y_3 + 18y_4 + \ldots)$ about 1 ord.
21. (*a*) 2·09440, (*b*) 1·97714, (*c*) 2·00139, (*d*) 2·00000 correct.
22. (*a*) 201·06, (*b*) 197·33, (*c*) 199·75, (*d*) 200·59, (*e*) 200·51.
23. (*a*) 141·42, (*b*) 158·93, (*c*) 158·02, (*d*) 157·08 correct.
25. (*a*) 23·37, (*b*) 11·36.
26. (*a*) 4·37, (*b*) 7·23, (*c*) 9·3 per cent.

Chapter 3

1. (*a*) 146, (*b*) 544.
2. 4·94 m.
3. 8550 ft^3, 6·82 ft^3/tonf.
5. 1·2 in. below keel.
6. 9 degrees.
7. 21 m.
8. 54,018 tonf, 13·41 ft below 1 WL (1331 Rule).
9. 39,030 tonf, 14·24 ft below 1 WL.
10. 0·27 m below $2\frac{1}{2}$ WL.
11. 12 ft $6\frac{3}{4}$ in., 18 ft $6\frac{1}{2}$ in.
12. 19 ft 6 in. abaft amidships, 22 ft $0\frac{1}{4}$ in.
13. 4·39 m.
14. 4·21 m forward, 4·70 m aft.
15. 37·38 ft forward, 40·43 ft aft.
16. 1·34, 1·65 m.

17. 3·704 m, 2·744 m.
18. 373 tonf, 13 ft 4 in., 16 ft 7 in.
19. 1·95 hours, 0·92 m, 3·56 m.
20. (*a*) 2350 tonf, 11·01 ft aft of amidships, (*b*) 205 tonf, 11·75 ft.
21. 3·77 m fwd, 4·33 m aft.
22. 10 ft 6¾ in., 14 ft 6¾ in.
23. 30 ft 3 in. forward, 30 ft 5½ in. aft.
24. 8847 m^3, 2·76 m.
25. 4459·2 tonf.
26. 42·2 ft before amidships, 8 ft 11·9 in., 8 ft 8·8 in.

Chapter 4

1. —.
2. —.
3. —.
4. 1·972 m, 0·536 m, 1·047 m.
5. 3875 tonnef m, 1252 tonnef m, 15·6 degrees.
6. 2·05 m, 1·99 m.
7. 2·1 m, 15·4 m.
8. 1·13 in.
9. 0·97*T*.
10. 1° 27′.
11. 4·388 m.
12. —.
13. 7° 7′.
14. 90 degrees.
15. 2·39 m, 27·13 m.
16. ·1·25 m.
17. 3·04 m.
18. 1·44 m.
19. 1° 22′.
20. 0·70 ft.
21. 0·28 m.
22. 3·28 m.
23. 45 degrees.
24. —.
25. 0·37 m.
26. (*a*) 70·4°, (*b*) 84·9°, (*c*) 87·1°, (*d*) 95·5°.
27. 74 MJ.
28. 2·60, 1·035 : 1.
29. 0·67 ft.
30. 1·54 ft.
31. 2·53 ft, 2·82 ft, 0·09 ft, 4·57 ft, 75·5°.
32. 0·509 m, 0·49 m.
33. − 1·425 ft.
34. —.

Chapter 5

1. 1·73 m port, 2·99 m starboard.
2. 13 ft 3 in., 2 ft 11 in.
3. 288 tonnef, 4·70 m, 5·83 m; 6·23 m, 5·81 m.
4. 20,740 tonnef m, 10° 37’, 93·31 m (added weight), 91·58 m (lost buoyancy).

5. (i) 167 tonf, (ii) 2·07 ft, (iii) 1·95 ft.
6. (*a*) 13·41 ft, 11·23 ft, (*b*) 2·65 ft.
7. 7·06 ft, 4·92 ft forward; 4·33 ft, 2·20 ft aft.
8. 1·70 m.
9. 9·70 ft, +0·98 ft.
10. 40 tonnef, 94 tonnef, 22 cm.
11. 10¾ degrees.
12. 1·8 degrees.
13. 53 tonnef, 1½ degrees.
14. 13·97 ft.
15. 4·63 min.
16. Starboard hit 40 per cent, 22½ per cent; Port hit 40 per cent, 10 per cent.

Chapter 6

1. 117 ft from fore end, 1569 tonf ft.
2. SF, $0{\cdot}21L$ from end = $W/10{\cdot}4$; BM amidships = $WL/32$.
3. Max. SF., 30·75 tonnef; max BM, 132·6 tonnef m,.
4. SF, 48 tonf. 30 ft from ends; BM 960 tonf ft amidships.
5. Max. SF, 690 tonnef; max. BM ,13,000 tonnef m.
6. 0·315 in.
7. SF, 714 tonnef at 3/4; BM 24,500 tonnef m at 5/6.
8. 113·5 pascals.
9. 3·58 and 4·24 tonf/in².
10. σ_D, 6·07 tonf/in², σ_K 3·67 tonf/in².
11. 35.
12. z_D, 8090 in² ft; z_K, 6220 in² ft; σ_D, 4·62 tonf/in²; σ_K 6·00 tonf/in².
13. σ_D, 2·62 tonf/in²; σ_K 3·78 tonf/in².
14. 52 cm².
15. 57·5, 42·2, 19·7 MPa.
16. 42·9, 42·9, 33·2 MPa.
17. $\delta M/WL$ = 0·144 and 0·115.
18. 3·10 in².
19. 19 mm.
20. 8½ tonf/in², 0·3 degrees.
21. 55 mm approx.

Chapter 7

1. 260 MPa.
2. 7700 cm³, 14,800 cm³.
3. (*a*) 1:2·92, (*b*) 1:2·20.
4. (*a*) 64 in⁴, (*b*) 0·29 in.
5. 18 mm.
6. 4·38 tonf.
7. 7·4 cm.
8. (*a*) 0·80 in., (*b*) 0·33 in.
9. (*a*) 0·112 MPa, (*b*) 0·035 MPa, (*c*) 5·7 mm.
10. (*a*) 16·3, (*b*) 106, (*c*) 32, (*d*) 58 lbf/in².
11. (*a*) 2·0, (*b*) 1·15.
12. 90 MPa.

13. 2·7 tonf/in^2, in rib at deck edge.
14. $M_{EA}=420$, $M_{AE}=840$, $M_{BA}=27{,}300$, $M_{BF}=8480$, $M_{BC}=35{,}760$, $M_{FB}=2120$ Nm.
15. $M_{AB}=2{\cdot}5$, $M_{BC}=9{\cdot}9$, $M_{CB}=39{\cdot}9$, $M_{CG}=6{\cdot}7$, $M_{CD}=46{\cdot}6$, $M_{GC}=3{\cdot}3$ tonf ft.
16. 0·03 MN.
17. Torque, 51·4 tonf ft; BM, 764 tonf ft.
18. 21·8 in., 15·5 hp.
19. 15·4 tonf/in^2.

Chapter 8

1. 75·5 m, 7·7 MN, 0·19 N/mm^2.
2. 18·36, 14·31 fwd, 22·41 aft tonnef/m^2.
3. —7·81 ft forward, 22·67 ft aft.
4. 28·5 ft.
5. (*a*) 97 m, (*b*) 1,480 tonnef, (*c*) 41,000 tonnef m.
6. (*a*) lift at 144 m (*b*) max. force on FP 7,950 tonnef, (*c*) min tip mt. 700,000 tonnef m, (*d*) floats off.
7. Max. BM, 55,250 tonf ft; max. SF, 432 tonf.
8. (*a*) 54 ft, (*b*) 1500 tonf, (*c*) 180,000 tonf ft.

Chapter 9

1. —.
2. 12·5 m/s, 8·0 s, 28·5 m.
3. 4·85 m, 3·03 m, 6·19 m.
4. $p(x)=\frac{1}{1{\cdot}925}\exp\left\{-\frac{(x-2{\cdot}19)^2}{1{\cdot}18}\right\}$,

   $p(x)=\frac{x}{5{\cdot}38}\exp\left\{\frac{-x^2}{10{\cdot}75}\right\}$.
5. 0·16 MN, 41 knots, 1° 35′.
6. 1·64 m, 2·40 m, 3·00 m.
7. 1·56, 1·56; 6·2, 3·1; 25, 6·2; 56, 9·4; 100, 12·5; 156, 15·6; 306, 21·8; 506, 28·1; 625, 31·2.
8. 24·3 knots, 16·96 m.
9. —.
10. 57·5 c.p.m.

Index